Advances in High-speed Rail Technology

"Advances in High-speed Rail Technology" presents the latest and most cutting-edge theories, techniques, and methodologies in the multidisciplinary field of high-speed railways, focusing on advances and findings from China. This series includes monographs, professional books and edited volumes from dedicated conferences and workshops. All volumes are authored or edited by established experts in their fields and undergo rigorous peer review, based on the editors' preview and selection and refereeing by independent experts. The intended audience includes researchers, engineers, industrial practitioners, graduate students, and professionals. Topics of interest in "Advances in High-speed Rail Technology" include, but are not limited to: Infrastructure, Mobile Equipment, Communication & Signal, Traction Power Supply, Operation Organization, etc.

Zhiqiang Long · Zhiqiang Wang · Mingda Zhai · Xiaolong Li

High-Speed Maglev Train's Levitation and Guidance Control

The Key Technologies

Springer

Zhiqiang Long
College of Intelligence Science
and Technology
National University of Defense Technology
Changsha, Hunan, China

Zhiqiang Wang
College of Intelligence Science
and Technology
National University of Defense Technology
Changsha, Hunan, China

Mingda Zhai
College of Intelligence Science
and Technology
National University of Defense Technology
Changsha, Hunan, China

Xiaolong Li
College of Intelligence Science
and Technology
National University of Defense Technology
Changsha, Hunan, China

ISSN 2363-5010 ISSN 2363-5029 (electronic)
Advances in High-speed Rail Technology
ISBN 978-981-97-2308-9 ISBN 978-981-97-2309-6 (eBook)
https://doi.org/10.1007/978-981-97-2309-6

This Springer imprint is published by the registered company Springer Nature Singapore Pte Ltd.
The registered company address is: 152 Beach Road, #21-01/04 Gateway East, Singapore 189721, Singapore

Preface

Maglev trains represent a novel approach to rail transit that use electromagnetic force to levitate the vehicle body above the guideway and utilize the linear motor for propulsion operation, which have the advantages of high travel speed, no friction loss, low noise, flexible line selection, and less maintenance workload at the later stage. Germany, Japan, China, the United States and South Korea have researched on maglev train technology, and the application areas include urban maglev transit with a maximum speed of 160 km/h, high-speed maglev transit with a speed of 400 km/h or higher, and ultra-high speed low-vacuum tube maglev system with a speed of 800 km/h.

Since 1980, the National University of Defense Technology, the Southwest Jiaotong University, the Institute of Electrical Engineering of Chinese Academy of Sciences, and other institutes in China have successively carried out research on maglev train technology. Under the leadership of Profs. Wensen Chang, Jisan Lian, Xiangming Wu, and Academician Luguang Yan etc., China has overcome a number of key technologies of urban maglev trains and high-speed maglev trains such as levitation and guidance control, which has promoted the progress of engineering application of China's EMS-type maglev train technology. China has successively built urban maglev operation lines in Beijing, Changsha, Fenghuang in Hunan, and Qingyuan in Guangdong. In December 2002, China and Germany cooperated to build an EMS-type high-speed maglev operation demonstration line in Pudong, Shanghai. On this basis, China researched on the localization and technological innovation of EMS-type high-speed maglev transit technology, and developed an EMS-type high-speed maglev test vehicle and a PEM hybrid levitation based high-speed maglev test vehicle. In July 2021, CRRC Qingdao Sifang Co., Ltd. led the development and test of a high-speed maglev train with a maximum design speed of 600 km/h.

Based on the academic researches, engineering applications, and technical innovations of high-speed maglev trains carried out by the maglev team of the National University of Defense Technology, this book summarizes the technical achievements in the field of levitation and guidance control technology of high-speed maglev train. The specific contents are as follows.

In Chap. 1, the definition and types of maglev transit systems in the world are introduced, and then the three stages of development of EMS-type high-speed maglev train technology are analyzed. Finally, a discussion on the current research status and principal achievements in the field of high-speed maglev trains is conducted.

In Chap. 2, the basic structure of the high-speed maglev transit is introduced, and then the technical analysis is carried out for the complex high-speed maglev vehicle system, and the composition of each subsystem is analyzed. Finally, for the key technical problems of the high-speed maglev train, the basic structures, basic requirements and control schemes of the levitation system and the guidance system are analyzed respectively.

In Chap. 3, the levitation control strategy of high-speed maglev train is mainly studied. Through the analysis of the levitation system model of high-speed maglev train, the nominal controller design based on linear quadratic regulator and the controller design based on active disturbance rejection control are carried out. Then, the designed control schemes are verified by the experiments of high-speed maglev train levitation double maglev bogies and the whole vehicle.

In Chap. 4, the guidance system of high-speed maglev train is taken as the research object, and the guidance system model considering disturbance and fault is analyzed in combination with practical engineering problems. The research of the nominal controller design and the robust controller design are carried out, and relevant verification experiments are carried out using single bogie and whole vehicle of high-speed maglev train.

In Chap. 5, aiming at the control performance evaluation of the levitation system after the design of the nominal controller is completed, the nominal performance and nominal performance index of the levitation system are introduced, then the performance index and performance evaluation method of the levitation system under various working conditions are analyzed respectively.

In Chap. 6, aiming at the control performance evaluation of the guidance system after the design of the nominal controller is completed, a quantifiable evaluation index is proposed based on a fuzzy comprehensive evaluation method by using the state information of guidance gap, current and acceleration, then the performance evaluation of the guidance system under different curve radii, running speeds and payloads are analyzed respectively.

In Chap. 7, the levitation control technology for permanent and electric magnet (PEM) hybrid levitation type high-speed maglev train is mainly analyzed. Firstly, the PEM hybrid levitation technology and the structural design scheme of the PEM hybrid levitation electromagnet are analyzed. Then, the PEM hybrid levitation system is modeled and analyzed. Finally, the zero-power controller based on state feedback is designed, and test research is carried out on the double maglev bogies and the whole vehicle.

In Chap. 8, aiming at the possible faults of the levitation system of the high-speed PEM hybrid levitation train, the sensor fault diagnosis problem is studied, and the fault diagnosis methods based on Kalman filter, data-driven, and signal analysis are analyzed, respectively.

In Chap. 9, the fault-tolerant control of the levitation system of the high-speed PEM hybrid levitation train is studied. Firstly, the hierarchical fault-tolerant control framework of the levitation system is introduced and analyzed. Then, the fault-tolerant control based on the online optimization of Youla parameters is analyzed. Finally, the active fault-tolerant control method based on signal reconstruction and control law switching is studied.

The whole book was edited and unified by Prof. Zhiqiang Long of the National University of Defense Technology. Among them, Chaps. 1–4 were written by Prof. Zhiqiang Long, Chaps. 5 and 6 were written by Dr. Mingda Zhai, Chap. 7 was written by Prof. Xiaolong Li and Zhiqiang Long, Chaps. 8 and 9 were written by Dr. Zhiqiang Wang. Professor Wensen Chang, and Yungang Li, Longhua She, Jie Li, Jun Wu, Zhisu Zhao, Shaoke Liu, Guirong Chen, Kun Luo, Ning Wang, Aming Hao, Hu Cheng, Huixing Chen and more than 10 other doctors have directly participated in the research work related to the book. Numerous high-level references from around the globe were consulted during project research and manuscript writing.

This work is supported by the National Natural Science Foundation of China—"Key Theories and Technologies for Performance Evaluation and Collaborative Optimization of Normal-Conductor High-Speed Maglev at 600 km/h Level" (Grant No.52332011). And the Tongji University National Maglev Technology Research Center, the Shanghai Maglev Transportation Development Co., Ltd., CRRC Qingdao Sifang Co., Ltd., and the National University of Defense Technology have given long-term support to this research. We would like to express our sincere gratitude.

Changsha, China
June 2024

Zhiqiang Long

Contents

Chapter 1
Introduction

Maglev trains use maglev technology to achieve support and guidance between vehicle and track, and realize propulsion and electric braking through linear motors. Its application scenarios include the urban maglev transit, the high-speed maglev transit, and the ultra-high-speed low vacuum tube maglev transit. This chapter gives the definition of maglev trains, classifies maglev trains from the perspective of levitation principle and propulsion principle, and describes the electromagnetic suspension (EMS)-type system, the electrodynamic suspension (EDS)-type system and the high-temperature superconducting (HTS) flux-pinning maglev system in detail. What's more, this chapter reviews the development history and technical evolution of maglev train technology in terms of structure schemes for levitation systems, and introduces the three technological stages of the rigid body vehicle scheme, the magnetic wheel structure scheme and the modular structure scheme that the EMS-type maglev train has gone through. Finally, this chapter discusses the research and application results of high-speed maglev based on the relevant technology researches in Germany, Japan, and China.

1.1 Definition and Classification

The maglev technology has been studied for more than a century. Currently, many countries are still conducting extensive researches on maglev technology, which mainly includes the urban maglev trains, the high-speed maglev trains used in the intercity rail transit, the maglev momentum flywheels, the maglev reaction flywheels, the maglev energy storage flywheels, the maglev gyro torques used in satellites and space stations, and the high-speed rotating machinery supported by different kinds of magnetic bearings such as various types of pumps, centrifugal machinery, grinders, etc. Besides, some high-precision and high-sensitivity instruments such as maglev

Z. Long et al., *High-Speed Maglev Train's Levitation and Guidance Control*, Advances in High-speed Rail Technology, https://doi.org/10.1007/978-981-97-2309-6_1

gyros and maglev gravimeters, also adopt maglev technology. Among all kinds of applications, the field of maglev transit is attracting more and more attention.

Maglev trains use maglev technology to achieve the support and guidance between vehicle and track, and linear motors to achieve propulsion and electric braking.

According to the different principles of maglev forces, maglev systems can be mainly classified into EMS-type maglev system, EDS-type maglev system, and HTS flux-pinning maglev system.

According to the speed range, maglev systems can be further divided into the urban maglev (maximum speed of 160 km/h), the high-speed maglev (maximum speed of 600 km/h), and the ultra-high-speed maglev (maximum speed of 800 km/h or higher, using low vacuum tube technology, currently in the exploration stage).

In terms of propulsion principles, maglev systems can be divided into the long-stator synchronous linear motor propulsion and the short-stator asynchronous linear motor propulsion. The high-speed and urban maglev systems mainly utilize the long-stator synchronous linear motor propulsion, while the urban maglev systems and some urban maglev systems equip with the short-stator asynchronous linear motor propulsion.

(1) EMS-type maglev system

EMS-type maglev trains utilize actively controlled electromagnetic force between the electromagnet and the track to levitate the vehicle. At present, Germany, Japan, China, and Korea are the representatives, and research primarily focuses on the development of the urban maglev trains and high-speed EMS-type maglev trains. Among them, the urban maglev trains mainly solve the problem of intra-city transit. Its cost is comparable to light rail, and the biggest advantage of this type of train is that it does not pollute the environment. The high-speed maglev trains are suitable for inter-city transit. Although there are major differences in propulsion, power supply and operation control between the urban maglev and the high-speed EMS-type maglev trains, the levitation principles are the same, using the active electromagnetic levitation principle, and the levitation gap is about 8~10 mm. Therefore, there are many similarities in the design of the levitation control systems for both high-speed EMS-type maglev trains and urban maglev trains. The schematic diagram is shown in Fig. 1.1.

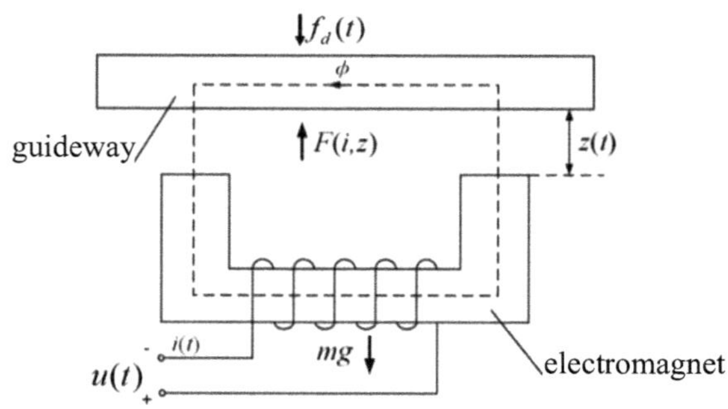

Fig. 1.1 Levitation schematic diagram

In Fig. 1.1, $f_{\mathrm{d}}(t)$ is the external disturbance force, $F(i, z)$ is the equivalent levitation force, mg is the gravitational force, $z(t)$ is the levitation gap, $u(t)$ is the voltage of the electromagnet coil, and $i(t)$ is the current of the electromagnet coil.

In the field of urban maglev trains, many countries have built the test line and operating line. In May 2003, Japan firstly completed the Nagoya East Hills operating line. In March 2016, Korea completed the Incheon Airport operating line. Germany proposed an embedded urban maglev scheme and built a test line. Since 1980, China has started the research on maglev transit technology, which has been commercialized or engineered so far. Since June 2016, China has started the construction of the urban maglev operating lines or maglev tourist lines in Beijing, Changsha, Qingyuan, and Fenghuang. Meanwhile, China has carried out innovative research on permanent and electric magnet (PEM) hybrid levitation technology and developed the experimental vehicles.

In the field of high-speed maglev trains, in December 2002, China and Germany cooperated to build a high-speed maglev commercial operating demonstration line from Shanghai Longyang Road to Pudong Airport. With the support of the 863 Program of the 10th Five-Year Plan, and the Science and Technology Support Program of the 11th Five-Year Plan and 12th Five-Year Plan, China has conducted the localization and innovative research on the high-speed maglev, and has developed single maglev bogie, double maglev bogies and a two-vehicle high-speed maglev test train based on the PEM hybrid levitation. The technology of the high-speed maglev trains has been rapidly promoted.

(2) EDS-type maglev system

For the EDS-type maglev systems, the levitation derives from the repulsive force generated by the interaction of source magnetic field and induced current, which is produced by the relative motion between the permanent magnet/superconducting magnet and the rail. According to the source magnetic field, the EDS-type systems can be divided into superconducting magnet type and permanent magnet type.

The EDS-type maglev technology is represented by the low-temperature superconducting magnet high-speed maglev trains of Japan and the permanent magnet low-speed maglev trains of USA. Initially, the research of the EDS-type maglev systems was mainly focused on superconducting magnets because the permanent magnet materials could not provide sufficient magnetic field.

With the adoption of permanent magnet materials with high remanent magnetic strength and the continuous application of Halbach-structured permanent magnet arrays in maglev technology, the permanent magnets can provide sufficient levitation force, which is gradually being introduced as an electrodynamic levitation technology based on the permanent magnet. Since the permanent magnet electrodynamic levitation technology does not require complex cooling devices, and it can achieve a high ratio of levitation to magnetic resistance at a certain speed, it becomes one of the maglev technologies with good development prospects.

In the USA, the applications of the permanent magnet electrodynamic levitation technology are typically represented by the Magplane maglev train, the GA Urban maglev train and the MIT Hyperloop. Professor Montgomery of MIT proposed the design of Magplane based on the permanent magnet electrodynamic scheme. Langley Aerothermodynamics Laboratory in the USA carried out the research of the permanent magnet electrodynamic levitation technology with the background of electromagnetic-assisted emission and developed the test vehicle for NASA's rocket boosting program.

In China, the Institute of Electrical Engineering of the Chinese Academy of Sciences, the National University of Defense Technology, the Southwest Jiaotong University, and the Naval Engineering University have carried out basic research work on the permanent magnet electrodynamic levitation technology one after another.

(3) HTS flux-pinning maglev system

The basic principle of HTS flux-pinning maglev system is to use the unique strong magnetic flux pinned ability of high-temperature superconductors in the external magnetic field, which makes it difficult for captured magnetic lines of force to escape from the pinned center, and difficult for uncaptured free magnetic lines of force to penetrate the superconductor. The pinned property enables the superconductor to induce a superconducting strong current that hinders this change in response to the change in the external magnetic field. This unique electromagnetic interaction achieves self-levitation and guidance of the levitating body at a macroscopic level without active control, and with zero magnetic resistance in the direction of operation, which in principle can achieve the high-speed stable operation. China, Germany, and Brazil have carried out relevant basic research and are currently in the verification stage of experimental line technology. In January 2021, the world's HTS flux-pinning high-speed maglev experimental prototype and the test line developed by the Southwest Jiaotong University was officially opened in Chengdu, Sichuan.

Among the various types of maglev technology, EMS is the only maglev transit technology that has entered commercial application stage. This book focuses on the EMS-type maglev transit system as the background and mainly studies the levitation and guidance control issues of the EMS-type high-speed maglev train.

1.2 The Development History of EMS-Type Maglev Trains Technology

The world's first maglev train levitation system was proposed by German engineer Kamper in 1922. In the 1970s, Germany began research on Transrapid (TR) maglev transit technology. And the initial TR01, TR02, and TR04 trains were also of its structure, advancing with the short-stator asynchronous linear motors for propulsion. But this structural solution was only suitable for the middle-low speed. Germany later abandoned Kamper's short stator and inverted U-shaped track scheme, and switched to the high-speed maglev technology with long-stator synchronous motor propulsion. Japan, Korea, and China insist on EMS technology, inverted U-shaped track structure, short-stator asynchronous linear propulsion, and outsourced vehicle structure.

In terms of the evolution of the structural scheme of the levitation system, the EMS-type maglev train has experienced the rigid body vehicle scheme, the magnetic wheel structure scheme and the modular structure scheme. From the view of the evolution of the structural scheme of the guidance system, the EMS-type maglev train mainly has passive guidance control scheme and active control guidance scheme. The evolution of the levitation system is described below.

1.2.1 Rigid Body Vehicle Scheme

The TR0l, TR02 and TR04, developed in Germany in the early years, have the following characteristics.

(1) A levitation system consisting of an inverted U-shaped track and a U-shaped electromagnet with lateral self-stabilization, eliminating the need for an additional active guidance system.
(2) Four electromagnets fixed directly on the carriage floor, which are a rigid structural constraint with no relative degrees of freedom of movement between the four electromagnets.
(3) A short-stator (on board) asynchronous motor is used for propulsion. The system structure is relatively simple and easy to implement. Later, Japan developed the HSST-01 maglev train based on the short-stator asynchronous motor.

In this period, it was thought that the concept of a rigid-free motion vehicle should be used to design the maglev train system by fixing four electromagnets directly on the carriage floor and using the concepts of yaw and pitch to describe and control the motion of the maglev train, such as TR01, TR02, TR04, MBB of Krauss-Maffei in Germany.

For lower operating speed, the contradictions of this structure are not prominent. However, when the speed is slightly increased, the problem becomes very serious. For example, Germany's TR04, originally designed to target 250 km/h, but in the experiment, serious vibration and swaying occurred with a speed less than 200 km/h. In addition, when the vehicle was levitated statically on the steel track beam, vibration instability occurred.

The Japanese HSST-01 and HSST-02 both adopted this concept, and the HSST-01 had the same shape as an airplane. A similar phenomenon occurred with the HSST-01, which led to the gradual realization that the concept of understanding the maglev train as a rigid-body free-degree vehicle was wrong.

The mistake is that although there is no mechanical contact between the electromagnet and the track, the motion of the electromagnet is strictly constrained by the track due to the levitation control system. According to the test results of the low-speed maglev trains, the control stiffness between the electromagnet and the track is above tens of thousands of N/m. It is due to this constraint that the vehicle body, as a whole rigid, suffers from at least the following problems.

(1) The track cannot be absolutely flat. The unevenness error of any local processing on the track will cause the adjustment effect of the point levitation controller. The whole vehicle body is a rigid body again.
(2) The four groups of controllers are coupled together and adjusting the parameters of anyone controller will affect the movement of the other levitation points. Thus, the four controllers of the whole train must be adjusted at the same time, which is very difficult to debug.
(3) Failures of any of the electromagnets or controllers under the carriage floor will result in levitation failure of the entire vehicle, thus causing a vehicle-wide failure.
(4) The two tracks where the train crosses the curve are not coplanar, resulting in conflicting levitation control objectives at each point.

Ultimately, the problem is the decoupling between the levitation points. Once the problem was identified, a flexible carriage floor was used to reduce the coupling between the electromagnets.

Japan Airlines developed HSST-02 based on HSST-01, and it did not achieve the expected results. The carriage floor adopts a flexible structure, and the connection between the electromagnets is flexible and can have a certain elastic deformation. Therefore, the coupling between the relative motion of the electromagnets has weaken a lot, and the decoupling effect is fine at low-speed operation, but when the speed adds up to 100 km/h, there is a more obvious vibration.

In addition, the structure of the maglev train at the 640 m line at Birmingham Airport in the UK in 1984 was similar to that of HSST-02, with electromagnets fixed to a carriage floor with a flexible base plate. The performance was moderate as the maximum speed of operation was only 40 km/h.

Since the carriage floor is also used to carry passengers or cargo, there are certain requirements for its overall stiffness, there is a limit to the use of flexible carriage floor decoupling structures. No application has been seen other than a demonstration of small maglev model vehicle with very low speed.

The small maglev experimental vehicle developed by the National University of Defense Technology in 1989 used the structure with a decoupled flexible carriage floor.

1.2.2 Magnetic Wheel Structure Scheme

After the TR04 failed to achieve the expected results, the German companies Krauss-Maffei and MBB Aerospace decided to join forces at the end of 1974 to revisit the concept of maglev trains. They abandoned the Krauss-Maffei structure in favor of the MBB structure with active lateral control and proposed a hierarchical progressive system concept based on the "magnetic wheel" concept (Fig. 1.2). To test whether this levitation system structure could meet the requirements of high-speed operation, they developed the test system Komet, which was propelled by rocket technology

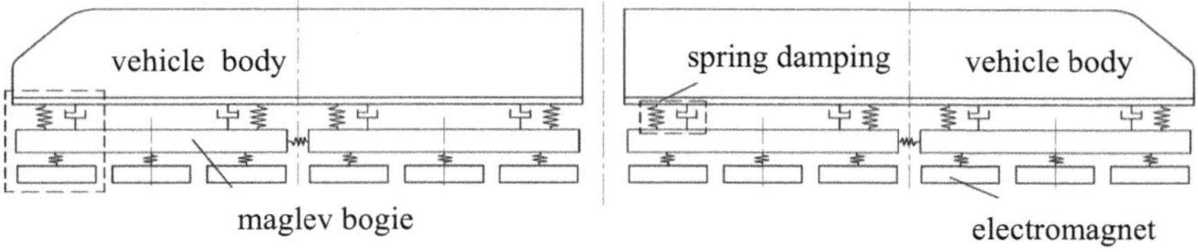

Fig. 1.2 Hierarchical progressive structure based on the "magnetic wheel" concept

and could reach a speed of 401.3 km/h (in 1976). The test achieved the expected results, which confirmed the necessity and feasibility of this mechanical decoupling system.

In this structure, each electromagnet (levitation or guidance) is individually connected to the maglev bogie by a spring damping, and the electromagnet has control freedom in only one direction with respect to the bogie (Fig. 1.3). Since each electromagnet acts like a wheel for support, it is called a "magnetic wheel". Each maglev bogie has three independent (motion and control) electromagnets on each side, allowing for lateral and yaw-wise movements relative to one another. This structure completely ensures decoupling of the motion between the electromagnets.

In parallel with the development of the hierarchical progressive control structure, Thyssen developed the HMB2 maglev train with the long-stator synchronous motor propulsion in 1976, which is 5 m, weighs 2.5 t, and can operate at speeds up to 36 km/h.

This was followed by Germany's combination of the levitation and guidance structure of the hierarchical progressive "magnetic wheel" concept and long-stator propulsion technology, resulting in the TR05 system.

The "magnetic wheel" structure completely ensures the decoupling of the motion between the electromagnets and the ability of the train to pass through the curves. It can be considered that the "magnetic wheel" concept has gone from one extreme

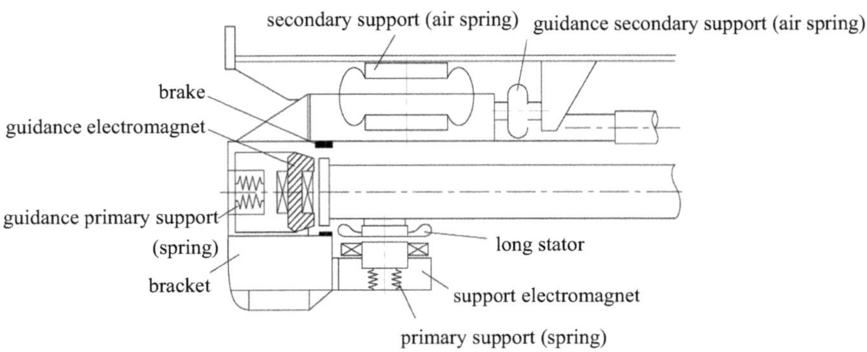

Fig. 1.3 Principle of levitation and guidance in the "magnetic wheel" structure

to another after the "rigid body vehicle" concept hit a wall. Although it solves the problem of mechanical decoupling very well, it brings some new problems as follows.

(1) As each "magnetic wheel" independently support the vehicle body, although the whole vehicle will not occur the phenomenon of overall vibration, the individual "magnetic wheel" is susceptible to levitation instability phenomenon.
(2) The failure of any one of the electromagnets or controllers of the "magnetic wheel" will cause the levitation and guidance to fail, and there is no redundant relationship between the "magnetic wheel" and the "magnetic wheel", thus causing a total vehicle failure.
(3) The vehicle based on the "magnetic wheel" concept is not compact enough structurally and is prone to mechanical failure during the high-speed operation, making it difficult to engineer.

1.2.3 Modular Structure Scheme

After theoretical analysis and engineering tests of the "rigid body vehicle" concept and the "magnetic wheel" concept in Germany, a compromise between the two was made and an "electromagnetic module"-based vehicle structure concept was gradually established. This concept was adopted in both TR07 and TR08. Figure 1.4 is a schematic diagram of the vehicle structure based on the "electromagnetic module" (the part circled with a dashed line in the figure) concept, without considering the joint structure between the two vehicles, each high-speed maglev vehicle has seven sets of independent electromagnetic modules on each side, which is a mechatronic functional structure formed by the combination of bracket arms, levitation electromagnets, guidance electromagnets, levitation sensors, guidance sensors, connecting parts, etc.

With the "electromagnetic module" concept, the vehicle system has the following advantages.

(1) The mechanical decoupling of each control point can be achieved by the displacement deformation of the primary hinge or primary spring and the torsional stiffness of the maglev bogie itself.

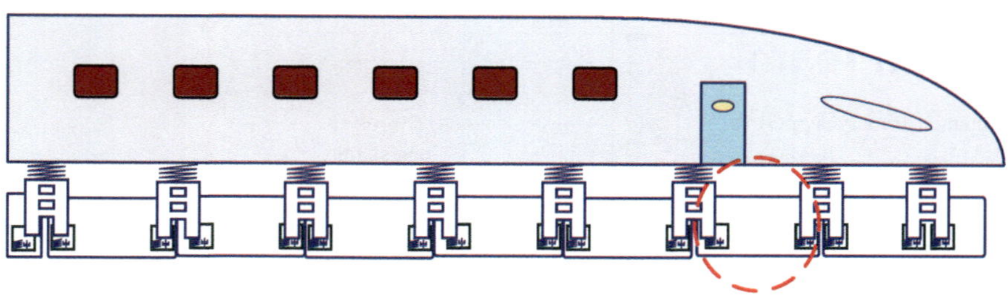

Fig. 1.4 Side view of the high-speed maglev vehicle based on "electromagnet module + joint structure"

(2) The levitation function and guidance function corresponding to each bracket arm are redundant, ensuring the functional safety of the system.
(3) The vehicle substructure is compact, which is conducive to the high-speed operation and easy to install and maintain. Also, it meets the needs of engineering and practical application.
(4) Due to the multiple degrees of freedom of movement provided by the connecting parts between the electromagnet and the maglev bogie, the vehicle is ensured to have a good curve passing capability.

The successful operations of the TR07 and TR08 have demonstrated that the "electromagnetic module" concept is mature and viable.

1.3 Research Status and Main Achievements of High-Speed Maglev

The main countries engaged in the research of maglev transit technology are Germany, Japan, China, UK, Korea, and USA. The research and application fields mainly relate to the urban maglev, the high-speed, and the ultra-high-speed maglev transit. The staggered development of the urban maglev technology and the high-speed maglev technology drives the exploration and research of ultra-high-speed pipeline maglev.

The urban maglev train adopts non-contact operation mode. Compared with ordinary inner-city wheel-track trains, it has the advantages of low noise, low vibration, small turning radius, relaxed line laying conditions, low construction cost, easy implementation and easy maintenance, etc. Moreover, because its propulsion force is not affected by the adhesion coefficient between wheels and rails, the urban maglev train also has the characteristics of strong climbing ability.

The high-speed maglev transit system mainly solves the problems of transit between big cities. And now Germany, Japan, and China are the main countries to develop the high-speed maglev technology.

1.3.1 Progress of High-Speed Maglev Research in Germany

The German high-speed maglev transit system uses EMS technology. In 1971, the first German maglev prototype vehicle was tested on a 660 m-long test line. In 1975, Thyssen Henschel pioneered the operation of line side long-stator linear synchronous motor driven maglev train on the HMB1 test line at the Kassel plant. In 1976, Thyssen Henschel carried out the operation of a manned long-stator test vehicle on the HMB2 test line.

In 1979, the TR05 maglev railroad demonstration line with a track length of 900 m and a maximum designed speed of 100 km/h was exhibited at the Hamburg

International Transport Fair for passenger operation, and the demonstration speed reached 75 km/h.

The German industrial circles formed the Transrapid consortium to build the Transrapid test line (TVE) in Emsland, northwestern Germany. The first phase included a 21.5 km-long test line, test center, and test vehicle TR06. The first phase was completed from 1979 to 1984. And the first section of the test line was officially put into operation on June, 1983.

From 1986 to 1989, Thyssen led the development of the application-oriented maglev train TR07. An independent, comprehensive evaluation and qualification of the Transrapid system was led by the Federal Railroad Central Bureau. After nearly two years of evaluation and qualification, it was concluded at the end of 1991 that the system was technically mature for its application. At this point, it can be said that the technology of high-speed EMS maglev train is mature enough to enter the line selection and construction phase of commercial operation line project. In 1993, the TR07 maglev train reached a maximum speed of 450 km/h on the TVE test line.

In September 1999, the TR08 (in three-section formation) was delivered to the test line and began tests in October, reaching a maximum speed of more than 400 km/h.

1.3.2 Progress of High-Speed Maglev Research in Japan

Japan began research on EDS-type high-speed maglev technology in 1962. The first vehicle model, ML100, was successfully verified in 1972, while a 7 km-long test line was opened in Miyazaki City. In 1979, the ML-500 set a speed record of 517 km/h. Subsequently, two test vehicle models, MLU001 and MLU002N, were tested in succession.

In 1991, the Yamanashi Line began to construct. The vehicle model of this test line was MLX001, which left the factory in 1995. The line, also completed in 1997, is 18.4 km long, with a maximum grade of 4%, and a minimum turning radius of 8000 m, and the tunnel length accounts for approximately 80% of the total length, all of which are substantial improvements over the Miyazaki City test line.

The first maglev train was successfully tested on the Yamanashi line in 1997, and the maximum speed was then continuously increased, achieving 550 km/h by the end of that year. At the same time, the multi-vehicle train tests and control tests have started one after another. In 2000, an evaluation by the Japanese Ministry of Transportation stated that superconducting maglev technology could already be applied to high-volume transit systems.

To get into service as soon as possible, research since 2000 has focused on reliability verification, system durability, cost control, and improvements in body streamlining. A new group of vehicle models MLX01-901 was put on Yamanashi Prefecture in 2002. The speed records were continuously broken in the following years, and the tests also focused on testing the reliability of the system and the vehicle over long periods and long distances.

From 2006 to the present, this phase began to focus on the commercial application of superconducting maglev, and the design of the vehicle models began to consider the transport of passengers. In 2010, the L0 was confirmed as the final vehicle model, after which L0 was put into mass production. At the end of 2011, the technology standard of superconducting maglev was also adopted by the competent authority.

In 2013, the Yamanashi Prefecture test line was opened with a total of 42.8 km and started running L0 series vehicles. On December 17, 2014, the world's first superconducting maglev high-speed railroad with a maximum speed of 505 km/h officially started construction. On April 16, 2015, at the Yamanashi Maglev test line, a high-speed operation test was conducted using the "L0 series" superconducting maglev train, which reached the world's highest speed of 590 km/h for manned travel. In 2015, Japan conducted a high-speed maglev operation test with a maximum speed of 603 km/h, setting another world record in railroad history. In December 2014, the Chuo Shinkansen, which used L0 series maglev technology, began construction and is scheduled to open the Tokyo-Nagoya section in 2027 and the Tokyo-Osaka section by 2045.

1.3.3 Progress of High-Speed Maglev Research in China

In December 2000, China decided to build a high-speed maglev transit line from Longyang Road Metro Station to Pudong International Airport in Pudong, Shanghai. Construction officially started in March 2001, and a three-section high-speed maglev train was tested on a single line in January 2003. In September 2003, the entire line was opened, and on December 31, 2003, the entire line was completed for examination and acceptance. The line is about 30 km, using the German TR08 high-speed maglev train technology, and has been operating safely for nearly 20 years, as shown in Fig. 1.5.

With funding from the 863 Program of the 10th Five-Year Plan and the Science and Technology Support Program of the 11th Five-Year Plan and the 12th Five-Year Plan, the Tongji University National Maglev Engineering Research Center has made significant studies in the localization of high-speed maglev trains. During

Fig. 1.5 Shanghai high-speed maglev train

Fig. 1.6 Tongji University high-speed maglev test line and PEM hybrid maglev train

this period, the National University of Defense Technology has carried out research on PEM hybrid levitation system, and gradually completed the development and commissioning of PEM hybrid levitation control system of single bogie, double bogie, single vehicle and two-section high-speed maglev test vehicles. In November 2017, the project task of the National Science and Technology Support Program of the 12th Five-Year Plan, "Research on levitation, guidance and eddy current brake controllers of the PEM hybrid high-speed maglev train", successfully passed the acceptance of the project. The localized PEM hybrid high-speed maglev train (as shown in Fig. 1.6) has completed the basic function tests such as levitation and guidance, and completed the propulsion operation test on the 1.5 km-long high-speed maglev test line in Tongji University. Due to the limitation of the length of the test line, the speed-related operational tests are not sufficient, so the main task in the next phase will be to improve the research on the levitation system of high-speed maglev trains based on the existing research and to promote the commercial operation of localized high-speed maglev trains.

With the support of the National Key Research and Development Program during the 13th Five-Year Plan, CRRC Qingdao Sifang Co., Ltd. took the lead in developing the first engineering prototype of a high-speed maglev train with a design speed of 600 km/h on May 23, 2020. On July 20, 2021, a five-carriage set of high-speed maglev train with a design speed of 600 km/h was officially unveiled and successfully underwent integrated testing on a 665 m test line in Qingdao. And the National University of Defense Technology provided the technology of levitation and guidance control and positioning test.

Bibliography

1. Chang W, She L (2003) Maglev technology development and automatic control. In: The 22nd Chinese control conference, vol 8, pp 27–30
2. Ding S (2022) 600 km/h high-speed maglev transportation system. Shanghai Scientific and Technical Publishers, Shanghai
3. Han H-S, Kim D-S (2016) Magnetic levitation. Springer, Berlin
4. Huaqing Liu Compile (1995) The German maglev train-Transrapid. University of Electronic Science and Technology Press, Sichuang
5. Lin G, Liu W, Xu J et al (2023) Opportunities and challenges for the development of high-speed maglev transportation in China. Sci Technol Foresight 2(4):7–18
6. Liu Z, Long Z, Li X (2015) Maglev trains. Springer, Berlin
7. Long Z, Li X, Cheng H (2023a) Research on permanent-electromagnetic levitation technology and application. Shanghai Scientific and Technical Publishers, Shanghai
8. Long Z, Zhai M, Wang Z (2023b) Condition monitoring, fault diagnosis and fault tolerant control of the maglev train. Shanghai Scientific and Technical Publishers, Shanghai
9. Masuda K, Fujie J, Kato J (1992) Maglev technology for railways. Japan OEM Company
10. Sawada K (1996) Development of magnetically levitated high speed transport system in Japan. IEEE Trans Magn 32(4):2230–2235
11. Sinha PK (1987) Electronmagnetically suspension dynamics control. Peter Peregrinus Ltd., London
12. Tandan GK, Sen PK, Sahu G et al (2015) A review on development and analysis of maglev train. Int J Res Advent Technol 3(12):14–17
13. Wang Y, Zhang K, Zhang H et al (2023) Characteristics and adaptability of different types of high-speed maglev transportation systems. Sci Technol Foresight 2(4):19–30
14. Wei Q, Kong Y, Shi J (2010) System and technology for maglev transit. China Science and Technology Press, Beijing
15. Wu X (2003) Maglev train. Shanghai Scientific and Technical Publishers, Shanghai
16. Yan L (2002) Consideration on the development strategy of high-speed maglev in China. Eng Sci 4(12):40–46
17. Zhang K (2021) High-speed maglev railway technology. China Railway Publishing House, Beijing
18. Zheng Q, Ni F (2023) High-speed maglev transportation: from Germany to China. Sci Technol Foresight 2(4):31–39

Chapter 2
Technology Analysis of EMS High-Speed Maglev System

The EMS-type high-speed maglev system consists of four major components, including maglev line, vehicle system, propulsion and power supply system, and operation control system. The maglev line is one of the main components of the maglev train that utilizes electromagnetic force to achieve support and guidance functions. The vehicle system is a complex system that includes several subsystems, including the vehicle body, maglev bogie, levitation and guidance, power supply, operation control and fault diagnosis, safety brake, and positioning speed measurement. Among them, the levitation and guidance system provide the levitation and guidance functions for the maglev train to achieve its safe and reliable operation. Taking the EMS-type high-speed maglev system as the object, this chapter analyzes the basic components of maglev line, vehicle system, propulsion and power supply system, and operation control system in the high-speed maglev transit system; gives the basic structure of the high-speed maglev vehicle system, analyzes the functional characteristics of components of the high-speed maglev vehicle system; focuses on the basic structure, technical requirements and control scheme of the levitation and guidance system of the high-speed maglev train, respectively.

2.1 Basic Components of EMS-Type High-Speed Maglev System

The EMS-type high-speed maglev system consists of four major parts, including maglev line, vehicle system, propulsion and power supply system, and operation control system. The analysis for each part is presented in detail as follows.

Z. Long et al., *High-Speed Maglev Train's Levitation and Guidance Control*, Advances in High-speed Rail Technology, https://doi.org/10.1007/978-981-97-2309-6_2

2.1.1 Maglev Line

Maglev trains mainly rely on electromagnetic force to achieve support, guidance, and propulsion functions, and one of the main components to achieve this function is the maglev line. The maglev line consists of two parts, one is the guideway supporting beam, which is the load-bearing structure, and the other is the guideway equipment (mainly functional components) attached to the beam.

To ensure the safety and comfort of maglev trains during high-speed operation, the high-speed maglev system imposes high design and manufacturing requirements on the guideway structure.

The deformation requirements: the deformation and deflection of the guideway structure should be controlled within a small range under the action of the train load and external environmental influences.

The dynamic characteristics requirements of guideway supporting beam: the vehicle-track coupling resonance may occur when the high-speed maglev train is running at high speed, passing at low speed or levitating statically. Therefore, the parameters related to the dynamic characteristics of the guideway beam should be controlled. Whether it is a wheel-track railway bridges or a maglev guideway beam, the first-order modal frequency of the beam body is strictly controlled in the design.

The accuracy requirements of guideway functional plane: to ensure that the maglev train runs smoothly with a high speed along the guideway, the manufacturing accuracy of the line is necessary to meet the technical requirements of electromechanical products. The unevenness of the maglev line is mainly reflected in the geometric tolerance of the functional plane. Geometric tolerance mainly includes two aspects: one is the tolerance that can be directly measured, such as position deviation, mutual misalignment, mutual gap, guideway width, clamp distance, deflection angle, etc. The other is the tolerance derived from the data obtained from the measurement and certain processing calculation, such as long-wave deviation, short-wave deviation, standard rate of slope change, etc.

(1) Guideway supporting beam

The guideway beam guides the train in the forward direction, bears the train load, and transmits it to the pier foundation, and has functional components such as stator cores and lateral guide planes installed on its upper side. The high-speed maglev line has strict requirements on various indexes of the guideway beam structure.

The main types of guideways supporting beam include low-set integral guideway supporting beam, elevated integral guideway supporting beam, composite beam, maintenance base steel beam, and bridge guideway supporting beam.

When the line runs across large rivers or other large plane obstacles, the guideway supporting beam used on the general section can be erected on large bridge structures to solve the problem of implementing large span structures that cannot be crossed by conventional length guideway supporting beam along the line, as shown in Fig. 2.1.

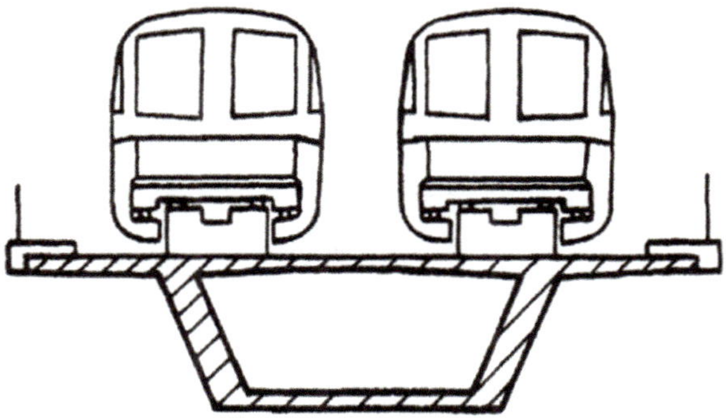

Fig. 2.1 Cross section of guideway supporting beam on bridge over general section

(2) Functional component

The functional components are arranged at the upper two ends of the guideway supporting beam to realize the functions of supporting, guiding and driving of the maglev train. They include top sliding plane, side guidance plane, stator core, power supply rail, etc., as shown in Fig. 2.2.

(3) Switch

The connection equipment for vehicles to turn from one guideway or cross to another is called switch, which is an indispensable device for trains to change forward direction. The high-speed maglev switch is a flexible bending beam segment divided into several sections, and the lateral moving power device is installed at its connection. Then the moving distance of each device is precisely controlled according to the switch rutting needs.

Figure 2.3 shows the schematic diagram of the double-open switch. There are 6 piers under the steel beam, among which the switch base is set on pier number zero and the moving crossbeam is set on pier number 1–5, which can make the switch move along the fixed slide of the crossbeam. Except for pier number zero, positioning and locking devices are set on the remaining supports to ensure that the switch steel beam can be bent to the designed position and can be fixed effectively.

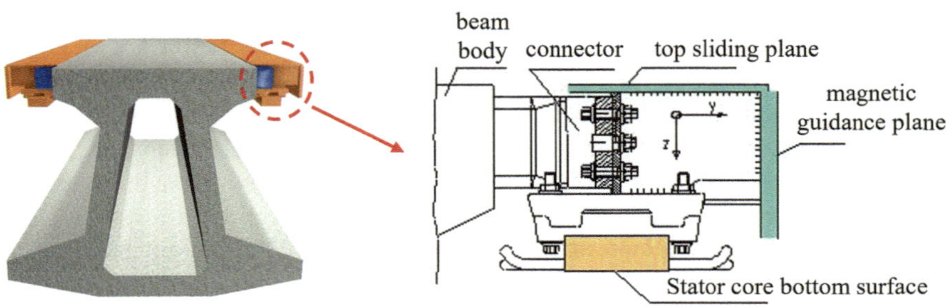

Fig. 2.2 Functional components

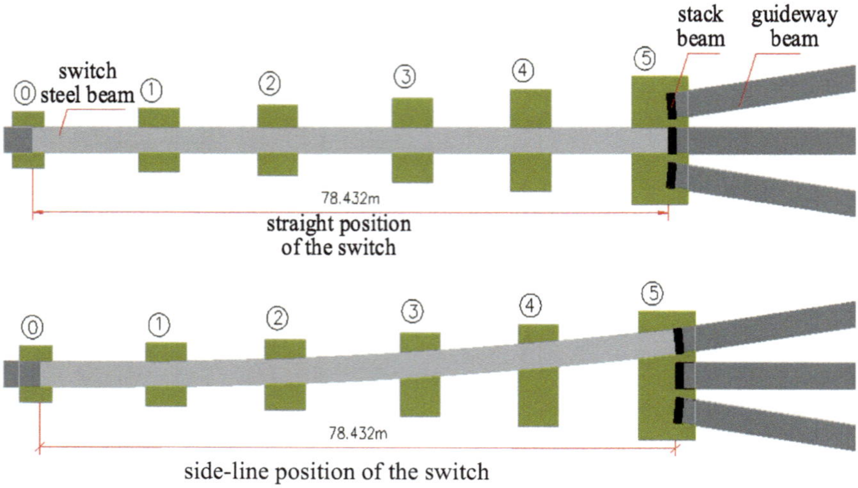

Fig. 2.3 Schematic diagram of the double-open switch structure

(4) Bridge infrastructure

Bridge infrastructure includes bridge piers and foundations. The pier structure of the elevated structure generally adopts one-column pier, two-column pier, portal pier, etc. The low-set structure generally adopts separated reinforced concrete column, wall column, etc.

2.1.2 Vehicle System

The high-speed maglev train is a complex system. This section first introduces the overall structure of the vehicle and then introduces each main subsystem of the vehicle according to its structure and functions.

Taking German TR08 high-speed maglev train as an example, the overall structure of the vehicle system of the single-section train is shown in Fig. 2.4. The main structure of the vehicle system includes carriage, secondary levitation device, bogie, primary levitation device, and electromagnet. Among them, the carriage and the bogie are connected by the secondary levitation device. The electromagnet and the bogie are connected by the primary levitation device, and the electromagnet and its corresponding primary levitation device are divided into the levitation, the guidance, and the eddy current brake.

The vehicle system adopts the assembly principle of bottom-up hierarchical levitation. The bogie is the installation and positioning component of levitation/guidance/brake electromagnets and secondary levitation device. Each electromagnet is connected to the bracket arm of the bogie through two sets of front and rear primary

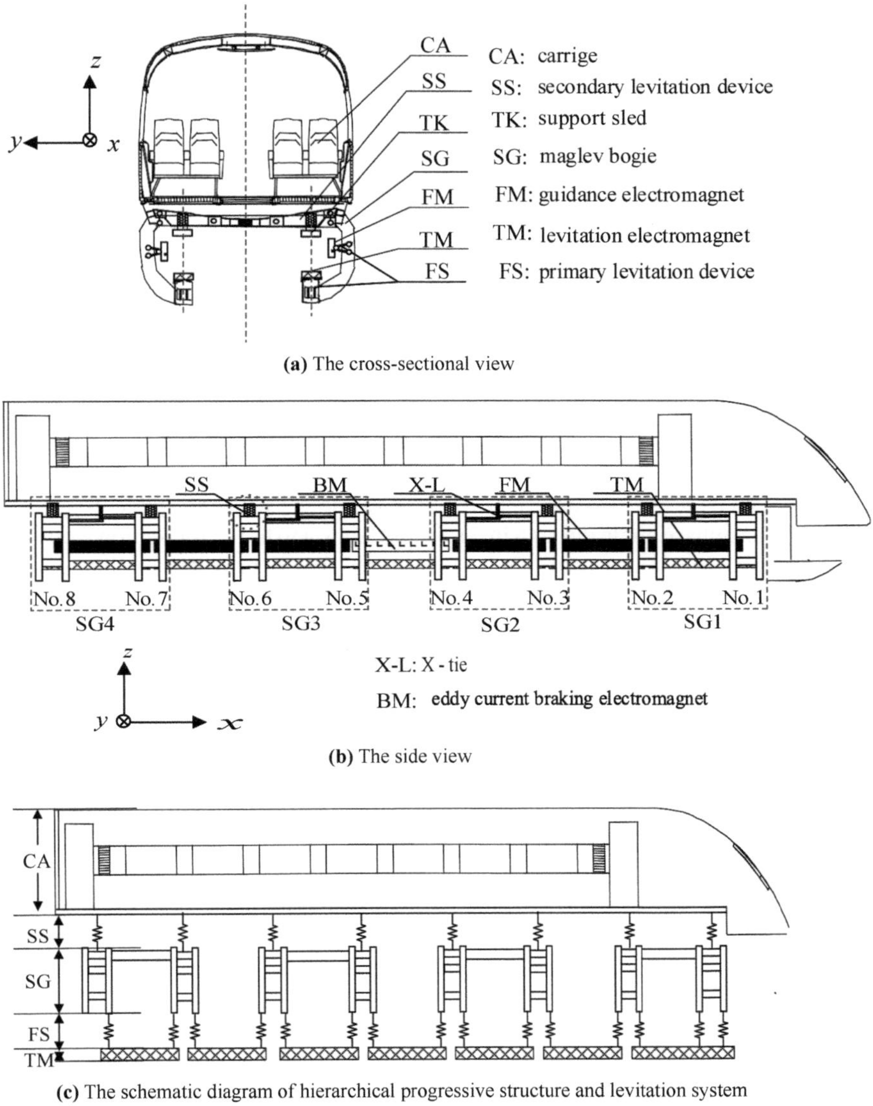

Fig. 2.4 Schematic diagram of the overall structure of TR08 high-speed maglev train

levitation devices, which is used to realize the transfer of force and compensation of relative motion between the electromagnet and the bogie. Each bogie is connected to the carriage floor through four sets of secondary levitation devices, which is used to realize the transfer of levitation and guidance force between the carriage and the bogie and the compensation of relative motion, inhibit the rolling of the carriage relative to the bogie, and play the role of vibration damping and mitigation for vertical and lateral vibration and impact. In addition, each bogie is connected to the carriage floor by one X-tie, which is used to transfer propulsion and brake force between the carriage and the bogie.

2.1.3 Propulsion and Power Supply System

(1) Power supply system

After receiving power through the high-voltage substation, the power supply system provides propulsion and power to the power supply equipment. The high-voltage substation reduces the 110 kV or 220 kV power supply to 20 kV medium voltage power supply through the main transformer, and then supplies power to the propulsion and power supply system and the auxiliary power supply system. The propulsion and power supply system rectifies 20 kV medium voltage into DC power after voltage reduction, and then converts it into three-phase AC power with adjustable frequency, amplitude, and phase angle through inverter, and feeds it to the stator section along the line for train propulsion or brake. The power supply system provides auxiliary power supply for the propulsion system, and sends the 20 kV power supply of the high-voltage substation to the track-side transformer substation along the line through the 20 kV ring network cable. The track-side transformer substation will supply power to the wireless communication equipment along the line, the auxiliary power supply of the stator switch station, the switch, and the power rail along the line after 20 kV voltage reduction or rectification.

Divided from the equipment structure unit, the high-voltage substation includes the 110 kV circuit breaker, the main transformer, the 20 kV feeder circuit-breaker, the filter, and the dynamic reactive power compensation device. The propulsion and power supply system is composed of the propulsion substation, the feeder cable, the stator switch station and the stator section coil. The power supply system is composed of the station transformer substation, the track-side transformer substation, the medium voltage ring network cable, and the power rail system.

(2) Propulsion system

The propulsion system mainly includes the high-power converter system and the propulsion control system. The task of the propulsion system is threefold, first is to control the amplitude, frequency and phase of the output voltage and current of the high-power power electronic converter system in real time according to the requirements of the operation control system; the second is to supply power to the stator section of the long-stator linear motor where the maglev train is located in sections; and the third is to implement the propulsion and brake of the train. The structure of the propulsion system of the German TR series high-speed maglev transit adopted by the Shanghai high-speed maglev line has the following main characteristics.

① All system equipment are set on the ground. The energy conversion and control of the propulsion power are realized through the fixed equipment.
② To reduce the line loss, the power supply of the stator coil of the long-stator synchronous linear motor adopts subsection mode. Sectionalized switching is implemented through switches along the track-side of the line.
③ Through the substations at both ends of the line, the long stator of the motor on the main line is powered at the same time, that is, double feeding for reducing line loss.
④ According to the line conditions and the requirements of the departure interval of the operation organization, the whole line needs to be divided into a number of different propulsion and power supply segments. Each propulsion segment can only run one train, and the length of the propulsion segment varies from several kilometers to tens of kilometers. The propulsion substations are usually set at both ends of propulsion segments.
⑤ According to the requirements of the departure interval of the operation organization, the substation can be configured as I-type or II-type. The propulsion substation that can supply power to the left and right sides of the substation at the same time is II-type propulsion substation, and the propulsion substation that can only supply power to the left or right side of the substation alternately is I-type propulsion substation.
⑥ According to the power requirements of different propulsion segments, the converter equipment with different power levels shall be adopted. It includes the high-power power electronic converter equipment (up to tens of megawatts), the medium-power power electronic converter equipment (up to several megawatts), and the low-power power electronic converter equipment (hundreds (thousands) of watts).

The propulsion power module includes different number of converter units according to the different stator stepping methods adopted. The propulsion power module is usually composed of the input switch device, the input transformer, the converter unit and its cooling system, the output transformer, and the output switch device. The propulsion power module operates under the control of the propulsion control system, and the specific composition is shown in Fig. 2.5.

(3) Track-side facilities

The track-side facilities include the track-side switch station, the power rail, and the track-side feeder cable, etc.

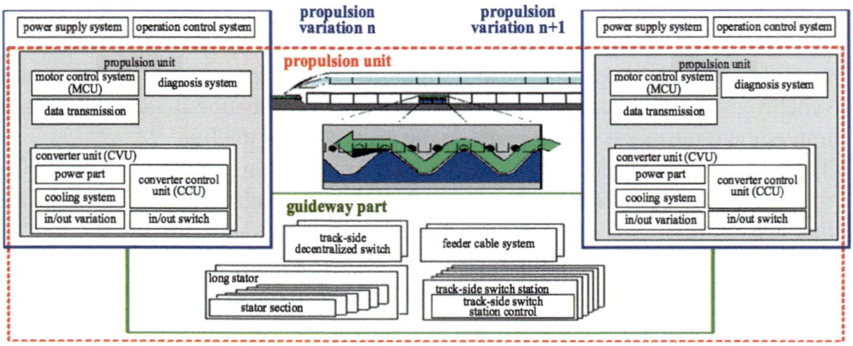

Fig. 2.5 Structure of the propulsion and power supply system

2.1.4 Operation Control System

The operation control system is connected to the vehicles, propulsion, line and switch equipment or systems of the maglev transit system through advanced technologies such as computer control, computer network, communication, and information processing, to complete the control, safety protection, automatic operation, and scheduling management of the train operation.

The operation control system of the high-speed maglev train can be composed of three-layer structure and four sets of systems. The three-layer structure refers to the centralized control layer, the decentralized control layer, and the vehicle control layer. The four sets of systems refer to the centralized control system, the decentralized control system, the vehicle control system, and the communication system, as shown in Fig. 2.6.

Taking the "Shanghai High-speed Maglev Demonstration Operation Line" as an example, the centralized control system mainly includes the automatic train operation system (ATO), the operator terminal system (OTS), and the diagnostic terminal system (DTS), etc.

The equipment of the decentralized control system mainly includes the decentralized control computer (DCC), the decentralized safety computer (DSC), the decentralized transmission computer (DTC), the decentralized propulsion shut-off computer, and the decentralized switch module (DSM), etc.

The vehicle control system is the operation control system located on the maglev train, which includes the vehicle safety/transmission computers (VSC1/VTC, VSC2) located at both ends of the train. Under the normal operation mode, the main functions of the vehicle system are performed by the vehicle safety computer 1 (VSC1), while the vehicle safety computer 2 (VSC2) is in the hot standby state.

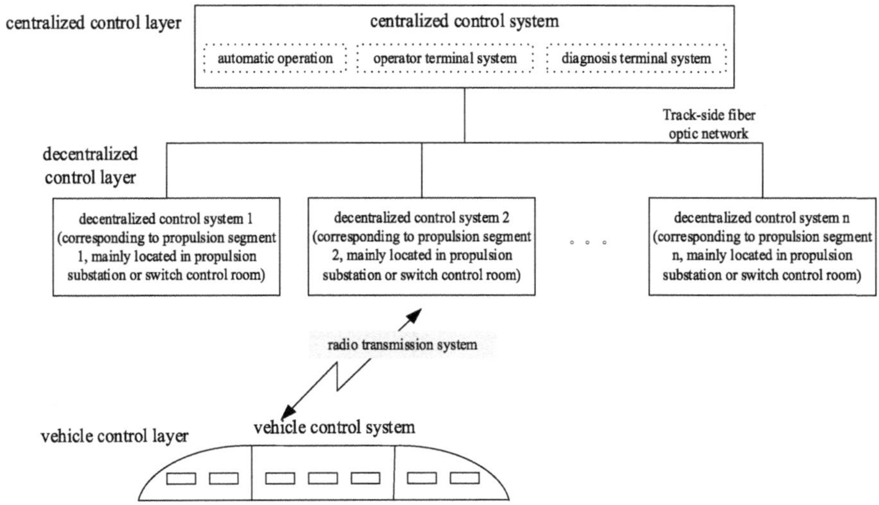

Fig. 2.6 Basic composition structure diagram of the operation control system

The communication system is a distributed and transparent information transmission network, which has the characteristics of safety, reliability, high speed, redundancy, and so on. The communication system is composed of three subcommunication networks, namely the operation control core network (also known as the wide area transmission network), the decentralized protection communication network (i.e., interlocking bus, IL-BUS), and the radio transmission network (i.e., 38 GHz radio transmission system).

The operation control core network is used for information transmission between the ground centralized control system and the decentralized control system, using the optical fiber transmission medium. The interlocking bus is included in the decentralized control system and adopts the optical fiber transmission medium.

The 38 GHz radio system includes the centralized radio system located in the control center (also known as the centralized radio control unit, CRCU), the decentralized radio system located in the decentralized operation control room (also known as the decentralized radio control unit, DRCU), the mobile radio system located on the maglev train (also known as the mobile radio control unit, MRCU), the radio base stations located at track-side, the optical fiber networks and other equipment connecting these base stations.

As a safety control and protection system, the operation control system shall comply with relevant safety principles and relevant international or national standards in the process of design, manufacture, use, and maintenance. One of the most basic safety principles followed by the operation control system provided by Siemens is the "fail-safe" principle, which requires the system to avoid the disastrous consequences of train operation caused by system failure within the normal operation life under the guidance of correct operation and maintenance specifications.

2.2 Technology Analysis of High-Speed Maglev Vehicle System

2.2.1 Vehicle Body Structure

There are two types of vehicle body of the maglev train: end vehicle body and intermediate vehicle body, which are roughly the same in structure. The vehicle body includes the carriage and the sandwich structure under the carriage. The carriage is mainly composed of several parts, such as underframe-side wall structure, roof, end wall structure and windows, designed to accommodate passenger seating, air conditioning units, and other equipment to provide the interior environment for carrying passengers. The sandwich is an aluminum structure with plug-in electrical equipment chassis, air compressors, controller and power cables, etc. The end vehicle includes the vehicle head structure.

2.2.2 Maglev Bogie

The maglev bogie is the support part of the entire vehicle, connecting the electromagnet module of the levitation, guidance, and safety brake system to the vehicle body, as shown in Fig. 2.7. It transmits the levitation, guidance, brake, and propulsion forces to the vehicle body part, and changes with the change of the guideway during operation. The maglev bogie largely determines the requirements of the line, the ability of the train to adapt to the line, and the comfort of the ride.

Structurally, the maglev bogie is mainly composed of the following parts: the secondary levitation device, the electromagnet connection device, and the skid device.

Functionally, the main function of the maglev bogie is to support the vehicle body and the electromagnet, and to transmit the vertical, longitudinal and lateral forces generated between the electromagnet and the guideway to the vehicle body.

2.2.3 Levitation and Guidance System

The levitation and guidance system of the high-speed maglev train allows for contactless levitation and guidance of the vehicle along the operating line. The levitation system provides the vertical force to support the vehicle levitating, while the guidance system provides the lateral force to align the vehicle centerline with the guideway centerline. To maintain a certain gap between the functional planes of the levitation and guidance electromagnets and the corresponding functional planes of the guideway, the magnitude of the levitation and guidance forces can be controlled by

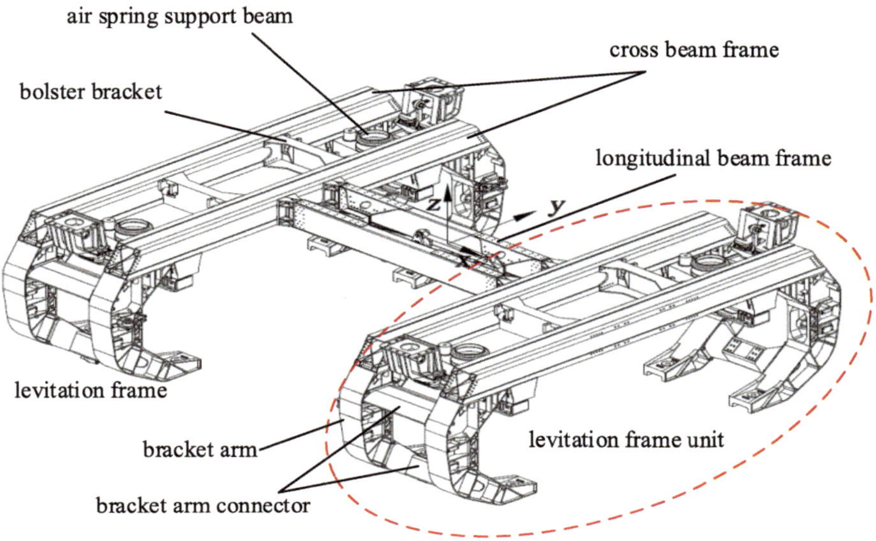

Fig. 2.7 Maglev bogie

adjusting the current in the excitation coils of the levitation and guidance electromagnets. The levitation and guidance electromagnets are arranged along the length of the whole train, and each vehicle has four bogies, and each bogie has four levitation frames, as shown in Fig. 2.4b. There are eight levitation frames on each side of a vehicle. The levitation and guidance electromagnets are arranged between two adjacent levitation frames, i.e., on two adjacent bracket arms of the bogie.

A levitation electromagnet and a guidance electromagnet mounted on the same two bracket arms form a basic electromagnet module unit. The schematic diagram of the joint-based levitation system is shown in Fig. 2.8a, and the schematic diagram of the guidance system is shown in Fig. 2.8b.

2.2.4 *Vehicle Power Supply System*

The main function is to provide electric power supply for the electric equipment on board, such as safe and reliable power supply for the vehicle control system, the levitation and guidance system, the brake system, the lighting, the air conditioners, and other equipment. When the train is running, the vehicle power grid provides the train with the electric power required for its normal operation and fault operation.

The maglev train has no contact with the ground when it is running at high speed. To provide electric energy to the train when it is running, the following power supply scheme is adopted: when the train is running at low speed, the ground power supply rail and the vehicle linear generator jointly supply power to the vehicle; when the

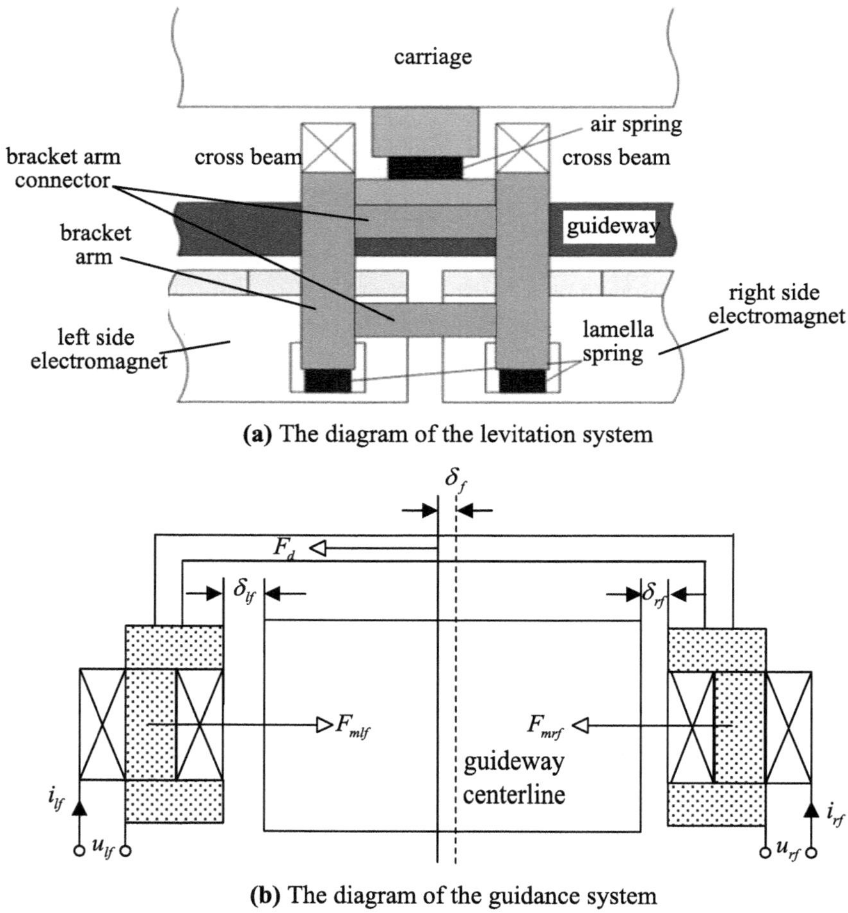

Fig. 2.8 Schematic diagram of the levitation and guidance system

train is running at a higher speed, the power is mainly supplied by the vehicle linear generator; when the train is in the case of emergency or failure, the vehicle battery is used as the power supply. When the eddy current brake is used for emergency brake, the electric energy of the high-speed operation section is provided by the linear generator. When the electric energy of the linear generator cannot meet the brake needs at a lower speed, the electric energy is provided by the battery and the linear generator jointly. And the electric energy of the power supply rail is not used during the emergency brake.

There are three main voltage levels of power grid on the maglev train, namely 440 V DC grid, 24 V DC grid, and 230 V AC grid. The train also uses several voltage levels of 28 V, 15 V, 12 V, and 5 V DC power supplies, which are converted by low-voltage switching power supplies.

2.2.5 *Vehicle Control and Diagnosis System*

The vehicle control is used to assign safety-related control commands from the vehicle operation control system to the corresponding train components and to transmit the safety-related monitoring signals generated by each vehicle component in the train back to the vehicle operation control system after certain processing. The safety-related functions include levitation, guidance, safety brake, opening/closing doors, vehicle power grid control, fire alarm, etc.

The vehicle diagnosis system implements train diagnostic functions and some non-safety related operation functions on the train. These functions are integrated in a two-level diagnostic network consisting of CAN fieldbus and Ethernet.

2.2.6 *Safety Brake System*

The main function is to achieve controlled and safe stopping of the train when the system fails, and it is difficult to operate and brake normally. The safety brake system is used to brake the vehicle in case of brake failure of the propulsion system or other emergencies, so that the vehicle can safely drive to the designated stopping area. The eddy current brake is used for safety brake when the train speed is above 10 km/h. And the brake of the train below 10 km/h is done by the skid.

Each vehicle has two eddy current brake electromagnets, which are located on the left and right sides of the middle of the vehicle. The total length of the brake electromagnet is about 3 m, of which the active section of the brake electromagnet is 2 m. Each brake electromagnet consists of 12 magnetic poles, which are fixed to the backrest of magnetic poles by fastening devices. The pole distance is 168 mm. And these magnetic poles are distributed along the running direction of the vehicle. The side of the eddy current brake electromagnet facing the lateral guide rail has a wear plate, which separates the eddy current brake electromagnet from the lateral guide rail, thus avoiding wear on the plane of the eddy current brake electromagnet.

The eddy current brake electromagnet is connected to the levitation frame by an articulation in the X-direction and a flexible articulation in the Y-direction. The eddy current brake electromagnet is connected with the levitation frame through a spring with certain prestress in the Y-direction. Therefore, when the eddy current brake electromagnet and the lateral guide rail have different attraction, the spring expansion will also change, so that the distance between the eddy current brake electromagnet and the lateral guide rail will change. As the attraction of both increases, the train

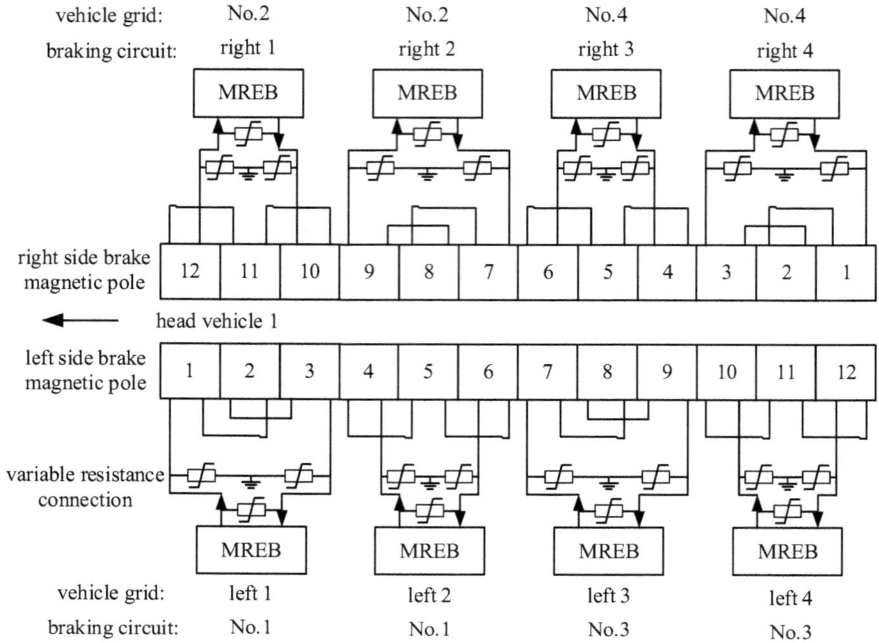

Fig. 2.9 Circuit wiring diagram of the eddy current brake

speed decreases, and the gap between the eddy current brake electromagnet and the lateral guide rail will be reduced. When the train speed is about 140–150 km/h, the wear plate of the brake electromagnet is in contact with the guide rail. The connection diagram of eddy current brake electromagnet and levitation frame is shown in Fig. 2.9.

The 12 magnetic poles of the brake electromagnet are divided into four groups. Three adjacent magnetic poles are a group, and the three magnetic poles are connected in series with each other. Each group corresponds to one set of eddy current brake control unit (MREB), so there are four sets of eddy current brake control units for each brake electromagnet and a total of eight sets of eddy current brake control units for each vehicle. The brake circuit of each group is independently supplied with DC power by the eddy current brake control unit. The failure of the brake circuit in one eddy current brake electromagnet will not affect the other brake circuits. The current supplied by the eddy current brake control units generates a magnetic field in the magnetic pole coils of the brake electromagnet during brake. The magnetic field cuts the lateral guide rail and generates an induced current (eddy current) in it when the vehicle is moving. The interaction between the magnetic field and the induced current generates a force that prevents the vehicle from moving. The magnitude of the force depends on the winding current of the electromagnet, the electromagnetic air gap between the electromagnet and the guide rail, and the driving speed of the vehicle. The magnitude of the eddy current brake force can be adjusted by the eddy current brake control unit.

2.2.7 Positioning and Speed Measurement System

The positioning and speed measurement system is applied to measure the position, speed, phase of propulsion linear motor and other parameters of the train on the line. The positioning system includes the absolute positioning and the relative positioning. The specific components include the absolute positioning sensors, the relative positioning sensors, the PRW electronic components, and the ORT positioning electronic components. The composition of the positioning and speed measurement system is shown in Fig. 2.10.

On the line, a set of coding board coded according to the absolute position of the line, also known as positioning sign board, is installed every 200–1200 m. When the train passes the board, the absolute position of the train is recorded and determined by reading the code on the coding board through the absolute positioning reader installed on the vehicle. This set of the absolute positioning device is installed according to the requirements of redundancy on the vehicle and line. In addition, the train is also equipped with an electromagnetic induction device to count the long-stator cogging on the line, which is used to determine the position of the train more accurately. The obtained position is the relative position based on the nearest coding board. The positioning accuracy is the width of one tooth or groove, i.e., 43 mm. The signals provided by the positioning system are also used to obtain the actual running speed and direction of the train.

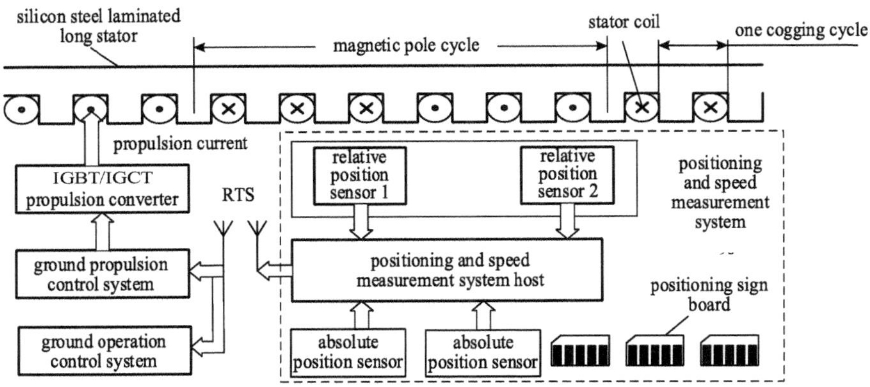

Fig. 2.10 Composition of the positioning and speed measurement system

The cogging count value and absolute positioning code value can also convert the relative position of the magnetic pole of the levitation electromagnet relative to the stator magnetic field of the long-stator linear synchronous motor, which is called the phase angle of the magnetic pole, for propulsion control of the linear motor at low speed (rotor flux linkage-oriented control).

All positioning data are first obtained from the vehicle, processed by the safety computer of the vehicle operation control system and transmitted via radio to the decentralized operation control system on the ground. The operation control center obtains the train position and speed data from the decentralized operation control system.

The PRW signals for propulsion motor control are not safety-related and are not processed by the vehicle safety computer, but are transmitted directly to the propulsion segments on the ground via the vehicle radio.

2.3 Analysis of the Levitation System of High-Speed Maglev Train

This section, together with Sect. 2.4, using the German TR08 EMS-type high-speed maglev train as the object, analyzes the basic structure, the basic design requirements, and the technical solutions of the levitation and guidance system.

2.3.1 Basic Structure of the Levitation System

German TR08 high-speed maglev trains generally use three to eight vehicles formation, each middle vehicle has 14 full section levitation electromagnets and 4 partial levitation electromagnets that span adjacent vehicles, and each end vehicle has 14 full section levitation electromagnets and 2 partial levitation electromagnets. Figure 2.11a, b give the side and top view of the structure of an end vehicle. Each high-speed maglev vehicle is supported by four bogies (SG1, SG2, SG3, SG4). Each bogie contains four joint structure levitation frames (JS). Each levitation frame provides a support point through an air spring. The levitation electromagnets are connected to each other by the levitation frames, forming a flexible levitation structure. The entire carriage is thus supported by sixteen levitation frames. The sixteen levitation frames are decoupled from each other by adoption of a mechanical structure. In high-speed maglev trains, the levitation system consisting of a levitation controller chassis, two sets of levitation sensors and the corresponding electromagnets and guideway is referred as a single-point levitation system. Each electromagnet located at the end of the vehicle body corresponds to three levitation points, and each electromagnet in the remaining positions corresponds to two levitation points. There are 16 levitation points on one side of the whole vehicle. The electromagnet at the rear of the carriage is used to connect the other carriage.

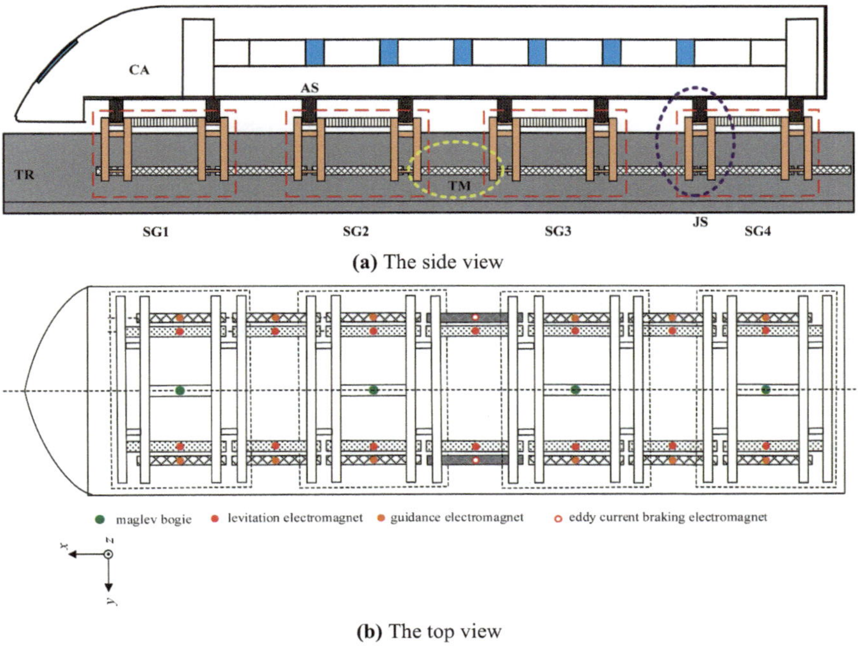

Fig. 2.11 Structural diagram of the German TR08 high-speed maglev train

Figure 2.12 shows a partial enlarged schematic of the joint structure. The electromagnet is connected to the bottom of the bracket arm in the levitation frame by a metal rubber spring, which is different from the direct solid connection of the electromagnet to the bracket arm in the urban maglev train. As can be seen in this diagram, the two joint structures corresponding to the end electromagnet are different. The left side is a general joint structure where there is a limited degree of freedom between the two electromagnets due to the metal rubber spring. The electromagnet corresponding to the two levitation points in the right joint structure is fixedly connected to an electromagnet back beam. This joint structure is called end joint structure.

2.3.2 Basic Requirements for the Levitation System

(1) Load bearing index of the levitation system

It is assumed that the maximum rated load of the middle vehicle is 64.5 t, the maximum rated lifting air gap is 31 mm, the maximum rated load of the end vehicle is 62.0 t, and the maximum rated lifting air gap is 31 mm. Through calculation, the requirements for the levitation forces of levitation electromagnet and levitation magnetic pole can be obtained, as shown in Table 2.1.

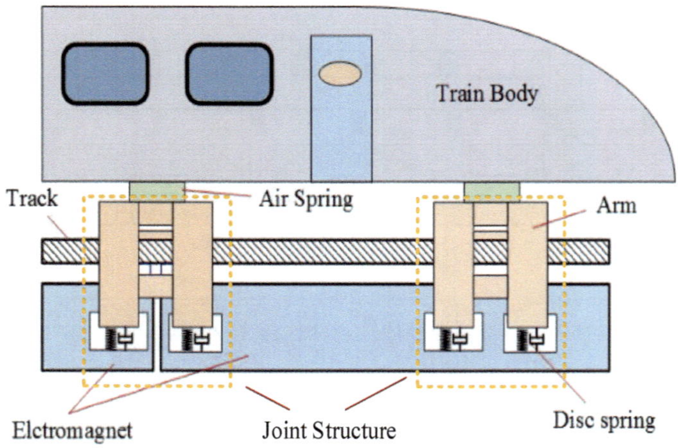

Fig. 2.12 Schematic diagram of the levitation joint structure of the high-speed maglev trains

Table 2.1 Vehicle static levitation force index

	Static levitation force (ton)	Remarks
Non end levitation electromagnet	4.031	The air gap is 31 mm
End levitation electromagnet	4.764	The air gap is 31 mm
Main levitation magnetic pole	0.367	The air gap is 31 mm
End levitation magnetic pole	0.184	The air gap is 31 mm
Middle vehicle	64.5	The air gap is 31 mm
End vehicle	62.0	The air gap is 31 mm

(2) Guideway dynamic tracking performance index of the levitation system

Although the vehicle runs non-contact along the guideway, the vehicle must track the guideway levitation according to certain requirements. When the vehicle passes the line section composed of straight section, curve section and slope section at a certain speed, the levitation system must have sufficient rapid response capacity. The evaluation index of this rapid response ability is the so-called guideway dynamic tracking performance index of the levitation system.

(3) Levitating-to-weight ratio index of the levitation system

Suppose that the sum of the self-weight of the levitation system composed of levitation controller (including chopper, etc.), levitation electromagnet, levitation sensor, etc., is m_{lev}, and the total weight that the levitation system can levitate is m_{total}, then the levitating-to-weight ratio W_{r} of the levitation system is defined as:

$$W_{\text{r}} = m_{\text{lev}}/m_{\text{total}} \tag{2.1}$$

A lower levitating-to-weight ratio can increase the payload of the system. Generally, W_r is less than 1/7.5.

(4) Power consumption index of the levitation system

The power consumption of the levitation system is mainly composed of the switching loss of the levitation chopper, copper loss, and iron loss of the electromagnet, etc. Under the rated levitation air gap, let the total power consumed by the levitation system be P_{total}, then the power consumption index P_r of the levitation system is defined as:

$$P_r = P_{total}/m_{total} \tag{2.2}$$

When the levitation air gap is 10 mm, it is generally required that P_r is not greater than 1 t/kW.

2.3.3 Levitation System Control Scheme

During the operation of high-speed maglev trains, the failures of individual levitation points may result in the train not being able to achieve full levitation. This will have a significant impact on the safety of the train operation. To reduce the chance of this occurring, the levitation joint structure is designed to solve this problem. A joint structure consists of two levitation control units, which are jointly responsible for the levitation of the same levitation frame.

There are three possible working scenarios of the levitation joint structure: both levitation control units work normally, one of the levitation control units loses levitation capacity, both levitation control units fail and lost levitation capacity. These three states are referred as normal working state, redundant working state and failure state respectively. The levitation joint structure in the failure state is not controllable and the levitation system cannot function. Thus, the normal working state and the redundant working state are considered in the design of the levitation control system. The control technology for the joint structure needs to achieve stable levitation in the normal working state and the redundant working state, and ensure that the joint structure is in the normal working state as much as possible.

For the high-speed maglev trains, the levitation system mainly consists of levitation electromagnets, levitation sensors, and levitation controllers.

(1) Levitation electromagnet

The levitation electromagnet is also the excitation pole of the propulsion linear motor. A standard full-section levitation electromagnet consists of two partial levitation electromagnets. The two levitation electromagnets on both sides of the head of the end vehicle are divided into three parts. Each partial levitation electromagnet corresponds to a pair of levitation sensors and a set of levitation controller. A linear generator coil is also embedded in the magnetic pole of the levitation electromagnet. The levitation electromagnet has three functions: levitation excitation, propulsion linear motor excitation, and power generation.

The German high-speed maglev train has two types of electromagnets with different numbers of magnetic poles. The two levitation electromagnets at the head of the end vehicle each have 14 magnetic poles, including 2 end magnetic poles, 6 high core magnetic poles, and 6 main magnetic poles, as shown in Fig. 2.13. The remaining levitation electromagnets each have 12 magnetic poles, including 2 end magnetic poles, 4 high core magnetic poles, and 6 main magnetic poles, as shown in Fig. 2.14. The high core magnetic poles are functionally the same as the main magnetic poles, except that the core is taller, leaving some space for the levitation gap measurement unit. Each levitation electromagnet is about 3 m in length, and the parts on the levitation electromagnet are connected by the back beam of the levitation electromagnet to form a whole.

(2) Levitation sensor

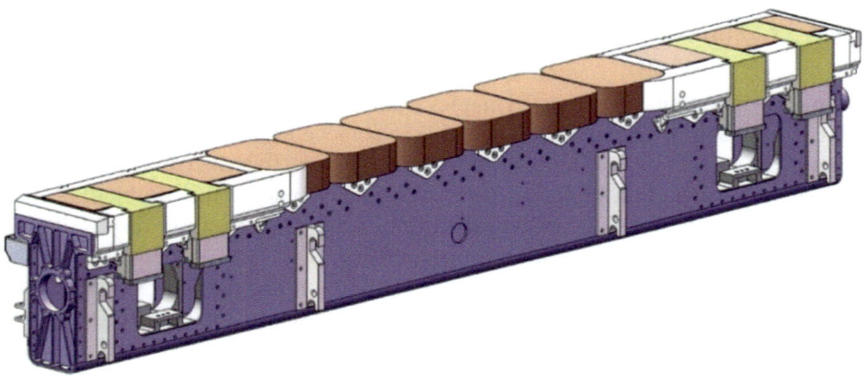

Fig. 2.13 Intermediate levitation electromagnet

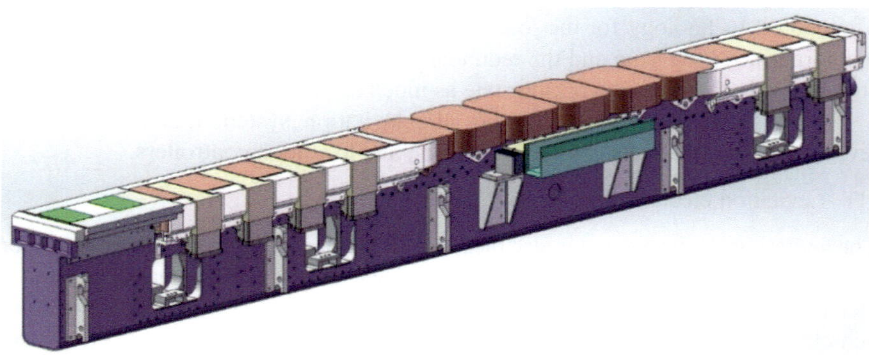

Fig. 2.14 End levitation electromagnet

Each vehicle (including end vehicle and middle vehicle) has 64 levitation sensors installed on the levitation electromagnet, corresponding to 32 levitation control systems. Each levitation sensor consists of a measurement coil, a signal conversion and transmission circuit board (hereinafter referred to as circuit board), an aluminum envelope for the measurement coil, an accelerometer, and a cable connector.

There are three independent coil group systems on the measurement coil: the gap measurement coil group of channel A, the gap measurement coil group of channel B, and the measurement coil group of vehicle running speed.

The circuit board has four functional modules: GapA gap measurement, GapB gap measurement, Acc acceleration measurement, and Vel velocity measurement.

The accelerometer is fixed in the inner cavity of the levitation sensor.

The cable connector is a 26-pin socket, which is the sensor power input and measurement signal output interface. In addition, there are 2 board compression strips in the levitation sensor to hold the board in place and to make the board resistant to shock and vibration.

(3) Levitation controller

There are 32 sets of levitation controllers for each vehicle (including end vehicle and middle vehicle), and each levitation electromagnet corresponds to 2–3 sets of levitation controllers, where the end electromagnet corresponds to 3 sets of levitation controllers.

The basic function principle of the levitation controller is to obtain the signals from the levitation sensor and the adjacent levitation electromagnet control unit, then carry out certain calculations in the levitation controller through the software, and finally realize the control of the levitation force and air gap by controlling the excitation current of the levitation electromagnet.

The internal structure of each levitation controller can be roughly divided into the levitation chopper part, the digital control part, and the signal conditioning part by function.

The levitation chopper part is mainly used to adjust the levitation electromagnet current by controlling the switching tubes in the chopper. The digital control part is the core of the whole levitation controller, which mainly analyzes the speed, gap, acceleration and other signals collected by the sensors to get the voltage given by the chopper. The signal conditioning part performs the pre-processing (e.g. filtering) on the signals collected by the sensors.

2.4 Analysis of the Guidance System of High-Speed Maglev Train

2.4.1 Basic Structure of the Guidance System

The guidance system of TR08 EMS-type high-speed maglev train is composed of guidance electromagnets and connectors on both the left and right sides. Each vehicle has 6 pairs of guidance electromagnets, which are installed on the bracket arm of the bogie between two adjacent levitation frames (except between the No. 4 and No. 5 levitation frames, eddy current brake electromagnets are installed between the two levitation frames). It can be seen from Fig. 2.15 that the guidance electromagnets also form a joint structure similar to that in the levitation system. The high-speed maglev train realizes the mechanical decoupling and guidance fault-tolerant design between bogies through the joint structure design and decomposes the guidance system of the whole vehicle into basic guidance units as shown in Fig. 2.15. The current is applied to the guidance electromagnets on both the left and right sides through the chopper, and the guidance force generated by the mutual attraction between the electromagnet and the side rail is transmitted to the whole carriage through the bogie, to realize the guidance function of the high-speed maglev train.

The guidance electromagnet is divided into the end part and the non-end part in terms of the installation position. In Fig. 2.15, there is no guidance electromagnet installed between No. 4 and No. 5 levitation frames, which is also equivalent to the end part. Therefore, if only a single carriage is considered, the end guidance electromagnets are installed between No. 1 and No. 2 levitation frames, No. 3 and No. 4 levitation frames, No. 5 and No. 6 levitation frames, and No. 7 and No. 8 levitation frames. The non-end guidance electromagnets are installed between the remaining No. 2 and No. 3 levitation frames, No. 6 and No. 7 levitation frames. Each end guidance electromagnet corresponds to two sets of guidance control units, and when one of them fails, the other guidance control unit assumes the guidance function of the frame. Each non-end guidance electromagnet corresponds to one set of guidance control unit, which forms a redundant structure with one set of guidance control unit corresponding to another non-end guidance electromagnet in the same levitation frame.

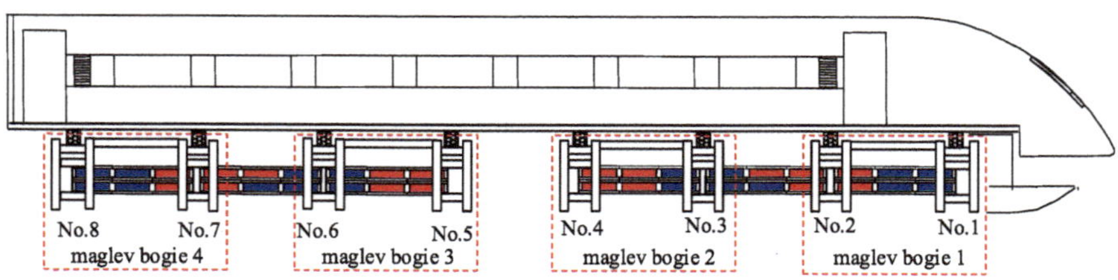

Fig. 2.15 Grouping of each guidance electromagnet in the single-section train system

2.4.2 Basic Requirements for the Guidance Systems

(1) Static guidance force performance index

The basis for determining the static guidance force performance index of the vehicle is: when the vehicle is statically levitated on the guideway under the conditions of maximum rated load and maximum line cross-slope angle, the vehicle can maintain static balance in the Y-direction.

Assume that the maximum rated load of a vehicle is 64.5 t, the maximum line cross-slope angle is 16°, then the following indicators can be obtained by calculation.

- The maximum static guidance force required of one vehicle is 17.8 t.
- The maximum static guidance force of a guidance electromagnet is 2.97 t.
- The maximum static guiding force of a guidance magnetic pole is 0.50 t.

2.4.3 Guidance System Control Scheme

Each guidance electromagnet contains six magnetic poles, which correspond to two different control schemes according to the end and non-end cases of the guidance electromagnet. The basic structure of the guidance electromagnet is shown in Fig. 2.16.

When the guidance electromagnet in Fig. 2.16 is installed at non-end, the six guidance magnetic poles are divided into two groups for control, with poles No. 1, No. 2 and No. 3 as the first group and poles No. 4, No. 5 and No. 6 as the second group. The two first groups of guidance magnetic poles on both the left and right sides are controlled by two controllers with the same structural parameters, and form a

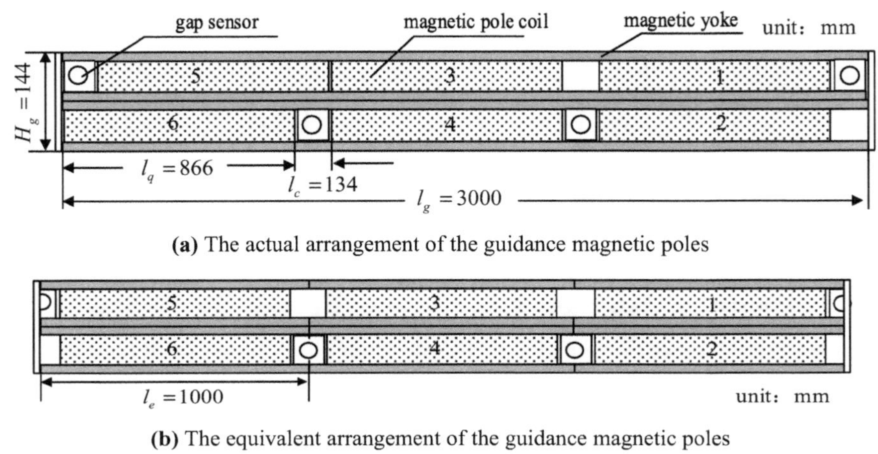

Fig. 2.16 Geometric arrangement of the magnetic poles in a guidance electromagnet

joint structure together with the two second groups of the preceding pair of guidance electromagnets. Similarly, the two second groups of guidance magnetic poles on both the left and right sides are also controlled by two controllers with the same structural parameters, and form a joint structure together with the two first groups of the latter pair of guidance electromagnets. The control scheme of the guidance joint structure is shown in Fig. 2.17. The voltage outputs from the four controllers are added to the respective electromagnets through four same choppers on the left and right, which are equivalent to four control units with the same structural parameters. To keep the train running in the center of the guideway, the gap sensor signals of the two electromagnets on the same side of the joint structure are summed and averaged, and the difference with the average value of the gap sensor signals of the two electromagnets on the opposite side is fed to the controller to perform derivate control.

When the electromagnet in Fig. 2.16 is installed at the end, the six magnetic poles used for guidance are divided into three groups for control, with poles No. 1 and No. 2 as the first group, poles No. 4 and No. 5 as the second group, and poles No. 3 and No. 6 as the third group. Similar to the non-end control scheme, the three groups of guidance magnetic poles on the both left and right sides are each controlled by two controllers with the same parameters, also divided into three control groups. The purpose of this is for redundancy. The first group of magnetic poles and another group of magnetic poles of the non-end guidance electromagnet form a joint structure, and the second group of magnetic poles and the third group of magnetic poles are jointly responsible for the guidance function of the end, forming a joint structure of the end (Fig. 2.18).

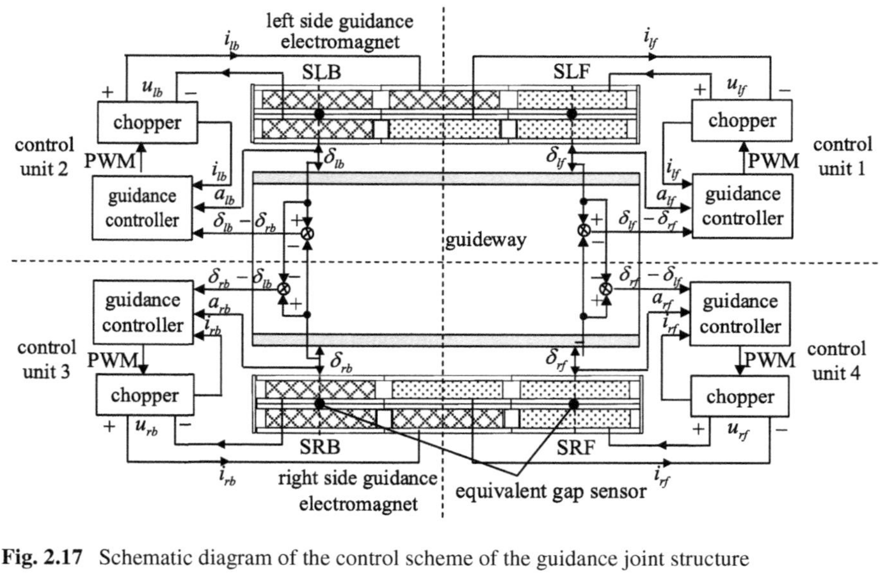

Fig. 2.17 Schematic diagram of the control scheme of the guidance joint structure

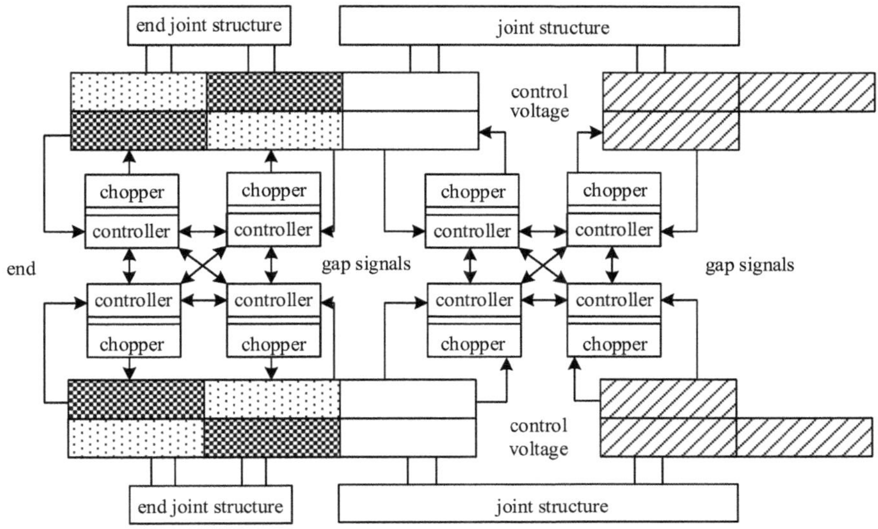

Fig. 2.18 Schematic diagram of control scheme of the end guidance system

The guidance system is mainly composed of the guidance electromagnet, the guidance sensor, and the guidance controller.

(1) Guidance electromagnet

The guidance electromagnet consists of magnetic poles, magnetic yoke, box beam body, sensor, and articulation mechanism, as shown in Fig. 2.19. Each guidance electromagnet contains six magnetic poles, which are divided into two rows along the vertical direction, with three coils in each row evenly arranged and symmetrical at the top and bottom. Each group has a pair of guidance sensors and a set of guidance controller. The plane of the guidance electromagnet is equipped with a magnetically conductive anti-collision plate.

According to the different installation positions of the guidance electromagnet in the vehicle, the six magnetic poles are divided into two or three groups, corresponding to two or three redundant control circuits, as shown in Fig. 2.20. When a control circuit fails, the guidance of the train can still be ensured through the mechanical connection device between the levitation frames.

(2) Guidance sensor

There are 64 guidance sensors installed on the guidance electromagnets of each vehicle (including end vehicle and middle vehicle). The four guidance electromagnets at both ends of each vehicle and the four guidance electromagnets near the brake electromagnet in the middle of the vehicle have six guidance sensors respectively, and the other four guidance electromagnets have four guidance sensors respectively.

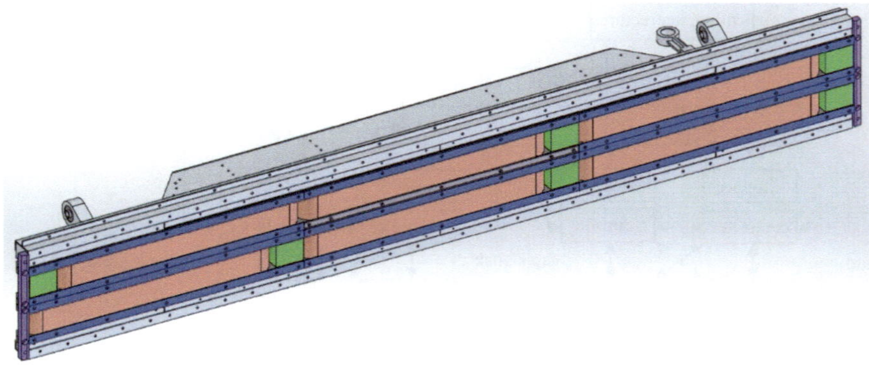

Fig. 2.19 Structure composition of the guidance electromagnet

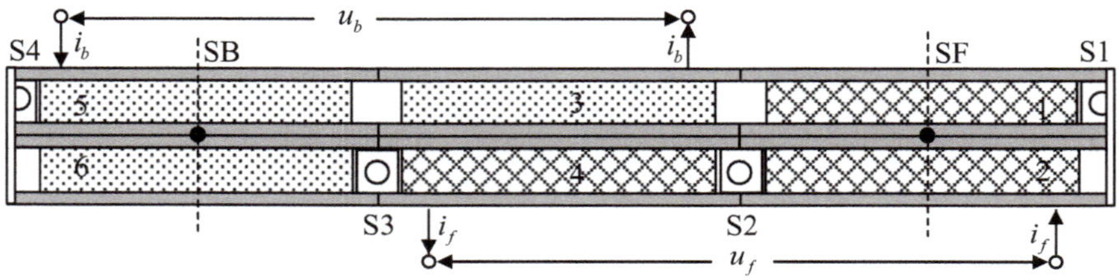

(a) The two-group division method

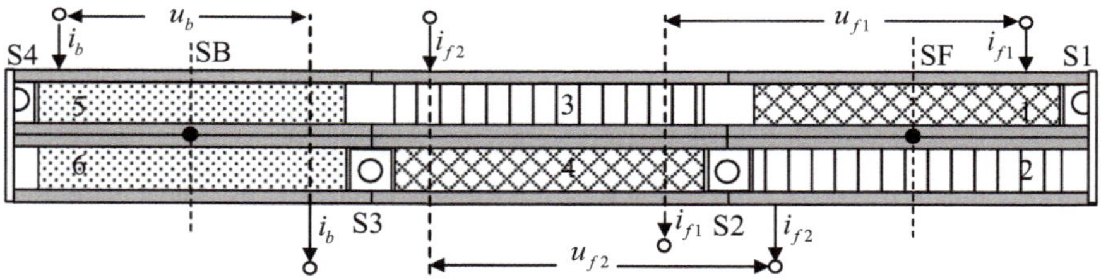

(b) The three-group division method

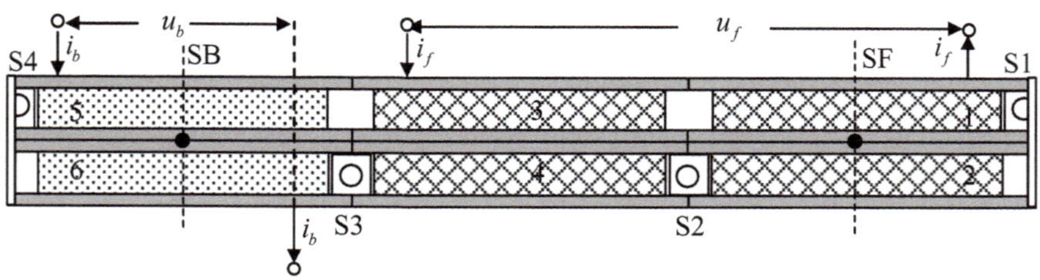

(c) The equivalent method of three-group division

Fig. 2.20 Grouping control method for magnetic poles in the guidance electromagnet

Each guidance sensor consists of a measurement coil, a signal conversion and transmission circuit board (hereinafter referred to as the circuit board), an aluminum envelope for the measurement coil, an accelerometer, and a cable connector.

There are two independent coil group systems on the measurement coil, which are the gap measurement coil group of channel A and the gap measurement coil group of channel B. The circuit board has four functional modules: the GapA gap measurement, the GapB gap measurement, the Acc acceleration measurement, and the Vel velocity measurement. The accelerometer is fixed in the inner cavity of the guidance sensor. The cable connector is a 26-pin socket, which is the sensor power input and measurement signal output interface.

In addition, there are 2 board compression strips in the guidance sensor to hold the board in place and to make the board resistant to shock and vibration.

(3) Guidance controller

There are 32 sets of guidance controllers in each vehicle (including end vehicle and middle vehicle), and each guidance electromagnet corresponds to 2–3 sets of guidance controllers. The four guidance electromagnets at both ends of a vehicle and the four guidance electromagnets close to the brake electromagnet have three sets of guidance controllers respectively, and the other four have two sets of guidance controllers respectively. The structure of the guidance controller is basically similar to that of the levitation controller, which is composed of three parts: the guidance chopper part, the digital control part, and the signal conditioning part.

2.5 Conclusion

Taking the EMS-type high-speed maglev system as the object, this chapter firstly analyzes the basic components of the high-speed maglev transit system, and introduces the maglev line, the vehicle system, the propulsion and power supply system, and operation control system respectively. Then, for the seven structural components in high-speed maglev vehicle system, including the vehicle body structure, the maglev bogie, the levitation and guidance system, the vehicle power supply system, the vehicle control and diagnosis system, the safety brake system, and the positioning speed measurement system, this chapter carries out a technical analysis. Finally, the basic structures, basic requirements, and the corresponding control schemes of the levitation system and the guidance system are elaborated.

Bibliography

1. Ding S (2022) 600 km/h high-speed maglev transportation system. Shanghai Scientific and Technical Publishers, Shanghai
2. Ding S, Fu S, Liang X (2023) Engineering practice and prospect of high-speed maglev transportation in China. Sci Technol Foresight 2(4):40–48

3. Ge Q, Zhang B, Wei R et al (2023) Present situation and prospect of traction power supply and control technologies for high-speed maglev transportation. Sci Technol Foresight 2(4):89–95
4. Li H, Shi J, Li X et al (2023) Current status and reflection on the development of high-speed maglev transportation. Railw Sci 2(3):327–335
5. Lin G, Sheng X (2018) Application and further development of maglev transportation in China. Innov Transp Syst Technol 4(3):36–43
6. Liu Z, Long Z, Li X (2015) Maglev trains. Springer, Berlin
7. Long Z, Zhai M, Wang Z (2023) Condition monitoring, fault diagnosis and fault tolerant control of the maglev train. Shanghai Scientific and Technical Publishers, Shanghai
8. Lu Q, Fang Y (2023) Development of driving technology for high-speed maglev transportation. Sci Technol Foresight 2(4):96–104
9. Lv G, Zhang Z, Liu Y et al (2021) Characteristics analysis of linear synchronous motor integrated with propulsion, levitation, and guidance in high-speed maglev system. IEEE Trans Transp Electrif 7(4):3185–3193
10. Wu X (2003) Maglev train. Shanghai Scientific and Technical Publishers, Shanghai
11. Wu W, Lu X, Liu G et al (2019) Operation control system of 600 km/h high-speed maglev levitated transport system in China. In: 2019 IEEE vehicle power and propulsion conference (VPPC), Hanoi, Vietnam, pp 1–5
12. Wu D, Wang Y, Zhang Z (2020) Analysis of influence of step-changing method on propulsion characteristics of high-speed maglev vehicle. Electr Drive Locomot 6:10–14
13. Wu D, Tian Y, Chen S et al (2021) Assistant stop area configuration method for high-speed maglev system based on segmented optimization. J China Railw Soc 43(9):86–92
14. Xu H, Zheng W, Liu X et al (2007) Design and realization of decentralized operation control system of high-speed maglev train. Adv Commun Technol 17(9):958–961
15. Zhu J, Ge Q, Sun P et al (2020) Traction-system research of high-speed maglev based on active disturbance rejection control. Trans China Electrotech Soc 35(5):1065–1074

Chapter 3
Controller Design of High-Speed Maglev Train Levitation System

In this chapter, a mathematical model of the levitation system based on the joint structure is firstly established and analyzed. After simplifying the model, a nominal controller of the levitation system is designed by adoption of the linear quadratic regulator (LQR). For issues of the track irregularities and propulsion disturbance, the active disturbance rejection control (ADRC) algorithm is proposed. A high-speed maglev train levitation control test platform is then built, and the high-speed maglev hardware and software systems are designed and implemented. The levitation system performance experiments are conducted to verify the effectiveness of the proposed levitation control algorithms.

3.1 Modeling and Analysis of the Levitation System

The levitation system mainly consists of the levitation electromagnets, the levitation gap measurement units, and the levitation control units. In this section, a detailed system model is given and its characteristics are analyzed.

3.1.1 Modeling of the Joint Structure-Based Levitation System

The levitation system is a typical electromechanical motion system, and the model of the joint structure-based levitation system can be modeled based on the electromagnetic principle and the theorem of dynamics. The motion mechanism of the joint structure-based levitation system covers: generating the coil current by adding voltage to the electromagnet coil, which in turn causes the electromagnetic force

Z. Long et al., *High-Speed Maglev Train's Levitation and Guidance Control*, Advances in High-speed Rail Technology, https://doi.org/10.1007/978-981-97-2309-6_3

generated by the electromagnet resulting in an upward force due to the electromagnet's generated electromagnetic force. To make the electromagnet steadily levitate with a fixed gap, the corresponding sensors are applied to measure the actual levitation gap. Meanwhile, the current in the electromagnet coil can be dynamically adjusted according to the difference between the actual levitation gap and the target gap. The two electromagnets in the joint structure support the train body through the levitation frame, and the interaction between the electromagnet and the levitation frame can be regarded as a spring-damper relationship. To obtain a complete model of the joint structure-based levitation system, the following three aspects need to be paid attention, including the dynamic relationship between the voltage input and the current generated by the electromagnet coil, the relationship between the electromagnetic force generated by the electromagnet and the coil current, and the dynamics equation of the joint structure-based levitation system under the influence of the electromagnetic force. The first two aspects are closely related to the specific structural form of the electromagnet. In the process of modeling the joint structure-based levitation system, the following assumptions are made: ignoring the influence of the leakage flux and the flux is all distributed in the closed circuit formed by the electromagnet and the guideway; ignoring the magnetic resistance of the core and the magnetic resistance is considered to exist mainly in the air gap and the permanent magnet.

The structure of electromagnet is shown in Fig. 3.1. Therein, δ is the distance between the upper surface of the electromagnet pole and the lower surface of the guideway (the levitation gap), S is the pole area. The mechanistic model of the joint structure-based levitation system can be established in the following steps.

First, the voltage–current relationship of electromagnet is given. Define N as the number of electromagnet coil turns, δ_0 as the levitation gap at the equilibrium point, i_0 as the levitation current at the equilibrium point, μ_0 as the vacuum permeability. According to the principle of electromagnetism, the process of generating current from voltage in a levitation system can be described as:

$$u = Ri + L_0\dot{i} - k_i\dot{\delta} \tag{3.1}$$

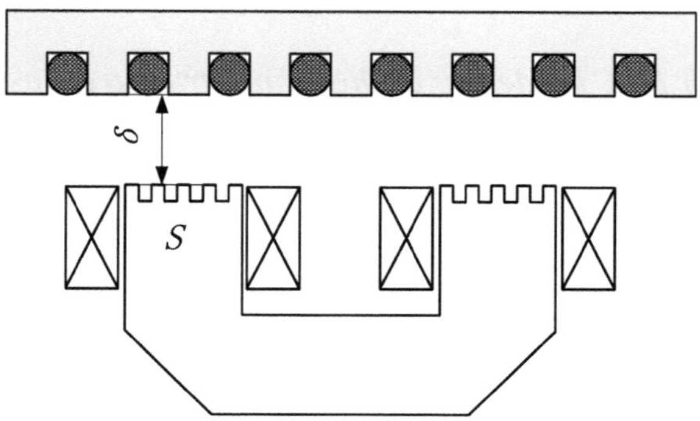

Fig. 3.1 Schematic of the high-speed maglev train electromagnet structure

where

u the value of voltage across the electromagnet coil
i the value of current generated by the electromagnet coil
R the value of resistance of the electromagnet coil
L_0 the value of the coil inductance

$$L_0 = \frac{\mu_0 N^2 S}{2\delta_0}$$

$$k_i = \frac{\mu_0 N^2 S i_0}{2\delta_0^2}$$

By using the subscripts l and r to refer to the physical quantities on the left and right sides, the corresponding voltage–current relationships can be expressed as:

$$u_l = R_l i_l + L_{l0}\dot{i}_l - k_{il}\dot{\delta}_l \tag{3.2}$$

$$u_r = R_r i_r + L_{r0}\dot{i}_r - k_{ir}\dot{\delta}_r \tag{3.3}$$

When the coil current is i and the levitation gap is δ, the corresponding electromagnetic force F_e is:

$$F_e = \frac{\mu_0 S N^2 i^2}{\delta^2} = k_e \frac{i^2}{\delta^2} \tag{3.4}$$

where $k_e = \mu_0 N^2 S/4$. Conducting linearization of F_e at equilibrium point (δ_0, i_0), the following can be obtained:

$$\Delta F_e = -2k_e \frac{i_0^2}{\delta_0^3}\Delta\delta + 2k_e \frac{i_0}{\delta_0^2}\Delta i \tag{3.5}$$

where $k_\delta = 2k_e i_0^2/\delta_0^3$ and $k_i = 2k_e i_0/\delta_0^2$ represent the influence degree of gap and current change on electromagnetic force, respectively. Thus, they are also referred as the gap coefficient and the current coefficient, respectively.

When the levitation gap for each side is δ_l and δ_r, respectively, and the corresponding electromagnet coil current for each side is i_l and i_r, the electromagnetic force on the left and right side is F_{el} and F_{er}, respectively, and they can be expressed as follows:

$$F_{el} = k_e \frac{i_l^2}{\delta_l^2} \tag{3.6}$$

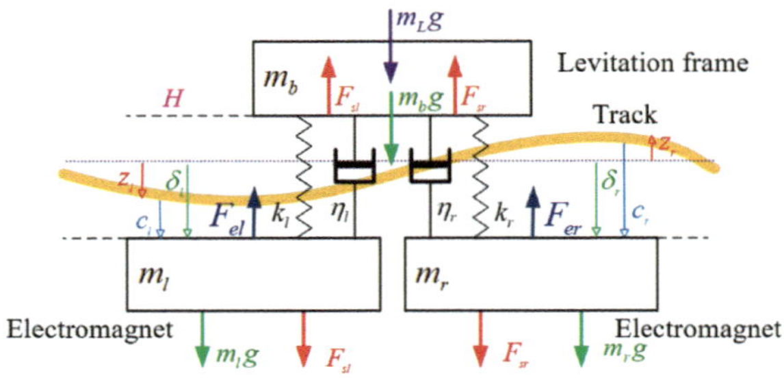

Fig. 3.2 Force analysis diagram of the joint structure-based levitation system

$$F_{\mathrm{er}} = k_{\mathrm{e}} \frac{i_{\mathrm{r}}^2}{\delta_{\mathrm{r}}^2} \tag{3.7}$$

As shown in Fig. 3.2, the voltage–current characteristics and force characteristics of the electromagnet can be described with Eqs. (3.2), (3.3), (3.6), and (3.7).

The dynamics of the joint structure-based levitation system then can be obtained by using Newton's laws of motion. Taking a general joint structure as an example, the force analysis of the structure can be shown in Fig. 3.2. Therein, δ_{l} and δ_{r} are the displacements of the left and right side electromagnet with respect to the reference plane. H is the displacement of the levitation frame. c_{l} and c_{r} are the gaps between the left and right-side electromagnet and guideway in case of guideway motion. z_{l} and z_{r} are the displacements of the left and right-side guideway in the case of guideway motion. The lower surface of the guideway is regarded as the reference plane for the motion of the electromagnet, and the vertically downward direction is taken as the positive direction, so the relative relationships can be expressed as:

$$\begin{cases} c_{\mathrm{l}} = \delta_{\mathrm{l}} - z_{\mathrm{l}} \\ c_{\mathrm{r}} = \delta_{\mathrm{r}} - z_{\mathrm{r}} \end{cases} \tag{3.8}$$

Under the condition of neglecting the guideway motion displacement, there exists a variable relationship of $\delta_{\mathrm{l}} = c_{\mathrm{l}}$, $\delta_{\mathrm{r}} = c_{\mathrm{r}}$.

In the force analysis of the levitation system, the pitching, side-rolling, and turning motions of the electromagnet and levitation frame are not considered. The mutual influence between two single levitation points in the electromagnet module is ignored. Only the motion of the electromagnet and levitation frame in the vertical direction is considered. As a result, the magnetic force generated by the electromagnet is reduced to a concentrated force with the point of action being its center of gravity. The force analysis of the left and right electromagnets and levitation frame can obtain their respective kinetic equations.

In this instance, for the left side electromagnet, the kinetic equation can be written as:

$$m_{\mathrm{l}} \ddot{\delta}_{\mathrm{l}} = m_{\mathrm{l}} g - F_{\mathrm{el}} + F_{\mathrm{sl}} \tag{3.9}$$

where $m_l g$ is gravity of the left-side electromagnet, F_{sl} is the pressure of the metal rubber spring, F_{el} is the electromagnetic force.

Then, the metal rubber spring can be considered as the spring-damper, and the corresponding pressure can be represented as:

$$F_{sl} = k_l(H - \delta_l) + \eta_l(\dot{H} - \dot{\delta}_l) \tag{3.10}$$

Similarly, the kinetic equation of the right-side electromagnet can be obtained as:

$$m_r\ddot{\delta}_r = m_r g - F_{er} + F_{sr} \tag{3.11}$$

$$F_{sr} = k_r(H - \delta_r) + \eta_r(\dot{H} - \dot{\delta}_r) \tag{3.12}$$

The kinetic equation of the levitation frame can be written as:

$$m_b\ddot{H} = m_b g - F_{sl} - F_{sr} + m_L g \tag{3.13}$$

where F_{sl} and F_{sr} are the forces of left and right-side spring, $m_b g$ is the gravity of the levitation frame, $m_L g$ is the downward force generated by the vehicle body through the air spring.

The properties of the joint structure-based levitation system can be described by Eqs. (3.2), (3.3), (3.6), (3.7), (3.9), (3.11), and (3.13).

3.1.2 Model Analysis of the Joint Structure-Based Levitation System

Assuming that the left and right-side single-point levitation systems in the joint structure have the same target levitation gaps, i.e., $\delta_{l0} = \delta_{r0} = \delta_0$. By setting the differential term in the kinetic equations of the system to zero, the equilibrium condition for a stable levitation system can be derived as follows:

$$\begin{cases} 0 = m_l g - F_{el} + F_{sl} \\ 0 = m_r g - F_{er} + F_{sr} \\ 0 = m_b g - F_{sl} - F_{sr} + m_L g \end{cases} \Rightarrow F_{el} + F_{er} = m_l g + m_r g + m_b g + m_L g \tag{3.14}$$

The small differences between the left and right-side single-point levitation systems in the joint structure are negligible, i.e., $m_l = m_r = m$, $R_l = R_r = R$, $k_l = k_r = k_s$, $\eta_l = \eta_r = \eta_s$. k_s is the equivalent stiffness of the metal rubber spring, η_s is the equivalent damping of the metal rubber spring. Setting the stable levitation gap as $\delta_0 = 12$ mm, and assuming that the electromagnets on the left and right side carry the same load during stable levitation, i.e., $i_{l0} = i_{r0} = i_0$, the value of current inside the electromagnet coils during steady-state levitation can be obtained as follows:

$$i_0 = \sqrt{\frac{2mg + m_\mathrm{b}g + m_\mathrm{L}g}{2k_\mathrm{e}}}\delta_0 \tag{3.15}$$

The system state variables, control variables, and output variables are taken as follows.

$$\begin{cases} \boldsymbol{x} = \left[\, x_1\ x_2\ x_3\ x_4\ x_5\ x_6\ x_7\ x_8 \,\right]^\mathrm{T} = \left[\, \delta_\mathrm{l}\ \dot{\delta}_\mathrm{l}\ i_\mathrm{l}\ \delta_\mathrm{r}\ \dot{\delta}_\mathrm{r}\ i_\mathrm{r}\ H\ \dot{H} \,\right]^\mathrm{T} \\ \boldsymbol{u} = \left[\, u_\mathrm{l}\ u_\mathrm{r} \,\right]^\mathrm{T},\ \boldsymbol{y} = \left[\, \delta_\mathrm{l}\ \delta_\mathrm{r} \,\right]^\mathrm{T} \end{cases} \tag{3.16}$$

where $k_i = \frac{2\mu_0 i_0 N^2 S}{\delta_0^2}$, $k_\delta = \frac{2\mu_0 i_0^2 N^2 S}{\delta_0^3}$, $L_0 = \frac{\mu_0 N^2 S}{\delta_0}$.

According to the Taylor expansion principle, the linearized model of the high-speed maglev train levitation system based on the joint structure can be obtained under the condition of neglecting the higher-order and infinitesimal terms as follows.

$$\begin{cases} \dot{\boldsymbol{x}}(t) = \boldsymbol{A}\boldsymbol{x}(t) + \boldsymbol{B}\boldsymbol{u}(t) \\ \boldsymbol{y}(t) = \boldsymbol{C}\boldsymbol{x}(t) + \boldsymbol{D}\boldsymbol{u}(t) \end{cases} \tag{3.17}$$

$$\boldsymbol{A} = \begin{bmatrix} 0 & 1 & 0 & 0 & 0 & 0 & 0 & 0 \\ \frac{k_\delta - k_\mathrm{s}}{m_\mathrm{l}} & 0 & -\frac{k_i}{m_\mathrm{l}} & 0 & 0 & 0 & \frac{k_\mathrm{s}}{m_\mathrm{l}} & 0 \\ 0 & \frac{k_i}{2L_0} & -\frac{R}{L_0} & 0 & 0 & 0 & 0 & 0 \\ 0 & 0 & 0 & 0 & 1 & 0 & 0 & 0 \\ 0 & 0 & 0 & \frac{k_\delta - k_\mathrm{s}}{m_\mathrm{r}} & 0 & -\frac{k_i}{m_\mathrm{r}} & \frac{k_\mathrm{s}}{m_\mathrm{r}} & 0 \\ 0 & 0 & 0 & 0 & \frac{k_i}{2L_0} & -\frac{R}{L_0} & 0 & 0 \\ 0 & 0 & 0 & 0 & 0 & 0 & 0 & 1 \\ \frac{k_\mathrm{s}}{m_\mathrm{h}} & 0 & 0 & \frac{k_\mathrm{s}}{m_\mathrm{h}} & 0 & 0 & -\frac{2k_\mathrm{s}}{m_\mathrm{h}} & 0 \end{bmatrix}$$

$$\boldsymbol{B} = \begin{bmatrix} 0\ 0\ \frac{1}{L_0}\ 0\ 0\ 0\ 0\ 0 \\ 0\ 0\ 0\ 0\ 0\ \frac{1}{L_0}\ 0\ 0 \end{bmatrix}^\mathrm{T}$$

$$\boldsymbol{C} = \begin{bmatrix} 1\,0\,0\,0\,0\,0\,0\,0 \\ 0\,0\,0\,1\,0\,0\,0\,0 \end{bmatrix}$$

$$\boldsymbol{D} = \begin{bmatrix} 0\ 0 \\ 0\ 0 \end{bmatrix}$$

Substituting the parameters of the joint structure-based levitation system into the model given by Eq. (3.17) and neglecting the damping coefficient η_s of the air spring, the controllability matrix $\boldsymbol{H}_\mathrm{ctrl} = \left[\, \boldsymbol{B}\ \boldsymbol{AB}\ \cdots\ \boldsymbol{A}^7\boldsymbol{B} \,\right]$ can be obtained with the rank in 8, which is full rank. From the linear system theory, it is known that the model of the joint structure-based levitation system is unstable but controllable. And the levitation system can be stabilized at the equilibrium point by designing the corresponding controller. Meanwhile, the rank of the observability matrix $\boldsymbol{H}_{\text{osb}} = \left[\boldsymbol{C}\ \boldsymbol{CA}\ \cdots\ \boldsymbol{CA}^7 \right]$ is 8, which is also full rank, indicating that the system variables can be observed.

3.2 Design of Nominal Controller for Levitation System

3.2.1 Simplification of the Levitation System Model

When the levitation gap fluctuates in a small range, the electromagnets can be regarded as a first-order inertia loop. And the inductance generated by the electromagnet coil makes the current change in the coil lag behind the voltage change. To improve the response speed of the electromagnet current, a common strategy is to implement current feedback to form a current loop to make the current value in the coil quickly track up to the current setting value (i.e., the control value). In this way, the current loop can be regarded as a proportional loop. In addition, by adjusting the current feedback coefficient, the current loop gain can be adjusted to a unit gain, which simplifies the subsequent controller design.

For the motion of the levitation frame, on the one hand, the motion of the levitation frame increases the complexity of the controller design. On the other hand, there is no sensor to obtain the displacement of the levitation frame directly, so it is necessary to design an observer to estimate the displacement of the levitation frame. Taking into account that the motion of the levitation frame is closely related to the state of the two electromagnets, the displacement of the levitation frame can be approximated as the average of the displacements of the left and right side electromagnets based on the initial deformation caused by the self-weight of the levitation frame $m_{\text{b}}g$ and the load $m_{\text{L}}g$, the displacement of the levitation frame H can be expressed as:

$$H = \frac{1}{2}(\delta_1 + \delta_2) + \frac{1}{2k_{\text{s}}}(m_{\text{b}} + m_{\text{L}})g \tag{3.18}$$

The equations of the joint structure system can be simplified as:

$$\begin{cases} m_{\text{l}}\ddot{\delta}_{\text{l}} = m_{\text{l}}g - F_{\text{el}} + \dfrac{1}{2}k_{\text{s}}(\delta_{\text{r}} - \delta_{\text{l}}) + \dfrac{1}{2}(m_{\text{b}} + m_{\text{L}})g \\ m_{\text{r}}\ddot{\delta}_{\text{r}} = m_{\text{r}}g - F_{\text{er}} + \dfrac{1}{2}k_{\text{s}}(\delta_{\text{l}} - \delta_{\text{r}}) + \dfrac{1}{2}(m_{\text{b}} + m_{\text{L}})g \end{cases} \tag{3.19}$$

where the new state variable $\boldsymbol{x}$ can be defined as $\boldsymbol{x} = \left[x_1\ x_2\ x_3\ x_4 \right]^{\text{T}} = \left[\delta_{\text{l}}\ \dot{\delta}_{\text{l}}\ \delta_{\text{r}}\ \dot{\delta}_{\text{r}} \right]^{\text{T}}$.

Then the simplified linear model of the joint structure-based levitation system can be expressed as:

$$\begin{cases} \begin{bmatrix} \dot{x}_1 \\ \dot{x}_2 \\ \dot{x}_3 \\ \dot{x}_4 \end{bmatrix} = \begin{bmatrix} 0 & 1 & 0 & 0 \\ \frac{2k_\delta - k_s}{2m} & 0 & \frac{k_s}{2m} & 0 \\ 0 & 0 & 0 & 1 \\ \frac{k_s}{2m} & 0 & \frac{2k_\delta - k_s}{2m} & 0 \end{bmatrix} \begin{bmatrix} x_1 \\ x_2 \\ x_3 \\ x_4 \end{bmatrix} + \begin{bmatrix} 0 & 0 \\ -\frac{k_i}{m} & 0 \\ 0 & 0 \\ 0 & -\frac{k_i}{m} \end{bmatrix} \begin{bmatrix} u_1 \\ u_2 \end{bmatrix} \\ \begin{bmatrix} \dot{y}_1 \\ \dot{y}_2 \end{bmatrix} = \begin{bmatrix} 1\,0\,0\,0 \\ 0\,0\,1\,0 \end{bmatrix} \begin{bmatrix} x_1 \\ x_2 \\ x_3 \\ x_4 \end{bmatrix} \end{cases} \tag{3.20}$$

The state variable $\begin{bmatrix} x_1 & x_2 \end{bmatrix}^{\mathrm{T}} = \begin{bmatrix} \delta & \dot{\delta} \end{bmatrix}^{\mathrm{T}}$ and the input variable $\begin{bmatrix} u_1 & u_2 \end{bmatrix}^{\mathrm{T}} = \begin{bmatrix} u_1 & u_r \end{bmatrix}^{\mathrm{T}}$ can be defined, and the system model can be linearized near the equilibrium point to obtain the mathematical model of the end-joint structure-based levitation system of the high-speed maglev train as shown in Eq. (3.21):

$$\begin{bmatrix} \dot{x}_1 \\ \dot{x}_2 \end{bmatrix} = \begin{bmatrix} 0 & 1 \\ 2k_e \frac{i_0^2}{m\delta_0^3} & 0 \end{bmatrix} \begin{bmatrix} x_1 \\ x_2 \end{bmatrix} + \begin{bmatrix} 0 & 0 \\ -k_e \frac{i_0}{m\delta_0^2} & -k_e \frac{i_0}{m\delta_0^2} \end{bmatrix} \begin{bmatrix} u_1 \\ u_2 \end{bmatrix} \tag{3.21}$$

When the relationship between the left and right control values is $u_l = u_r$, the system can be simplified to a SISO system which is equivalent to a single-point levitation system.

$$\begin{aligned} \begin{bmatrix} \dot{x}_1 \\ \dot{x}_2 \end{bmatrix} &= \begin{bmatrix} 0 & 1 \\ 2k_e \frac{i_0^2}{m\delta_0^3} & 0 \end{bmatrix} \begin{bmatrix} x_1 \\ x_2 \end{bmatrix} + \begin{bmatrix} 0 & 0 \\ 0 & -2k_e \frac{i_0}{m\delta_0^2} \end{bmatrix} \begin{bmatrix} 0 \\ u_2 \end{bmatrix} \\ &= \begin{bmatrix} 0 & 1 \\ 2k_e \frac{i_0^2}{m\delta_0^3} & 0 \end{bmatrix} \begin{bmatrix} x_1 \\ x_2 \end{bmatrix} + \begin{bmatrix} 0 \\ -2k_e \frac{i_0}{m\delta_0^2} \end{bmatrix} u_2 \end{aligned} \tag{3.22}$$

3.2.2 *Design and Analysis of Levitation System Controller*

To facilitate implementation and easy adjustment of the controller parameters, the controller in the form of linear feedback is used. Among the design methods for linear systems, the controller designed with LQR has simple and practical features, and there are many references available. A nominal controller using LQR for the joint structure-based levitation system is designed, and the simulation analysis is performed.

The levitation response curve of the joint structure-based levitation system is shown in Fig. 3.3. From the simulation results, the system response is fast and can reach stable levitation in 0.1 s with basically no overshoot, and the levitation gap is 12 mm at steady state with no static error. The simulation results show that the incorporation of the integration loop and the LQR-based controller can improve the performance of the system.

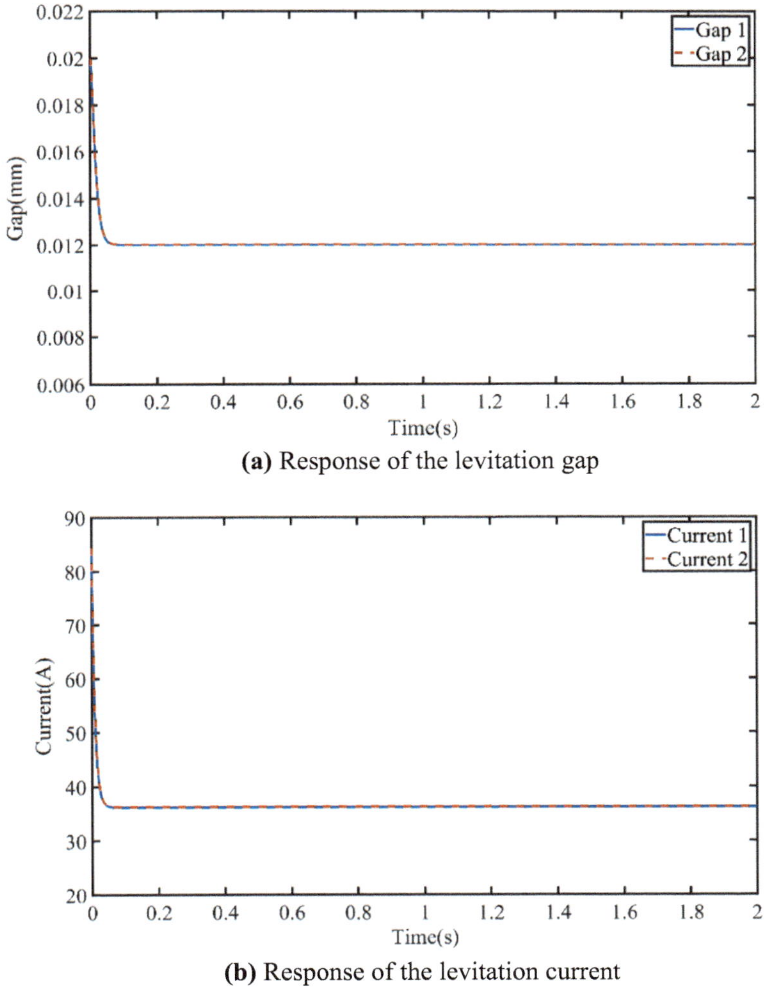

Fig. 3.3 Levitation response of the joint structure-based levitation system

3.3 Design of ADRC Controller for Levitation System

The levitation system is the core system of the high-speed maglev train. Its performance directly affects the safe and reliable operation of the high-speed maglev train. In the previous section, the modeling and design of the nominal system for the levitation system of the high-speed maglev train are introduced. In actual operation, the levitation system is subject to many external disturbances, resulting in large fluctuations of the levitation gap and even instability. To ensure the safe and reliable operation of the maglev train, it is very important to solve the impact of external disturbances on the levitation system. And it is also one of the research focuses in the field of the levitation control.

The main disturbance source in the operation of the high-speed maglev train comes from the track irregularity and the normal force of propulsion system. The vibration amplitude of the vehicle body will increase due to the track irregularity, which directly affects the levitation stability, running safety and riding comfort of the maglev train. In addition, the violent fluctuations of the normal force during the acceleration and deceleration of the maglev train will also cause the instability of the levitation system. Based on the joint structure model established in the previous section, in this section, the influence of the track irregularity and the fluctuations of the normal force on the levitation system are first analyzed. Then, the ADRC method is introduced, and an ADRC levitation controller is designed to solve the disturbance rejection problem of the system. Finally, through simulation experiments, it is proved that ADRC controller has better control performance and better disturbance rejection ability than PID controller.

3.3.1 Influence of Track Irregularity on Levitation System

Due to the foundation settlement, the processing, installation and measurement error of the track, the sleeper looseness, irregularities often occur in the track of the maglev train, which can be generally divided into three types: the short-wave irregularity, the long-wave irregularity and the railway step.

For the high-speed maglev trains, the short-wave irregularity affects the system in the form of high-frequency excitation. Away from the system cutoff frequency, it generally does not affect the operation and comfort of the maglev train. The levitation gap of the long-wave irregularity changes slowly, and the general controller can basically meet the stable levitation control. The railway step is a special kind of short-wave irregularity, and its levitation gap changes greatly in a moment. This change has a great impact on the levitation control system, mainly in the following three aspects: the large change in model parameters, the unmodeled parts appear in the model, and the fluctuations of the normal force. If the amplitude of the railway step is large, the maglev train will even smash the track.

(1) Variation of model parameters

Through the analysis of the joint structure model (shown in Eq. 3.20), it can be seen that the values of model parameters k_δ and k_i in the linearized model near the equilibrium point are related to the equilibrium gap. The equilibrium point δ_0 of the levitation control system can be considered constant when the fluctuation of the levitation gap is small. However, when the maglev train crosses the railway step, the equilibrium point of the system changes, and the variations of k_δ and k_i at the new equilibrium point are Δk_δ and Δk_i. At this time, the differential equations of x_2 and x_4 in the model become:

$$\begin{cases} \dot{x}_2 = \frac{2k_\delta - k_s}{2m}x_1 + \frac{k_s}{2m}x_3 + \frac{-k_i}{m}u_1 + \Delta K_1 \\ \dot{x}_4 = \frac{k_s}{2m}x_1 + \frac{2k_\delta - k_s}{2m}x_3 + \frac{-k_i}{m}u_2 + \Delta K_2 \\ \Delta K_1 = \frac{\Delta k_\delta}{m}x_1 + \frac{-\Delta k_i}{m}u_1 \\ \Delta K_2 = \frac{\Delta k_\delta}{m}x_3 + \frac{-\Delta k_i}{m}u_2 \end{cases} \tag{3.23}$$

The variation of the model parameters caused by the deviation from the equilibrium point can be regarded as an internal disturbance of the model, which is related to the levitation gap of the equilibrium point. In case of the railway step, a step change will occur in the levitation gap, resulting in the change of the equilibrium point. Therefore, the disturbance caused by the change of model parameters can be regarded as a step disturbance.

The change of the equilibrium point except for k_i and k_δ will also change the inductance of the electromagnet. According to Eq. (3.24), the inductance of the electromagnet is related to the levitation gap.

$$L(\delta) = 2k_e/\delta \tag{3.24}$$

When the levitation gap fluctuates little, the inductance of the electromagnet is regarded as a constant ($L_0 = 2k_e/\delta_0$), and the electrical equation can be simplified as:

$$U_{in} = Ri + L_0\dot{i} \tag{3.25}$$

Equation (3.25) is regarded as a proportional loop by designing the current loop. The current loop adopts the current feedback method to design the time constant of the first-order inertia loop as $T = 0.0025$, and the proportional coefficient is 1. The transfer function of the voltage–current of the electromagnet after the design of the current loop is:

$$G(s) = \frac{k_{c1}/(R + k_{c2}k_{c1})}{\frac{2k_e}{\delta_0(R + k_{c2}k_{c1})}s + 1} \tag{3.26}$$

When k_{c2} and k_{c1} are determined, the only variable that affects the time constant of the current loop is equilibrium point gap (δ_0). When the gap is increased a times, i.e., $\delta = a\delta_0$, time constant $\overline{T} = aT$. In general, the maximum value of a does not exceed 2. When $a = 2$, $\overline{T} = 2T$. In actual engineering, $T \approx 0.002$, when the equilibrium gap is doubled, the current loop time constant is 0.004 s. On the other hand, the equilibrium gap has no effect on the proportional characteristics of the current loop. Therefore, the influence of the change of the electromagnet inductance caused by the railway step on the levitation control can be ignored.

(2) Unmodeled part

The electromagnetic force equation of the electromagnet is:

$$F_e = k_e\left(\frac{i}{\delta}\right)^2 \tag{3.27}$$

The premise of using the minimum deviation method to linearize the nonlinear mathematical model is that the deviation of the variables from the equilibrium point is very small. According to the series theory, if all derivatives of the variable exist in the given interval, the nonlinear function can be expanded into Taylor series in the neighborhood of the given equilibrium point, as shown in Eq. (3.28). When the range of deviation is very small, the higher-order terms of deviation in the series can be ignored, and the linear equation containing only the first-order deviation term can be obtained.

$$\begin{aligned} F_{el} = {} & k_e\left(\frac{i_0}{\delta_0}\right)^2 + \frac{2k_e i_0}{\delta_0^2}\Delta i - \frac{2k_e i_0^2}{\delta_0^3}\Delta\delta \\ & + \frac{1}{2}\left(\frac{2k_e}{\delta_0^2}\Delta i^2 - \frac{4k_e i_0}{\delta_0^3}\Delta i\Delta\delta + \frac{6k_e i_0^2}{\delta_0^4}\Delta\delta^2\right) + \cdots \end{aligned} \tag{3.28}$$

When the state variable is far from the equilibrium point, ignoring the higher-order terms in the Taylor series will cause an unmodeled part of the model. When designing the controller, the unmodeled part will impact the performance of the controller. Particularly, if the deviation from the equilibrium point is large, the stability of the controller will decrease.

It can be seen from Eq. (3.28) that the higher-order terms of Taylor series of the electromagnetic force is a function of the deviation value of the equilibrium point and is proportional to the deviation value. When considering the model inaccuracy caused by the railway step, it is necessary to add a nonlinear disturbance related to the levitation gap into the linearized model, which donate as $g(\delta)$. The disturbance is reflected in the model of the controlled object as shown in Eq. (3.29).

$$\begin{cases} \dot{x}_2 = \frac{2k_\delta - k_s}{2m} x_1 + \frac{k_s}{2m} x_3 + \frac{-k_i}{m} u_1 + g(\delta) \\ \dot{x}_4 = \frac{k_s}{2m} x_1 + \frac{2k_\delta - k_s}{2m} x_3 + \frac{-k_i}{m} u_2 + g(\delta) \end{cases} \tag{3.29}$$

To sum up, the model parameters changes and the unmodeled part of the joint structure can be regarded as the internal disturbance of the system in case of the railway step, and both are added to the levitation control model in the form of step disturbance.

3.3.2 Design of ADRC Controller

Through the analysis of the influence of the railway step and the normal force on the levitation system, the model of the original joint structure can be rewritten as:

$$\begin{cases} \dot{x}_1 = x_2 \\ \dot{x}_2 = \dfrac{2k_\delta - k_s}{2m} x_1 + \dfrac{k_s}{2m} x_3 + \dfrac{-k_i}{m} u_1 + \Delta K_1 + \Delta F_{\text{ver}}(t) + g(\delta) \\ \dot{x}_3 = x_4 \\ \dot{x}_4 = \dfrac{k_s}{2m} x_1 + \dfrac{2k_\delta - k_s}{2m} x_3 + \dfrac{-k_i}{m} u_2 + \Delta K_2 + \Delta F_{\text{ver}}(t) + g(\delta) \\ y_1 = x_1 \\ y_2 = x_2 \end{cases} \tag{3.30}$$

It can be seen from Eq. (3.30) that the joint structure-based levitation system is a system with uncertain model parameters, unmodeled parts and the external torque fluctuation during the train operation, and the magnitude and frequency of the disturbances are unknown. For the torque fluctuation, the information of the torque disturbance can be obtained by adding the acceleration feedback in the project, but it is difficult to obtain the internal disturbance of the system by adding sensors.

The nominal controller mentioned above cannot observe and compensate the system disturbance. Therefore, it is necessary to introduce a new control method to solve the problem of the system's uncertain disturbance.

ADRC is a very valuable control method proposed by Mr. Han in 1993. The core idea is to solve the internal and external disturbances of the system by observing the total disturbance of the system and eliminating it by the control law. To solve the instability problem of the levitation controller caused by the railway step and the normal force, it has certain research value and engineering application prospect to apply the ADRC method to the design of the levitation controller of the high-speed maglev train.

The ADRC consists of a tracking differentiator (TD), an extended state observer (ESO), and a nonlinear state error feedback (NLSEF). The control block diagram is shown in Fig. 3.4. The control principle of ADRC is to design an ESO to estimate the system's total disturbance through the input and output characteristics of the model

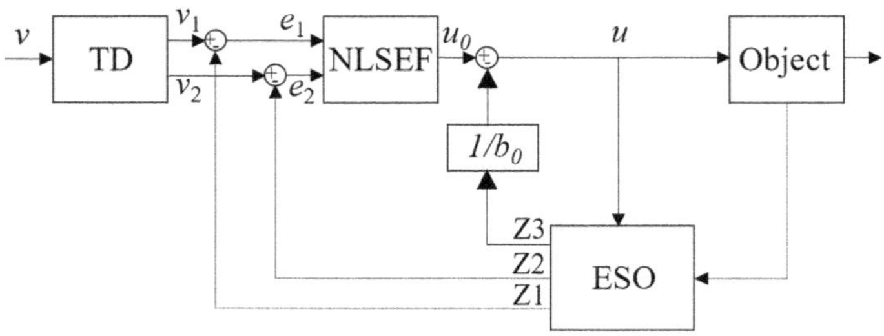

Fig. 3.4 Block diagram of the ADRC controller

itself. And the obtained system's total disturbance is feedback to the control law to cancel the disturbance in the model to achieve the purpose of disturbance rejection.

In this section, a nonlinear extended state observer (NESO) is used to observe the total disturbance. The observer can be expressed by the following equation:

$$\begin{cases} e_1 = z_1 - x_1, e_2 = z_2 - x_3 \\ \dot{z}_1 = z_2 - \beta_1 |e_1|^{a_1} \text{sign}(e_1) \\ \dot{z}_2 = \frac{2k_\delta - k_s}{2m} x_1 + \frac{k_s}{2m} x_3 + \frac{-k_i}{m} u_1 - \beta_2 \text{fal}(e_1, a_2, \sigma) \\ \dot{z}_3 = -\beta_3 \text{fal}(e_1, a_3, \sigma) \\ \dot{z}_4 = z_5 - \beta_1 |e_2|^{a_1} \text{sign}(e_2) \\ \dot{z}_5 = \frac{k_s}{2m} x_1 + \frac{2k_\delta - k_s}{2m} x_3 + \frac{-k_i}{m} u_2 - \beta_2 \text{fal}(e_2, a_2, \sigma) \\ \dot{z}_6 = -\beta_3 \text{fal}(e_2, a_3, \sigma) \end{cases} \tag{3.31}$$

where β_i are the gain of NESO, σ is boundary layer thickness of NESO, a_i are nonlinear feedback degree.

$\text{fal}(\cdot)$ is a nonlinear feedback function, which is used to solve the chattering problem of NESO in the tracking process. The expression is:

$$\text{fal}(e, a_i, \sigma) = \begin{cases} |e|^{a_i} \text{sign}(e), & |e| > \sigma \\ e/\sigma^{1-a_i}, & |e| \le \sigma \end{cases} \tag{3.32}$$

According to the theory of ADRC, by setting the appropriate NESO gain, z_1, z_2, z_4, z_5 can approximate x_1, x_2, x_3, x_4 in a limited time, and the errors are determined by β_i and a_i. Generally, β_i can be expressed using the bandwidth w_o. The expression is: $\beta_1 = 3 * w_o$, $\beta_2 = 3 * w_o^2/5$, $\beta_3 = w_o^3/9$.

Selecting the appropriate observation bandwidth can make the observer achieve the required accuracy. Thus, z_3 and z_6 can be approximately regarded as the total disturbance of the system. The total disturbance is feedback to the controller to obtain a new control value: $\overline{u}_1 = \frac{-k_i}{m} u_1 - z_3$, $\overline{u}_2 = \frac{-k_i}{m} u_2 - z_6$.

The most classical PID control is designed for u_1 and u_2, and the control values are:

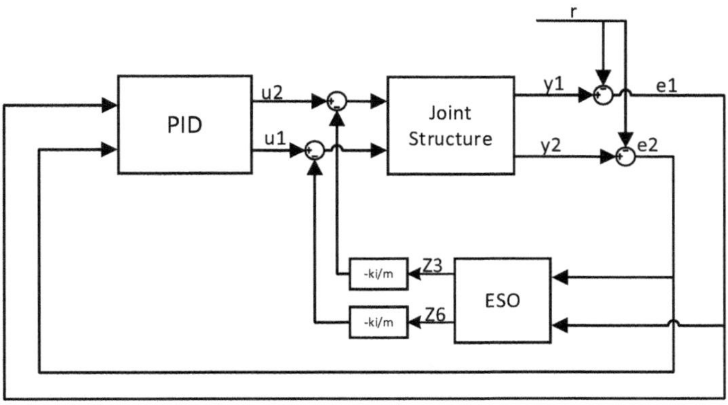

Fig. 3.5 Block diagram of the levitation controller based on ADRC

$$u_1 = K_p(x_1 - \delta_l) + K_d(\dot{x}_1 - \dot{\delta}_l) + K_i \int (x_1 - \delta_l)dt \tag{3.33}$$

$$u_2 = K_p(x_3 - \delta_r) + K_d(\dot{x}_3 - \dot{\delta}_r) + K_i \int (x_3 - \delta_r)dt \tag{3.34}$$

The block diagram of the levitation controller based on ADRC is shown in Fig. 3.5.

The characteristic of ADRC levitation controller is that it can observe the system torque disturbance, model uncertainty and model parameter change caused by the normal force and the railway step through NESO. By eliminating the disturbance, the model of the levitation control system can be "basically unchanged" under any working conditions. Therefore, the levitation control system based on ADRC can adapt to different system internal and external disturbances without parameter readjustment.

3.3.3 Simulation Analysis

(1) Simulation experiment of the railway step

Relevant parameters of NESO are set as: $w_o = 50$, $a_1 = 1$, $a_3 = 0.25$, $\sigma = 0.005$, $\varepsilon = 0.002$. PID parameters are set to $K_p = 6000$, $K_d = 35$, $K_i = 2000$. These parameters of PID are set by the method of trial and error. Simulation step $h = 0.001$.

When the maglev train crosses the railway step, three kinds of disturbances will occur: the variation of model parameters, the disturbance of the unmodeled part and the normal force disturbance. Therein, the variation of model parameters and the disturbance of the unmodeled part belong to the step disturbance, whose magnitude is related to the height of the railway step. The normal force disturbance belongs to the slope disturbance, and its change rate and amplitude are related to the speed of the maglev train and the height of the railway step, respectively.

The process of a high-speed maglev train passing through a railway step with a height of 6 mm and a length of 30.96 m at speed of 600 km/h is taken as the simulation object. Through calculation, it can be obtained that the interval time of disturbance generated by passing through the railway step is 186 ms. The amplitude of the step disturbance is designed to be 10 mm, which is added to the system at $T = 60$ ms and $T = 250$ ms respectively. The designed slope of the slope disturbance is 10, which increases from $T = 60$ ms to $T = 70$ ms, and then decreases to 0 with a slope of -10 after $T = 250$ ms. In addition, to simulate the changes of system parameters when passing through the railway step, the designed equilibrium point (d_0) becomes 18 mm at $T = 60$ ms, $d_0 = 12$ mm at $T = 250$ ms. The system parameters are calculated according to the changed equilibrium point and added to the simulation model.

Figure 3.6 shows the levitation gap curves of the PID levitation controller. The maximum fluctuation values of the levitation gap are 8.9 mm and 6.5 mm respectively when passing through the railway step twice. The maximum difference between the actual levitation gap and the target levitation gap is 5.5 mm. The joint structure takes 50 ms to return to the target levitation gap of 12 mm, and there is large vibration.

Figure 3.7 shows the levitation gap curves of the levitation controller based on ADRC. It can be seen that the levitation gap fluctuates only at the moment of passing through the railway step and can be restored to the target levitation gap in a very short time. The maximum fluctuation values of the gap are 9.5 mm and 9 mm respectively

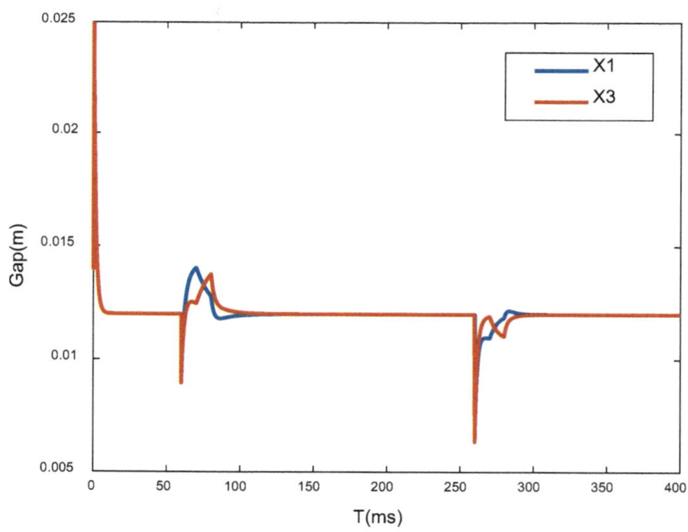

Fig. 3.6 Levitation gap curves of the PID controller in case of the railway step

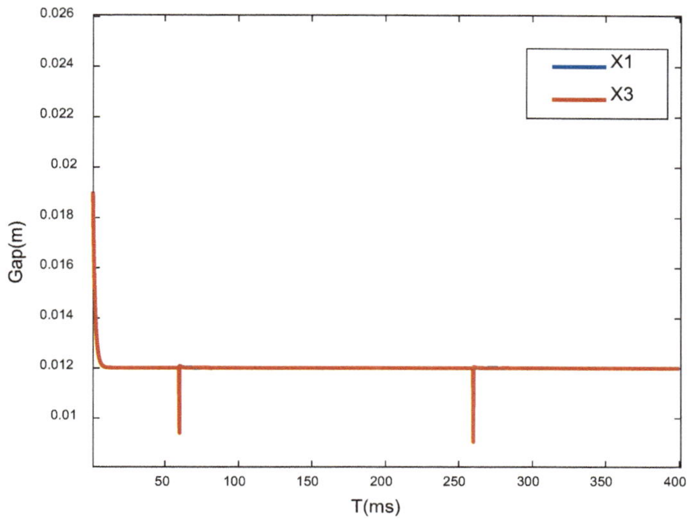

Fig. 3.7 Levitation gap curves of the ADRC controller in case of the railway step

when passing through the railway step twice. The maximum difference between the actual levitation gap and the target levitation gap is 3 mm.

Figure 3.8 shows the NESO's observation curves of disturbance caused by train passing through the railway step. It can be seen that z_3 and z_6 are superposition curves of square-wave disturbance and slope disturbance, and they are accurately tracked at the time when the designed disturbance occurs.

As shown in Fig. 3.9, the levitation gap curves of ADRC controller and PID controller are compared. ADRC can observe the total disturbance of the system in real time and compensate it through the control law, thus eliminating the influence of disturbance on the control system. Therefore, using ADRC in the design of the levitation controller can not only reduce the fluctuation of the levitation gap, but also improve the response speed of the system, and thus reduce the vibration and impact of the train when passing through the railway step.

(2) Simulation experiment of maglev train acceleration and deceleration process

During the acceleration and deceleration phase of the maglev train, the normal force fluctuates violently due to the change of the propulsion force. Through the analysis in Sect. 3.3.2, the normal force changes greatly during the acceleration process. When the train speed tends to be constant, the normal force also decreases and finally remains constant. When $T = 60$ ms, a slope disturbance (FV) with a slope of 200 is added, and it remains at -2000 at $T = 70$ ms. This stage simulates the increase of the normal force caused by the sudden increase of the propulsion force. When $T = 75$ ms, the slope of the slope disturbance is 20, and the disturbance stops changing and remains -400 at $T = 155$ ms. In this stage, the change of normal force in the process of propulsion force decreasing and tending to be constant is simulated.

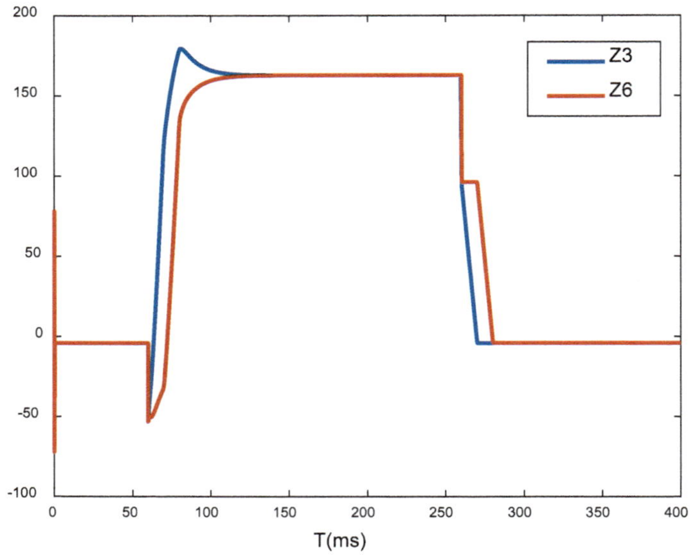

Fig. 3.8 NESO disturbance observation curve

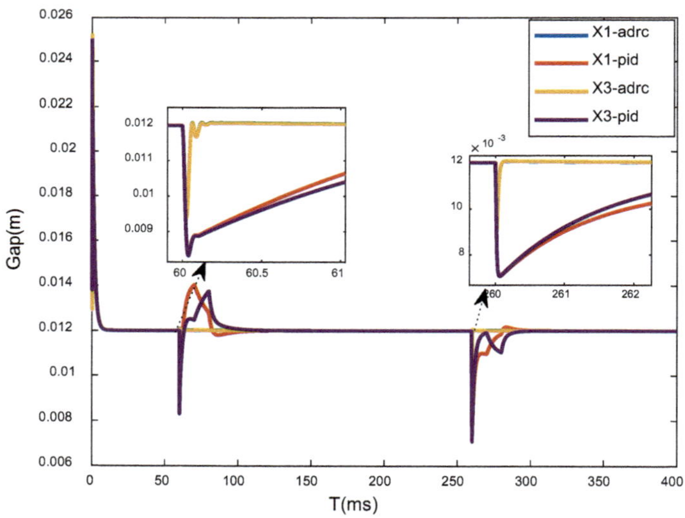

Fig. 3.9 Comparison curves of the levitation gaps between ADRC and PID controller in case of the railway step

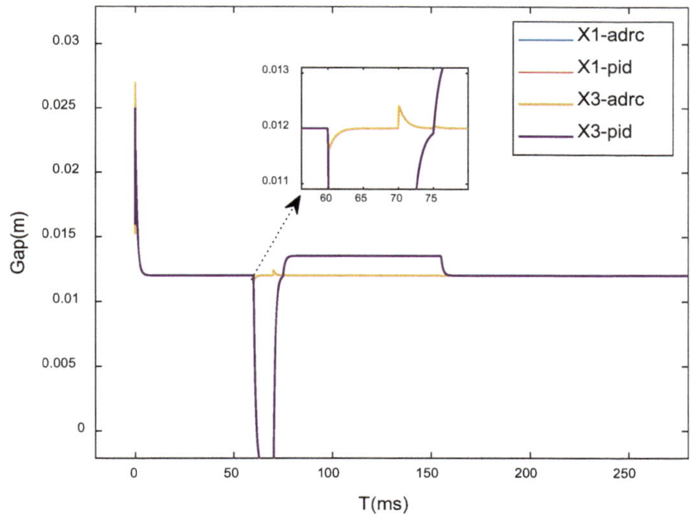

Fig. 3.10 Comparison curves of the levitation gaps between ADRC and PID controller when the normal force increases

As shown in Fig. 3.10, the variation curves of the levitation gaps between ADRC controller and PID controller are compared. The PID controller loses stability during the violent fluctuation of the normal force, and the levitation gap becomes 0. At this moment, the sliding block of the train rubs against the guideway, and the levitation fails. When the normal force fluctuation is small, the PID control restores the levitation gap to about 12 mm again, and it is stabilized at 12 mm after a period. The ADRC controller can observe and compensate the torque disturbance, so the levitation gap basically keeps unchanged during the violent fluctuation of the normal force. The ADRC controller can ensure the levitation stability of the train during acceleration and deceleration, and avoid the levitation failure.

As shown in Fig. 3.11, FV is the designed torque disturbance curve, and z_3 is the observation curve of disturbance. It can be seen that z_3 can track FV well, and the place with large error only occurs at the inflection point of the curve. This is because there is a certain phase lag in the tracking of NESO, but its influence on the slope disturbance can be ignored.

Through the above experiments, we can get the following conclusions: (1) the rejection disturbance ability of the levitation controller based on ADRC is much better than that of the PID controller. (2) Aiming at the disturbance problem caused by the normal force and the railway step during the operation of the high-speed maglev train, the ADRC control method can solve the influence of the internal and external disturbances of the system on the levitation stability, and ensure the stable operation of the maglev train.

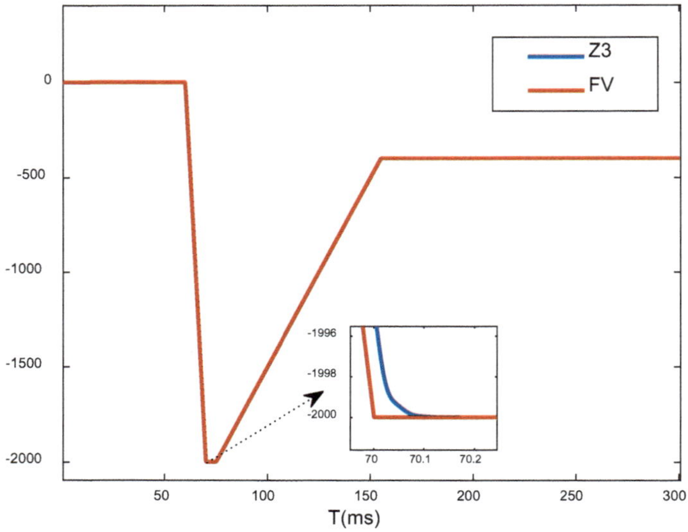

Fig. 3.11 NESO disturbance observation curve

3.4 Design and Implementation of Controller for Levitation System

In the engineering application of the high-speed maglev train levitation system, first, the main hardware circuit of the levitation controller and the hardware circuit of the kernel computer are designed. Secondly, aiming at the software system of the high-speed maglev train controller, the software architecture based on the dual DSP is realized so as to design and develop the high-speed maglev train controller. Finally, a high-speed maglev dual maglev bogie test platform is built, and a series of performance tests of the levitation system are carried out to verify the basic functions of the levitation controller.

3.4.1 Hardware Design of Levitation Controller

According to the main hardware circuit of the high-speed maglev train levitation controller shown in Fig. 3.12, the input power of the levitation controller is 440 V DC voltage. When the controller receives the levitation command, the auxiliary contactor is connected immediately and the 440 V DC power starts to charge the capacitor through the EMI filter. And the capacitor charging ends when the voltage across the capacitor is 440 V. At this moment, the main contractor in the levitation controller immediately turns on and the auxiliary contactor disconnects. The kernel computer in the levitation controller calculates the control value in real-time by

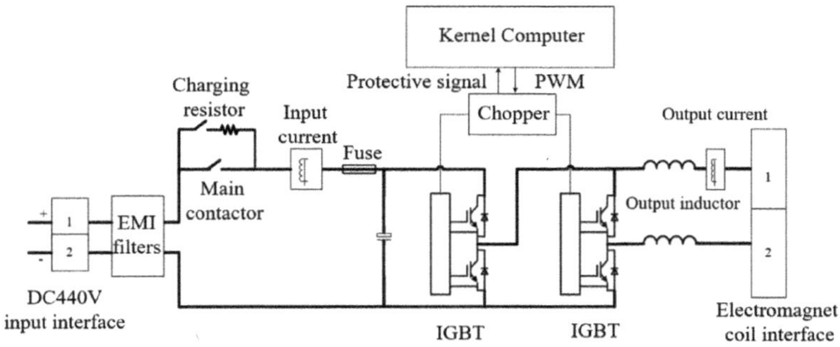

Fig. 3.12 Main circuit of the high-speed maglev train levitation controller

integrating the measured data of gap, acceleration, current, voltage, etc., from the sensors and outputs it to the chopper in the form of PWM wave. The chopper in the levitation controller consists of two sets of IGBT modules and starts to output current, which drives the electromagnet to generate levitation force to make the train levitate stably.

The external interface unit of the high-speed maglev train levitation controller mainly contains sensors as well as an online commissioning and monitoring system. The sensors include gap sensors, current transformers, accelerometers, etc. The sensors are mainly used to obtain the state information of the high-speed maglev train electromagnet in real-time. And the current transformers are mainly used for current feedback by detecting and measuring the current in the chopper. The kernel computer is composed of four circuit boards: the control board, the signal processing board, the chopper management board, and the interface board (Fig. 3.13).

3.4.2 *Software Design of Levitation Controller*

After the hardware circuit design of the high-speed maglev train levitation controller is completed, the corresponding software system needs to be developed to implement the above control algorithms, and the control circuit in the kernel computer adopts the mode of dual DSP core processing chips.

DSPA is mainly responsible for obtaining real-time data of the levitation control system. A large amount of data transmitted from the levitation sensor is processed by the signal processing circuit of the kernel computer and sent to DSPA. To enhance the rate of data transmission and storage and ensure real-time performance, DSPA completes the data acquisition by DMA in combination with McBSP. The software architecture of DSPA adopts the timing interrupt mode, and every 200 μs, DSPA reads and updates the data and writes them into the dual-port RAM.

The main function of DSPB is to implement the control algorithm and output the control value. When the program enters the main loop, the gap, current, voltage and

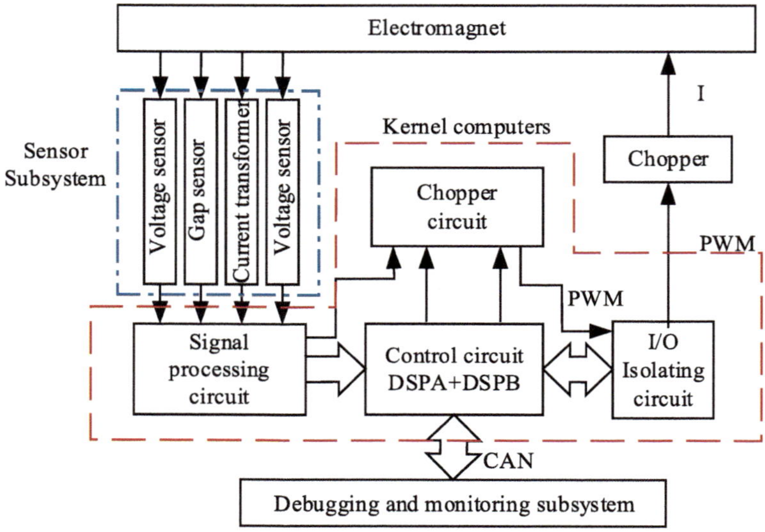

Fig. 3.13 Kernel computer of the high-speed maglev train levitation controller

some other state information of the levitation point will be transmitted to the host computer via CAN bus, and the commands from the host computer will be accepted at the same time. DSPB runs in interrupt mode synchronized with DSPA, that is, the control cycle and data acquisition cycle are kept in line.

3.4.3 Test Platform for Levitation System

To carry out the performance test of the high-speed maglev train levitation system and enrich the engineering application of the levitation controller, the dual maglev bogie test platform is built by using the levitation controllers, the sensors, and the electromagnets. The dual maglev bogie test platform is the smallest functional unit of the high-speed maglev train levitation and guidance system and can be used to carry out a series of performance tests. The dual maglev bogie test platform is shown in Fig. 3.14, where the weights on the dual maglev bogie test platform are used to simulate the load.

3.4.4 Performance Testing of Levitation System

To verify the basic functions of the levitation controller, a series of levitation system performance tests need to be conducted using the test platform shown in Fig. 3.14.

Fig. 3.14 High-speed maglev train dual maglev bogie test platform

The performance tests mainly include a series of tests on levitating, landing, static levitation and applied disturbance, etc. Through these tests, the performance of the levitation controller can be fully verified.

(1) Levitating test

The levitating test is the process to check the function of the levitation controller by realizing the levitation system from the initial state to the stable levitation state. The initial levitation gap is set at about 20 mm, and the levitation gap will change to the desired levitation gap of 12 mm after the levitating process is completed. During the design and commission of the actual levitation control system, to improve the comfort of the high-speed maglev train, a transition process is added into the input process, so that the entire levitating process is also relatively gentle.

(2) Landing test

The landing test is the process to check the function of the levitation controller by realizing the levitation system from the initial stable levitation state to the landing state. The stable levitation gap is set to about 12 mm during the landing process, and the levitation gap will become 20 mm after the landing process is completed. During the design and commission of the actual levitation control system, to improve the comfort of the high-speed maglev train, a transition process is added into the input process, so that the entire landing process is also relatively gentle.

(3) Static levitation test

The static levitation test is to check the function of the levitation controller by realizing that the vehicle is statically levitated on the guideway surface and has no horizontal movement relative to the guideway.

(4) Disturbance test

The disturbance test is to verify the performance of the levitation system under the disturbance by applying disturbance to the levitation system while it remains in levitation.

The performances of the levitation system are shown below:

1. Basic performance test of the levitation system

When the levitation system test platform receives the levitating command, the left and right-side electromagnets of the joint structure start to float, and the left and right-side electromagnets reach the given levitation gap and then levitate stably at the desired gap. The dynamic responses of the left- and right-side electromagnets during the levitating process are shown in Figs. 3.15 and 3.16.

When the levitation system test platform receives the landing command, the left and right-side electromagnets of the joint structure start to land and slowly begin to descend from the stable levitation state. The dynamic responses of the left- and right-side electromagnets during the landing process are shown in Figs. 3.17 and 3.18.

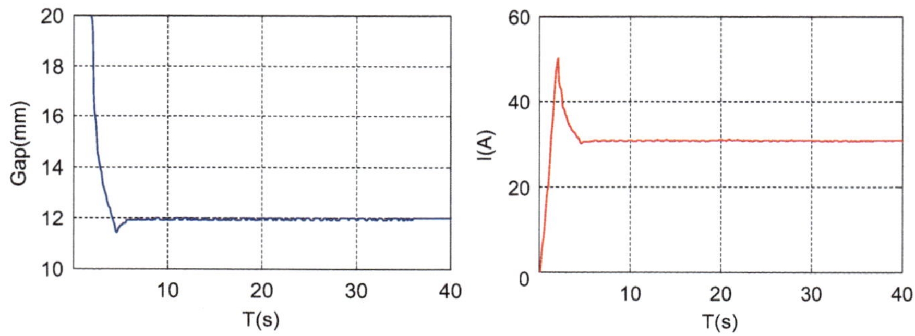

Fig. 3.15 Dynamic characteristic curves of the left-side electromagnet during the levitating process

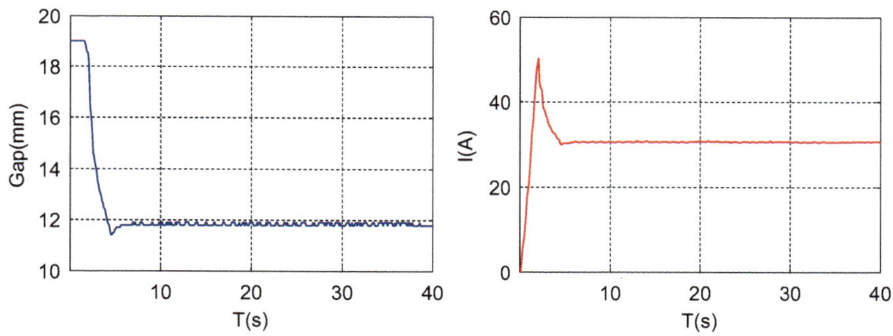

Fig. 3.16 Dynamic characteristic curves of the right-side electromagnet during the levitating process

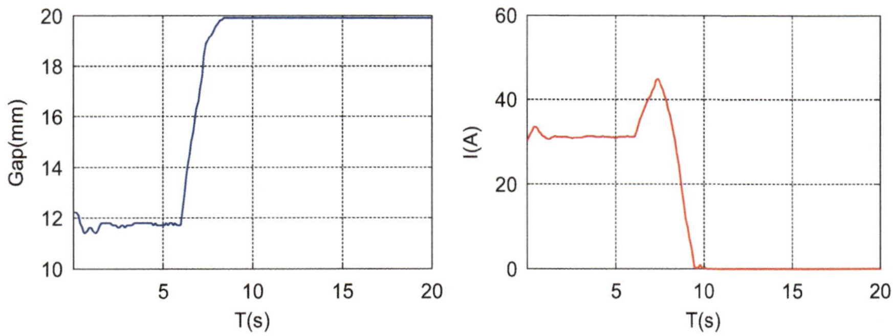

Fig. 3.17 Dynamic characteristic curves of the left-side electromagnet during the landing process

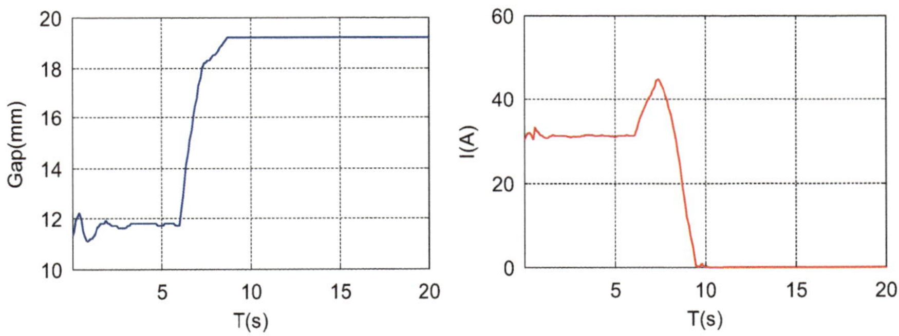

Fig. 3.18 Dynamic characteristic curves of the right-side electromagnet during the landing process

When the levitation system test platform is in static levitation, the left and right-side electromagnets of the joint structure are steadily levitated in the desired gap. The dynamic responses of the left- and right-side electromagnets during the static levitation are shown in Figs. 3.19 and 3.20.

As can be seen from Figs. 3.15, 3.16, 3.17, 3.18, 3.19 and 3.20, a series of basic levitation performance tests such as levitating, landing, and static levitation were completed, and the experimental data could verify that the developed levitation controller can achieve stable levitating, landing, and static levitation.

2. Performance test of levitation systems under disturbance

To further assess the performance of the levitation system under disturbance, the system is subjected to disturbance when it is in its normal levitation state. The dynamic responses of the left- and right-side electromagnets of the joint structure are shown in Figs. 3.21 and 3.22.

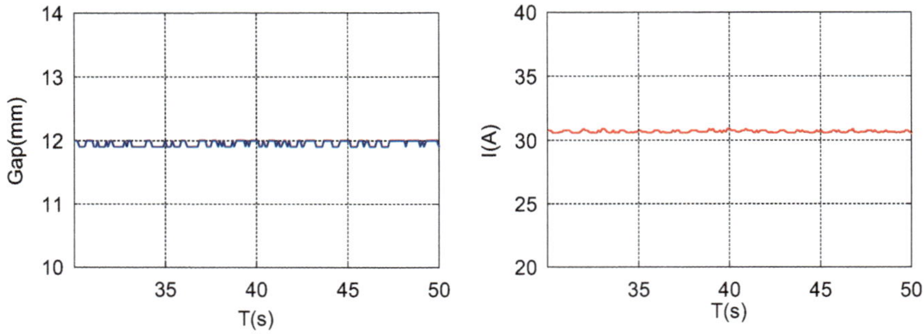

Fig. 3.19 Dynamic characteristic curves of the left-side electromagnet during the static levitation

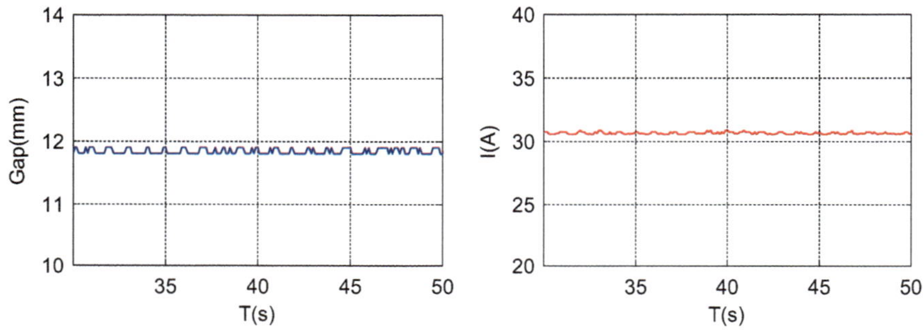

Fig. 3.20 Dynamic characteristic curves of the right-side electromagnet during the static levitation

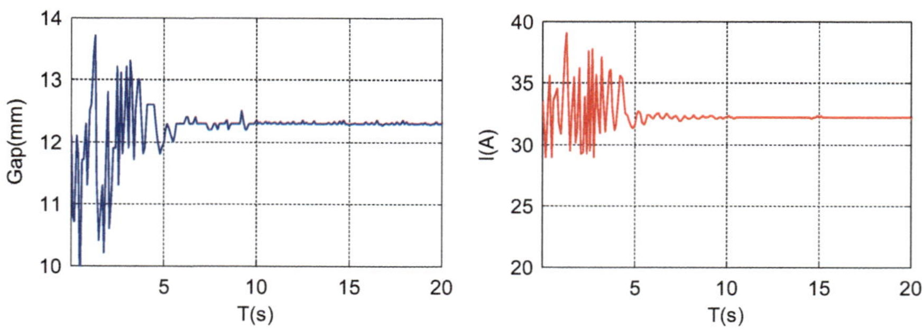

Fig. 3.21 Dynamic characteristic curves of the left-side electromagnet with disturbance

From Figs. 3.21 and 3.22, it can be seen that the current response and the gap response of the levitation system fluctuate when it is disturbed, but under the dynamic regulation of the levitation controller, the performance of the levitation system is optimized and recovered in 5 s. And the gap response and current response in the dynamic characteristic curve are quickly restored to the stable levitation state.

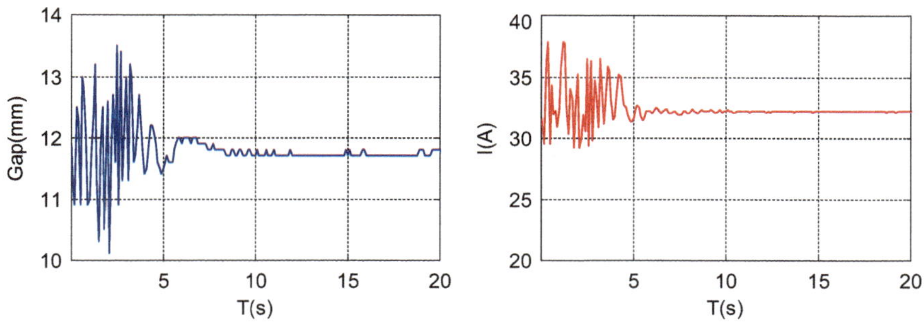

Fig. 3.22 Dynamic characteristic curves of the right-side electromagnet with disturbance

In summary, a series of performance tests of the levitation system, such as levitating, landing, static levitation, and applied disturbance, were completed by using the high-speed maglev dual maglev bogie test platform. It can be seen from the test results that the levitation controller can ensure the stable levitation and has good performance.

3.5 Conclusion

Firstly, the basic structure of the levitation system is analyzed, the force analysis of the electromagnet of the levitation system is carried out, the mathematical model of the levitation system based on the joint structure is established, and the linearization and stability analysis is carried out. In addition, for issues of track irregularities and propulsion disturbance, the ADRC for the levitation system is designed. Finally, the high-speed maglev dual maglev bogie test platform is built, and on this basis, a series of levitation system performance tests are carried out to verify the basic functions of the levitation controller. The results show that the designed levitation controller is effective and has good performance.

Bibliography

1. Chen C, Xu J, Lin G et al (2022) Model identification and nonlinear adaptive control of suspension system of high-speed maglev train. Veh Syst Dyn 60(3):884–905
2. Ding S (2022) 600 km/h high-speed maglev transportation system. Shanghai Scientific and Technical Publishers, Shanghai
3. Fu S, Wu D, Han W et al (2023) Modeling and analysis of high-speed maglev electromagnets based on nonlinear materials. J Southwest Jiaotong Univ 58(4):879–885
4. Hao A (2008) Research on key technologies of permanent magnet guideway high-speed maglev train suspension and guidance system. Ph.D. dissertation, National University of Defense Technology Doctor Degree, Changsha

5. Huaqing Liu Compile (1995) The German maglev train-Transrapid. University of Electronic Science and Technology Press, Sichuang
6. Li X (2009) Research on key technologies of suspension control and signal processing for high-speed maglev trains. Ph.D. dissertation, National University of Defense Technology Doctor Degree, Changsha
7. Li Y, Chang W (1999) Cascade control of an EMS maglev vehicle's levitation control system. Acta Autom Sin 25(2):247–251
8. Lin G, Liu W, Xu J et al (2023) Opportunities and challenges for the development of high-speed maglev transportation in China. Sci Technol Foresight 2(4):7–18
9. Liu Z, Long Z, Li X (2015) Maglev trains. Springer, Berlin
10. Long Z, Hao A, Chang W (2003a) Suspension controller design of maglev train considering the rail track periodical irregularity. J Natl Univ Defense Technol 25(2):84–89
11. Long Z, Hong H, Zhou X (2003b) Research of nonlinear control for maglev train. Control Theory Appl 20(3):399–402
12. Long Z, Hao A, Cao C (2004) Sensitivity constrained robust controller design of suspension controller for maglev train. Control Theory Appl 21(5):804–808
13. Long Z, Dou F, Wang Z et al (2023a) Current status and prospects of high-speed maglev suspension and guidance control technology. Sci Technol Foresight 2(4):78–88
14. Long Z, Li X, Cheng H (2023b) Research on permanent-electromagnetic levitation technology and application. Shanghai Scientific and Technical Publishers, Shanghai
15. Long Z, Zhai M, Wang Z (2023c) Condition monitoring, fault diagnosis and fault tolerant control of the maglev train. Shanghai Scientific and Technical Publishers, Shanghai
16. Shi X, Long Z (2009) Nonlinear vibration analysis of the maglev guideway-vehicle coupling control system. J China Railw Soc 31(4):38–42
17. Sun Y, Xu J, Qiang H et al (2019) Adaptive sliding mode control of maglev system based on RBF neural network minimum parameter learning method. Measurement 141:217–226
18. Wang Z (2019) Fault diagnosis and tolerant control for high speed maglev train suspension system. Ph.D. dissertation, National University of Defense Technology Doctor Degree, Changsha
19. Wu X (2003) Maglev train. Shanghai Scientific and Technical Publishers, Shanghai
20. Xie Y, Long Z (2009) A high-speed nonlinear discrete tracking-differentiator with high precision. Control Theory Appl 26(2):127–132
21. Yang Q (1989) The decoupled control system of the magnetic suspension test vehicle model. Acta Autom Sin 15(1):23–29
22. Zhai M (2019) Research on performance optimization of magnetic levitation system in high speed maglev train. Ph.D. dissertation, National University of Defense Technology Doctor Degree, Changsha
23. Zhang K (2021) High-speed maglev railway technology. China Railway Publishing House, Beijing
24. Zheng Q, Ni F (2023) High-speed maglev transportation: from Germany to China. Sci Technol Foresight 2(4):31–39
25. Zhou D, Li J, Yu P et al (2019) Analysis and adaptive control of the track induced self-excited vibration for the maglev transport. Acta Autom Sin 45(12):2328–2343

Chapter 4
Controller Design of High-Speed Maglev Train Guidance System

The guidance system of the high-speed maglev train enables the vehicle to realize non-contact guidance on the operating line. The guidance system provides the lateral force paralleling to align the vehicle centerline with the track centerline. By adjusting the current magnitude of the excitation coils in the guidance electromagnets, the magnitude of the guidance forces of the guidance system can be tuned so that the functional surfaces of both levitation and guidance electromagnets can maintain a certain gap from the functional surfaces of the track. In this chapter, by adoption of the nominal state, a joint structure-based mathematical model and a mathematical model of the guidance system under the consideration of disturbances and faults are established respectively. Then, the linear quadratic regulator (LQR) optimal control method is applied to design the nominal controller. A robust controller is proposed to guarantee good dynamic performance when the guidance system is disturbed by external disturbances. Finally, an experimental platform of the guidance system is set up and the performance test of the guidance control is carried out to verify the effectiveness of the designed guidance control algorithm.

4.1 Modeling and Analysis of Guidance System

The guidance system mainly consists of guidance electromagnets, guidance gap measurement units, and guidance electromagnet control units. The fundamental function of the guidance system is to realize lateral stability control of the train and maglev line tracking. There are 12 guidance electromagnets for each vehicle in total with 6 on each side. Each guidance electromagnet consists of 6 poles, and they are divided into two or three groups according to the different installation positions of electromagnets. Each group has a pair of guidance gap measurement units and a guidance electromagnet control unit. Based on the model of the guidance system, the design and simulation of the guidance controller are carried out.

Z. Long et al., *High-Speed Maglev Train's Levitation and Guidance Control*, Advances in High-speed Rail Technology, https://doi.org/10.1007/978-981-97-2309-6_4

Taking the non-end guidance unit as an object, the guidance system is reasonably assumed and simplified without affecting the design accuracy of the controller. The mathematical model of a single guidance unit based on the joint structure is established. The system is linearized at the equilibrium point so as to apply the linear control theory to simplify the analysis process. Finally, a model of the guidance system with disturbances and faults is established.

4.1.1 Force Analysis of Guidance Electromagnet

The guidance system only needs to overcome the torsional forces of the various levitation bogies against one another when the high-speed maglev trains are not disturbed from the outside. In this circumstance, the needed guidance force is almost zero value. The guidance system needs to provide a larger guidance force to counteract the wind and centrifugal forces when the train runs in windy conditions or through curves. The maximum guidance force that a guidance unit may provide when the train operates in normal weather is discussed below. The coordinate system is depicted in Fig. 4.1. A high-speed maglev line is allowed to have a minimum turning radius of 350 m and a maximum cross slope angle of 12°. When the train passes the curve at speed v, there exists an included angle between the levitation force and the y-axis direction due to the lateral deflection. The guidance force in this guidance unit is:

$$F_g = F_{g1} - F_{g2} = mg\sin\theta - (F_{l1} + F_{l2})\sin\alpha - m\frac{v^2}{r}\cos\theta - F_d \tag{4.1}$$

where F_{g1}, F_{g2}, F_{l1}, and F_{l2} are the left-side guidance force, the right-side guidance force, the left-side levitation force, and the right-side levitation force, respectively. m is the mass of the bogie, F_v is the centrifugal force, F_d is the disturbance force in the direction of the negative half-axis of the x-axis, θ and α are the angle of slope and deviation angle of the levitation force from the y-axis.

It is shown in Eq. (4.1) that when the train is laterally deflected, the force provided by the levitation force in the positive direction along the x-axis is opposite to the direction of deflection, and it can share part of the lateral force for the guidance system. It is also the principle that the passive guidance method is used by urban

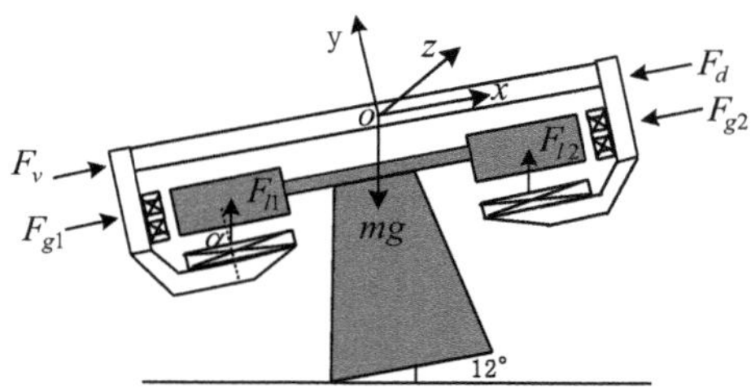

Fig. 4.1 Schematic of the basic guidance unit

maglev trains. Therefore, without the consideration of disturbance force outside, the levitation force is offset from the y-axis by 0° and the train speed is 0. When the train stops on the curve and is still at the center of the track, the guidance force supplied by the guidance system is the maximum. The magnitude of this maximum guidance force is the component force of the train gravity in the x-direction. The working load allowed for the non-end carriage of the new generation of high-speed maglev trains is 67 t, so a guidance electromagnet set has to carry approximately 1/16 of that weight. This implies a guidance force of roughly 8.5 kN.

4.1.2 Modeling of Guidance System Based on Joint Structure

The guidance electromagnet group of each guidance unit is composed of three electromagnets with identical parameters in series. This structure makes the mutual inductance of the three electromagnets in a group very small, and the gaps between them and the side rail are also approximately equal. Therefore, it can be equivalent to an electromagnet, and the influence of the other group of electromagnets in the same guidance unit is treated as an interference force.

Due to the complexity of the actual guidance system's components and working conditions, it is difficult to establish a mathematical model which is identical to the real system. Therefore, a few minor elements are ignored to simplify the design of the guidance controller, and the following assumptions are given.

(1) The guidance electromagnet is connected to the maglev bogie through the levitation device, the bracket arm, and the cross beam. Their elastic deformation can be ignored during modeling, and they can be regarded as rigid bodies.
(2) The deformation of the guideway is ignored, and the guideway is a flat rigid body.
(3) Only the lateral translation of the guidance electromagnet relative to the side rail is considered. Other motion directions, such as side-rolling, are not considered.
(4) The influence of the leakage flux is ignored. It is considered that the magnetic potential is uniformly distributed on the guidance gap, and the guidance force is uniformly concentrated on the center of a triangular structure's gravity of a group of guidance electromagnets.

Based on these assumptions, a basic guidance unit can be simplified to the structure shown in Fig. 4.2.

z_{l1}, z_{l2}, z_{r1}, and z_{r2} are the gaps from the left-side electromagnet 1, left-side electromagnet 2, right-side electromagnet 1, and right-side electromagnet 2 to the side rails, respectively. F_{l1}, F_{l2}, F_{r1}, and F_{r2} are the electromagnetic forces of the left-side electromagnet 1, left-side electromagnet 2, right-side electromagnet 1, and right-side electromagnet 2, respectively. F_g is the horizontal leftward interference force. u_{l1}, u_{l2}, u_{r1}, and u_{r2} are the voltages at the terminals of the left-side electromagnet 1, left-side electromagnet 2, right-side electromagnet 1, and right-side electromagnet 2, respectively. i_{l1}, i_{l2}, i_{r1}, and i_{r2} are the current of the left-side electromagnet 1,

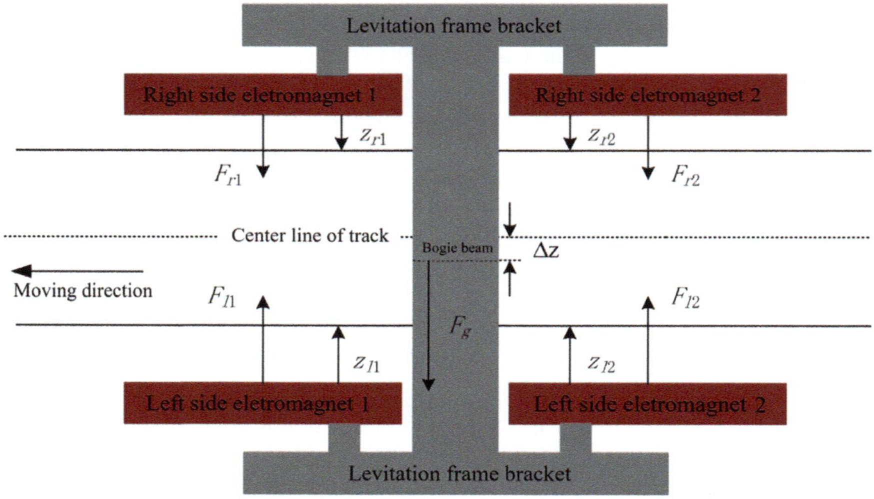

Fig. 4.2 Schematic of the guidance joint structure

left-side electromagnet 2, right-side electromagnet 1, and right-side electromagnet 2. $m_{l1}, m_{l2}, m_{r1}, m_{r2}$ are the qualities of the left-side electromagnet 1, left-side electromagnet 2, right-side electromagnet 1, and right-side electromagnet 2, respectively. m_d is the equivalent quality of the bogie. Δz is the gap deviation of the train from the center line of the track. N is the number of electromagnet coil turns. R_e is the resistance of the electromagnet coil. A_p is the polar area of one electromagnet group. μ_0 is the vacuum permeability. L is the equivalent inductance of the electromagnet coil during operation.

According to the model, the electromagnetic force equation, electrical equation, and kinetic equation are given respectively as follows:

$$\begin{cases} F_{l1} = \frac{\mu_0 N^2 A_p}{4} \cdot \left(\frac{i_{l1}}{z_{l1}}\right)^2 \\ F_{l2} = \frac{\mu_0 N^2 A_p}{4} \cdot \left(\frac{i_{l2}}{z_{l2}}\right)^2 \\ F_{r1} = \frac{\mu_0 N^2 A_p}{4} \cdot \left(\frac{i_{r1}}{z_{r1}}\right)^2 \\ F_{r2} = \frac{\mu_0 N^2 A_p}{4} \cdot \left(\frac{i_{r2}}{z_{r2}}\right)^2 \end{cases} \tag{4.2}$$

$$\begin{cases} u_{l1} = R_e i_{l1} + \dfrac{\mathrm{d}(L \cdot i_{l1})}{\mathrm{d}t} \\ u_{l2} = R_e i_{l2} + \dfrac{\mathrm{d}(L \cdot i_{l2})}{\mathrm{d}t} \\ u_{r1} = R_e i_{r1} + \dfrac{\mathrm{d}(L \cdot i_{r1})}{\mathrm{d}t} \\ u_{r2} = R_e i_{r2} + \dfrac{\mathrm{d}(L \cdot i_{r2})}{\mathrm{d}t} \end{cases} \tag{4.3}$$

$$(m_{l1} + m_{l2} + m_{r1} + m_{r2} + m_d)\ddot{z} = F_{r1} + F_{r2} - F_{l1} - F_{l2} + F_g \tag{4.4}$$

It can be seen from the above equations that the guidance system is a nonlinear system. Therefore, it is linearized near the equilibrium point, and the linear control theory can be adopted for system analysis and controller design. i_{l10}, i_{l20}, i_{r10}, and i_{r20} are the operating current of the left-side electromagnet 1, left-side electromagnet 2, right-side electromagnet 1, and right-side electromagnet 2 at the equilibrium point, respectively. Taking the right-side electromagnet 1 as an example, the electromagnetic force equation is expanded by first-order Taylor at the equilibrium point (i_{l10}, z_{r10}):

$$\begin{aligned} F_{r1}(i_{r1}, z_{r1}) &= F_{r1}(i_{r10}, z_{r10}) + \frac{\delta F_{r1}(i_{r1}, z_{r1})}{\delta i}\Big|_{i_{r1}=i_{r10}, z_{r1}=z_{r10}}(i_{r1} - i_{r10}) \\ &\quad + \frac{\delta F_{r1}(i_{r1}, z_{r1})}{\delta z}\Big|_{i_{r1}=i_{r10}, z_{r1}=z_{r10}}(z_{r1} - z_{r10}) \\ &= \frac{\mu_0 N^2 A_p}{4} \cdot \left(\frac{i_{r10}}{z_{r10}}\right)^2 + \frac{\mu_0 N^2 A_p i_{r10}}{2z_{r10}^2}\Delta i_{r1} - \frac{\mu_0 N^2 A_p i_{r10}^2}{2z_{r10}^3}\Delta z_{r1} \end{aligned} \tag{4.5}$$

The electrical equation's first-order Taylor expansion at the equilibrium point is:

$$\begin{aligned} u_{r1}\left(i_{r1}, \dot{i}_{r1}, z_{r1}, \dot{z}_{r1}\right) &= u_{r1}\left(i_{r10}, \dot{i}_{r10}, z_{r10}, \dot{z}_{r10}\right) + \frac{\delta u_{r1}}{\delta i_{r1}}\Big|_{i_{r10}, \dot{i}_{r10}, z_{r10}, \dot{z}_{r10}} \\ &\quad + \frac{\delta u_{r1}}{\delta z_{r1}}\Big|_{i_{r10}, \dot{i}_{r10}, z_{r10}, \dot{z}_{r10}} + \frac{\delta u_{r1}}{\delta \dot{i}_{r1}}\Big|_{i_{r10}, \dot{i}_{r10}, z_{r10}, \dot{z}_{r10}} \\ &\quad + \frac{\delta u_{r1}}{\delta \dot{z}_{r1}}\Big|_{i_{r10}, \dot{i}_{r10}, z_{r10}, \dot{z}_{r10}} \\ &= R_e i_{r10} + R_e \Delta i_{r1} + L_0 \Delta \dot{i}_{r1} - \frac{\mu_0 N^2 A_p i_{r10}}{2z_{r10}^2}\Delta \dot{z}_{r1} \end{aligned} \tag{4.6}$$

where $\dot{i}_{r10}$ and $\dot{z}_{r10}$ are equal to 0, and $L = L_0 = \frac{\mu_0 N^2 A_p}{2z_{r10}}$.

Considering that the gap fluctuation allowed by the guidance electromagnet is very small, $\Delta \dot{z}_{r1}$ in Eq. (4.6) can be ignored. Linearizing the equations of the other three electromagnets and the linearized equations for the system are as follows:

$$
\begin{cases}
\Delta u_{l1} = R_e \Delta i_{l1} + L_0 \Delta \dot{i}_{l1} \\
\Delta u_{l2} = R_e \Delta i_{l2} + L_0 \Delta \dot{i}_{l2} \\
\Delta u_{r1} = R_e \Delta i_{r1} + L_0 \Delta \dot{i}_{r1} \\
\Delta u_{r2} = R_e \Delta i_{r2} + L_0 \Delta \dot{i}_{r2} \\
\Delta F_{l1} = \frac{\mu_0 N^2 A_p i_{l10}}{2z_0^2} \Delta i_{l1} - \frac{\mu_0 N^2 A_p i_{l10}^2}{2z_0^3} \Delta z_{l1} \\
\Delta F_{l2} = \frac{\mu_0 N^2 A_p i_{l20}}{2z_0^2} \Delta i_{l2} - \frac{\mu_0 N^2 A_p i_{l20}^2}{2z_0^3} \Delta z_{l2} \\
\Delta F_{r1} = \frac{\mu_0 N^2 A_p i_{r10}}{2z_0^2} \Delta i_{r1} - \frac{\mu_0 N^2 A_p i_{r10}^2}{2z_0^3} \Delta z_{r1} \\
\Delta F_{r2} = \frac{\mu_0 N^2 A_p i_{r20}}{2z_0^2} \Delta i_{r2} - \frac{\mu_0 N^2 A_p i_{r20}^2}{2z_0^3} \Delta z_{r2} \\
(m_{l1} + m_{l2} + m_{r1} + m_{r2} + m_d)\Delta \ddot{z} = \Delta F_{r1} + \Delta F_{r2} - \Delta F_{l1} - \Delta F_{l2}
\end{cases}
\tag{4.7}
$$

where $\Delta z_{l1} = \Delta z_{l2} = \Delta z = -\Delta z_{r1} = -\Delta z_{r2}$.

The system state variables, control variables, and output variables are taken as follows:

$$
\begin{cases}
\boldsymbol{x} = \begin{bmatrix} x_1 & x_2 & x_3 & x_4 & x_5 & x_6 \end{bmatrix}^{\mathrm{T}} = \begin{bmatrix} \Delta z & \Delta \dot{z} & \Delta i_{l1} & \Delta i_{l2} & \Delta i_{r1} & \Delta i_{r2} \end{bmatrix}^{\mathrm{T}} \\
\boldsymbol{u} = \begin{bmatrix} \Delta u_{l1} & \Delta u_{l2} & \Delta u_{r1} & \Delta u_{r2} \end{bmatrix}^{\mathrm{T}}, y = \Delta z
\end{cases}
$$

The state space expression for the guidance system is shown in Eq. (4.8). k_e is defined as $k_e = \frac{\mu_0 N^2 A_p}{4}$, and M is defined as: $M = m_{l1} + m_{l2} + m_{r1} + m_{r2} + m_d$.

$$
\begin{cases}
\dot{\boldsymbol{x}} = \boldsymbol{A}\boldsymbol{x} + \boldsymbol{B}\boldsymbol{u} \\
y = \boldsymbol{C}\boldsymbol{x} + \boldsymbol{D}\boldsymbol{u}
\end{cases}
\tag{4.8}
$$

$$
\boldsymbol{A} = \begin{bmatrix}
0 & 1 & 0 & 0 & 0 & 0 \\
\frac{2k_e}{z_0^3 M}\left(i_{l10}^2 + i_{l20}^2 + i_{r10}^2 + i_{r20}^2\right) & 0 & -\frac{2k_e i_{l10}}{z_0^2 M} & -\frac{2k_e i_{l20}}{z_0^2 M} & \frac{2k_e i_{r10}}{z_0^2 M} & \frac{2k_e i_{r20}}{z_0^2 M} \\
0 & 0 & -\frac{R_e}{L_0} & 0 & 0 & 0 \\
0 & 0 & 0 & -\frac{R_e}{L_0} & 0 & 0 \\
0 & 0 & 0 & 0 & -\frac{R_e}{L_0} & 0 \\
0 & 0 & 0 & 0 & 0 & -\frac{R_e}{L_0}
\end{bmatrix}
$$

$$
\boldsymbol{B} = \begin{bmatrix}
0 & 0 & \frac{1}{L_0} & 0 & 0 & 0 \\
0 & 0 & 0 & \frac{1}{L_0} & 0 & 0 \\
0 & 0 & 0 & 0 & \frac{1}{L_0} & 0 \\
0 & 0 & 0 & 0 & 0 & \frac{1}{L_0}
\end{bmatrix}^{\mathrm{T}}
$$

$$
\boldsymbol{C} = \begin{bmatrix} 1 & 0 & 0 & 0 & 0 & 0 \end{bmatrix}
$$

$$
\boldsymbol{D} = \begin{bmatrix} 0 & 0 & 0 & 0 \end{bmatrix}
$$

Firstly, the eigenvalues of the system matrix $\boldsymbol{A}$ are obtained to analyze the established guidance model: $\lambda_{1,2} = \pm 24.98$, $\lambda_{1,2,3,4} = -19.90$. It can be seen that one of the poles is located to the right of the imaginary axis, demonstrating the guidance system model is unstable. Then, the ranks of the controllability matrix $\boldsymbol{Q_c} = [\boldsymbol{B}\ \boldsymbol{AB}\ \boldsymbol{A}^2\boldsymbol{B}\ \boldsymbol{A}^3\boldsymbol{B}\ \boldsymbol{A}^4\boldsymbol{B}\ \boldsymbol{A}^5\boldsymbol{B}]$ and observability matrix $\boldsymbol{Q_o} = [\boldsymbol{C}\ \boldsymbol{CA}\ \boldsymbol{CA}^2\ \boldsymbol{CA}^3\ \boldsymbol{CA}^4\ \boldsymbol{CA}^5]^{\mathrm{T}}$ are calculated to be 6 and 3, which means the system isn't completely observable but is completely controllable. Therefore, the required state variables are still available. It will be explained further in the section on the guidance controller's design.

4.1.3 Guidance System Model with Disturbances and Faults

Many factors during train operation have an impact on the guidance system directly or indirectly, including the strong wind and the guideway design is not enough to offset the partial force of the train's own gravity when passing a curve. There are also the changes in the gap between the guidance electromagnet and the side rail, such as the track irregularity, the joint-passing, and the roll and pitch of the electromagnet. Therefore, these conditions should be considered in advance. The changes in the electromagnetic force of the guidance electromagnet, caused by either the disturbance force applied by the external environment or the changes in the guidance gap, can be all converted to the lateral disturbance force on the guidance electromagnet. The guidance system's model with disturbances is established:

$$\begin{cases} \dot{\boldsymbol{x}} = \boldsymbol{Ax} + \boldsymbol{Bu} + \boldsymbol{E}_{\mathrm{d}}\boldsymbol{d} \\ \boldsymbol{y} = \boldsymbol{Cx} + \boldsymbol{Du} + \boldsymbol{F}_{d}\boldsymbol{d} \end{cases} \tag{4.9}$$

where the $\boldsymbol{A}$, $\boldsymbol{B}$, $\boldsymbol{C}$, and $\boldsymbol{D}$ matrices are provided before, $\boldsymbol{d}$ represents the deterministic unknown disturbance. $\boldsymbol{E}_{\mathrm{d}}$ and $\boldsymbol{F}_{\mathrm{d}}$ represent the application mode and the location of the disturbance respectively. For example, the disturbance force F_{g} acts directly on the guidance system horizontally to the left as shown in Fig. 4.2. At this moment, $\boldsymbol{E}_{\mathrm{d}} = \begin{bmatrix} 0 & 1/M & 0 & 0 & 0 & 0 \end{bmatrix}^{\mathrm{T}}$, and $\boldsymbol{F}_{\mathrm{d}} = 0$.

The fault of the guidance sensor f_{S}, the fault of the actuator f_{A} (including the fault of the guidance electromagnet and the fault of the chopper) and the fault of the process f_{P} are mainly considered in the fault model of the guidance system. The guidance system can be expressed:

$$\begin{cases} \dot{\boldsymbol{x}} = \boldsymbol{Ax} + \boldsymbol{Bu} + \boldsymbol{E}_{\mathrm{d}}\boldsymbol{d} + \boldsymbol{E}_{\mathrm{f}}\boldsymbol{f} \\ \boldsymbol{y} = \boldsymbol{Cx} + \boldsymbol{Du} + \boldsymbol{F}_{\mathrm{d}}\boldsymbol{d} + \boldsymbol{F}_{\mathrm{f}}\boldsymbol{f} \end{cases} \tag{4.10}$$

where $\boldsymbol{E}_{\mathrm{f}}$ and $\boldsymbol{F}_{\mathrm{f}}$ are the manner and the place where the faults occur. When $\boldsymbol{E}_{\mathrm{f}} = 0$, $\boldsymbol{F}_{\mathrm{f}} = \boldsymbol{I}$, the symbol $\boldsymbol{f}$ represents the fault of the sensor f_{S}. When $\boldsymbol{E}_{\mathrm{f}} = \boldsymbol{B}$, $\boldsymbol{F}_{\mathrm{f}} = \boldsymbol{D}$,

$\boldsymbol{f}$ represents the fault of the actuator f_A. When $\boldsymbol{E}_f = \boldsymbol{E}_0$, $\boldsymbol{F}_f = \boldsymbol{F}_0$, $\boldsymbol{f}$ represents the fault of the process f_P. Considering the disturbances and faults, the guidance model can be described as:

$$\begin{cases} \dot{x} = \boldsymbol{A}\boldsymbol{x} + \boldsymbol{B}\boldsymbol{u} + \boldsymbol{E}_d\boldsymbol{d} + \begin{bmatrix} \boldsymbol{0} & \boldsymbol{B} & \boldsymbol{E}_0 \end{bmatrix} \begin{bmatrix} f_S \\ f_A \\ f_P \end{bmatrix} \\ y = \boldsymbol{C}\boldsymbol{x} + \boldsymbol{D}\boldsymbol{u} + \boldsymbol{F}_d\boldsymbol{d} + \begin{bmatrix} \boldsymbol{I} & \boldsymbol{D} & \boldsymbol{F}_0 \end{bmatrix} \begin{bmatrix} f_S \\ f_A \\ f_P \end{bmatrix} \end{cases} \tag{4.11}$$

where $\boldsymbol{f} = \begin{bmatrix} f_S & f_A & f_P \end{bmatrix}^T$, $\boldsymbol{E}_f = \begin{bmatrix} 0 & \boldsymbol{B} & \boldsymbol{E}_0 \end{bmatrix}^T$, $\boldsymbol{F}_f = \begin{bmatrix} \boldsymbol{I} & \boldsymbol{D} & \boldsymbol{F}_0 \end{bmatrix}^T$.

4.2 Design of Nominal Controller for Guidance System

4.2.1 Simplification of the Guidance System Model

The more situations that are considered, the more accurate the model will be, which is closer to the real guidance system. However, it is preferable to ignore minor factors and keep the guidance system's order as low as possible while maintaining its fundamental characteristics when designing the guidance controller. The designed lower-order controller will then be used to control the higher-order guidance model and the actual guidance system. In this way, the guidance system model is appropriately simplified.

Each electromagnet group in the guidance joint structure depicted in Fig. 4.2 corresponds to a controller and a chopper. Firstly, the guidance gap of the two electromagnet groups on the same side of the guidance joint structure is always consistent under the condition that the track irregularity and the electromagnet roll measurement are not considered, and the connection between the bracket arm of the maglev bogie and the electromagnet has absolute rigidity. For greater precision, the gap values recorded by the two electromagnet guidance sensors are typically averaged. Take the right side as an example: $z_r = (z_{r1} + z_{r2})/2$. What's more, the initial currents feed into both electromagnets are also the same: $i_{r10} = i_{r20} = i_{r0}$, which means that the structure of the two electromagnets and the parameters of the designed controller on the same side are identical. Finally, with the same controller settings, the two electromagnet groups behave as two identical actuators. Therefore, the equivalent bogie mass of the load can be reduced by half, and the guidance controller design for the joint structure may be simplified as an electromagnet group on each side. The guidance model becomes:

$$\begin{cases} \begin{bmatrix} \Delta\dot{z} \\ \Delta\ddot{z} \\ \Delta\dot{i}_\mathrm{l} \\ \Delta\dot{i}_\mathrm{r} \end{bmatrix} = \begin{bmatrix} 0 & 1 & 0 & 0 \\ \frac{2k_\mathrm{e}}{z_0^3 M_0}\left(i_{\mathrm{l}0}^2 + i_{\mathrm{r}0}^2\right) & 0 & -\frac{2k_\mathrm{e} i_{\mathrm{l}0}}{z_0^2 M_0} & \frac{2k_\mathrm{e} i_{\mathrm{r}0}}{z_0^2 M_0} \\ 0 & 0 & -\frac{R_\mathrm{e}}{L_0} & 0 \\ 0 & 0 & 0 & -\frac{R_\mathrm{e}}{L_0} \end{bmatrix} \\ \qquad \begin{bmatrix} \Delta z \\ \Delta\dot{z} \\ \Delta i_\mathrm{l} \\ \Delta i_\mathrm{r} \end{bmatrix} + \begin{bmatrix} 0 & 0 \\ 0 & 0 \\ \frac{1}{L_0} & 0 \\ 0 & \frac{1}{L_0} \end{bmatrix} \begin{bmatrix} \Delta u_\mathrm{l} \\ \Delta u_\mathrm{r} \end{bmatrix} \\ y = \begin{bmatrix} 1\ 0\ 0\ 0 \end{bmatrix} \begin{bmatrix} \Delta z \\ \Delta\dot{z} \\ \Delta i_\mathrm{l} \\ \Delta i_\mathrm{r} \end{bmatrix} \end{cases} \tag{4.12}$$

where $M_0 = m_{\mathrm{l}1} + m_{\mathrm{r}1} + m_\mathrm{d}/2$. The system matrix and input matrix of the simplified guidance model demonstrate that the left and right-side controllers' inputs Δu_l and Δu_r to the guidance system have opposing impacts on the guidance system's acceleration of lateral movement. In the initial circumstance, the left and right-side guidance gaps ($z_{\mathrm{l}0}$ and $z_{\mathrm{r}0}$) are 10 mm, and both sides of the guidance electromagnets are operating at the same current. However, if the levitation gap of 10 mm is always used as the control goal, the presence of track irregularities would result in a situation where a particular track is wider than the others, reducing the guidance gap on both sides. The control voltage will be simultaneously decreased to widen the guidance gap when both sides are controlled independently, which enables the guidance system's motion unchanged. The existing gap error makes the controller output decrease until it ultimately brings the voltage across the electromagnet to zero. However, the chopper requires a certain opening voltage in practice, the above situation will result in a dead zone in the control process, severely reducing the stiffness of the guidance system and impairing the system's ability to operate safely and steadily. Therefore, the guidance control adopts a differential control scheme. In this scheme, the initial guidance gap and static current are the same on both sides, and the difference in the gap sensor data between the left and right sides $[(z_{\mathrm{l}0} + \Delta z) - (z_{\mathrm{r}0} - \Delta z)]/2 = \Delta z$ is taken as the feedback value for their control inputs. As a result, the voltage across the left electromagnet increases Δu, while the voltage on the right side decreases Δu. It is equivalent to the differential control composed of one guidance controller and two actuators. The model is further simplified as:

$$\begin{cases} \begin{bmatrix} \Delta\dot{z} \\ \Delta\ddot{z} \\ \Delta\dot{i} \end{bmatrix} = \begin{bmatrix} 0 & 1 & 0 \\ \frac{4i_{\mathrm{l}0}^2 k_\mathrm{e}}{z_0^3 M_0} & 0 & -\frac{4k_\mathrm{e} i_{\mathrm{l}0}}{z_0^2 M_0} \\ 0 & 0 & -\frac{R_\mathrm{e}}{L_0} \end{bmatrix} \begin{bmatrix} \Delta z \\ \Delta\dot{z} \\ \Delta i \end{bmatrix} + \begin{bmatrix} 0 \\ 0 \\ \frac{1}{L_0} \end{bmatrix} \Delta u \\ y = \begin{bmatrix} 1\ 0\ 0 \end{bmatrix} \begin{bmatrix} \Delta z \\ \Delta\dot{z} \\ \Delta i \end{bmatrix} \end{cases} \tag{4.13}$$

4.2.2 Design of Nominal Controller for Guidance System

In this section, the linear quadratic regulator (LQR) method is used to determine the corresponding state and input matrix in combination with the actual situation, so that the introduced optimal control performance index is optimal, and the optimal state feedback coefficient under this target can be obtained.

Substituting the corresponding actual physical parameters and transforming the simplified state space expression of the guidance system into a transfer function:

$$G(s) = \frac{z(s)}{u(s)} = \frac{-0.7521}{(s-10.21)(s+10.21)(s+19.97)} \tag{4.14}$$

From Eq. (4.14), it can be seen that the guidance system is a 0-type system. That is, the steady-state errors occur when the system stabilizes. When the train is operating under certain operating circumstances, the left and the right-side guidance electromagnets stabilize at various equilibrium points, respectively. As a result, the train deviates from the central line of the guideway all the time, which may lead to serious accidents. To make the closed-loop guidance system have suitable transition process characteristics and consistently tracking guideway input without steady-state errors, an integrating element is incorporated into the feedforward channel to convert the original system into a I-type system. It is shown in Fig. 4.3.

The deviation vector $\boldsymbol{e}(t) = \boldsymbol{y}(t) - \boldsymbol{r}(t)$ and the new state variable $\boldsymbol{x}_{\mathrm{e}}(t) = \int_0^t \boldsymbol{e}(\tau)\mathrm{d}\tau$ which is the integral of the error are defined, wherein $\dot{\boldsymbol{x}}_{\mathrm{e}}(t) = \boldsymbol{e}(t) = \boldsymbol{C}\boldsymbol{x}(t) - \boldsymbol{r}(t)$. The augmented guidance system can be expressed as:

$$\begin{cases} \begin{bmatrix} \dot{\boldsymbol{x}} \\ \dot{\boldsymbol{x}}_{\mathrm{e}} \end{bmatrix} = \begin{bmatrix} \boldsymbol{A} & 0 \\ \boldsymbol{C} & 0 \end{bmatrix} \begin{bmatrix} \boldsymbol{x} \\ \boldsymbol{x}_{\mathrm{e}} \end{bmatrix} + \begin{bmatrix} \boldsymbol{B} \\ 0 \end{bmatrix} \boldsymbol{u} + \begin{bmatrix} 0 \\ -1 \end{bmatrix} \boldsymbol{r} \\ \boldsymbol{y} = \begin{bmatrix} \boldsymbol{C} & 0 \end{bmatrix} \begin{bmatrix} \boldsymbol{x} \\ \boldsymbol{x}_{\mathrm{e}} \end{bmatrix} \end{cases} \tag{4.15}$$

The first step in designing a controller using LQR is to build an energy function. The optimal control trajectory should minimize the energy function if the system is completely controllable. By solving the observability and controllability matrixes of the augmented guidance system, we can get that the system is entirely observable and controllable. The quadratic performance metric is defined as:

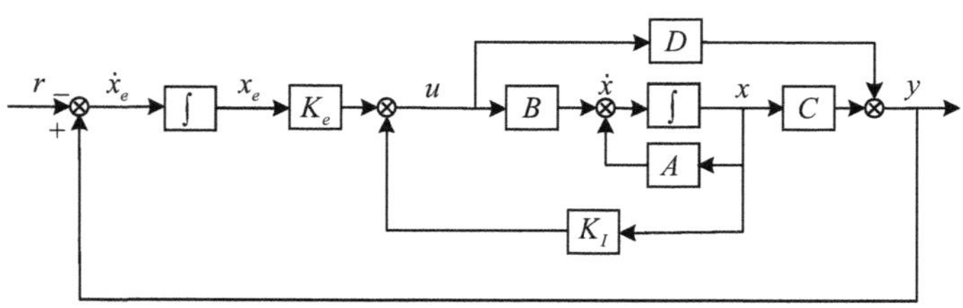

Fig. 4.3 Diagram of the augmented guidance system structure

$$J = \int_0^{\infty} \left(\tilde{\boldsymbol{x}}^{\mathrm{T}} \boldsymbol{Q} \tilde{\boldsymbol{x}} + \boldsymbol{u}^{\mathrm{T}} \boldsymbol{R} \boldsymbol{u}\right) \mathrm{d}t \tag{4.16}$$

Solving the Riccati equation:

$$\boldsymbol{P} = \boldsymbol{Q} + \tilde{\boldsymbol{A}}^{\mathrm{T}} \boldsymbol{P} \tilde{\boldsymbol{A}} - \tilde{\boldsymbol{A}}^{\mathrm{T}} \boldsymbol{P} \tilde{\boldsymbol{B}} \left(\boldsymbol{R} + \tilde{\boldsymbol{B}}^{\mathrm{T}} \boldsymbol{P} \tilde{\boldsymbol{B}}\right)^{-1} \tilde{\boldsymbol{B}}^{\mathrm{T}} \boldsymbol{P} \tilde{\boldsymbol{A}} \tag{4.17}$$

The state feedback matrix can be obtained:

$$\boldsymbol{K} = -\left(\boldsymbol{R} + \tilde{\boldsymbol{B}}^{\mathrm{T}} \boldsymbol{P} \tilde{\boldsymbol{B}}\right)^{-1} \tilde{\boldsymbol{B}}^{\mathrm{T}} \boldsymbol{P} \tilde{\boldsymbol{A}} = \left[\boldsymbol{K}_{\mathrm{I}}\ \boldsymbol{K}_{\mathrm{e}}\right] \tag{4.18}$$

$\tilde{\boldsymbol{x}}$ denotes the new state variable of the augmented guidance system in Eq. (4.16). $\tilde{\boldsymbol{A}}$ and $\tilde{\boldsymbol{B}}$ are new system matrix and input matrix respectively, where, $\tilde{\boldsymbol{A}} = \begin{bmatrix} \boldsymbol{A} & 0 \\ \boldsymbol{C} & 0 \end{bmatrix}$, $\tilde{\boldsymbol{B}} = \begin{bmatrix} \boldsymbol{B} \\ 0 \end{bmatrix}$. $\boldsymbol{Q}$ and $\boldsymbol{R}$ are weighting matrixes which are chosen as constant matrixes for the single input guidance systems in nominal conditions. The larger the $\boldsymbol{Q}$ value is, the smaller the state $\tilde{\boldsymbol{x}}$ is, which means that the poles of the closed-loop system are further far from the left of the s-plane, and the state $\tilde{\boldsymbol{x}}$ decays faster to 0. The larger the $\boldsymbol{R}$ value is, the smaller the input $\boldsymbol{u}$ is, which means that the optimal control can be obtained with minimal control cost, but the decay of the state $\tilde{\boldsymbol{x}}$ will be slower. To make sure that the designed closed-loop guidance system has sufficient stiffness, the gap could only vary within a small range. Therefore, the state variable Δz must correlate to a higher weight to provide a certain degree of interference protection. In consideration of the comfort and safety of the train, the lateral speed change must not be too fast and the corresponding weight for $\Delta \dot{z}$ must be chosen to be large. What's more, to guarantee that the train can track the input, maintain the nominal guidance gap without steady-state errors, and prevent the large overshoot, the error $\boldsymbol{x}_{\mathrm{e}}$ has to correspond to a large weight. During the operation of the guidance system, the operational current fluctuates quickly in a large range, hence the corresponding weight for the current variation Δi should be small. Finally, the appropriate weight $\boldsymbol{R}$ for the control voltage should likewise be small to ensure that the system responds rapidly. According to experience and experiment, the final weighting matrixes in the quadratic performance index are selected as: $\boldsymbol{Q} = \mathrm{diag}\left(\left[2 \times 10^{7}\ 2 \times 10^{6}\ 10^{3}\ 3 \times 10^{10}\right]\right)$, $\boldsymbol{R} = 1$. The state feedback matrix $\boldsymbol{K} = \left[\boldsymbol{K}_{\mathrm{I}}\ \boldsymbol{K}_{\mathrm{e}}\right] = \left[-90{,}748\ -5655.8\ -33.8\ -173{,}210\right]$ is calculated, and the input $\boldsymbol{u} = \boldsymbol{K}\tilde{\boldsymbol{x}} = \boldsymbol{K}_{\mathrm{I}}\boldsymbol{x} + \boldsymbol{K}_{\mathrm{e}}\boldsymbol{x}_{\mathrm{e}}$ can also be obtained.

4.2.3 Analysis of Guidance System Simulation

To validate the accuracy of the previously established model and the designed controller, simulations of the guidance system are performed. The simulations are carried out under the nominal state and introducing the external disturbances state. The closed-loop guidance system's characteristics under the control of the nominal controller are analyzed.

The linear model of the guidance system in Eq. (4.13) is first constructed and the linear controller using LQR in the preceding section is incorporated to form a closed-loop system. To preliminary verify the performance of the designed guidance controller, the step signal of the guidance gap and the disturbance force are given to imitate the track conditions and disturbance force changes during the train operation. The simulation results of the system responses are shown in Fig. 4.4.

It can be seen from the simulation results that the linearized closed-loop guidance system is capable of tracking gap changes fast without overshoot or steady-state error. The gap widens briefly and then quickly returns to the nominal gap when a continuous static disturbance force enters the system. It indicates that the disturbance force is successfully suppressed, and the controller's effectiveness is preliminarily confirmed.

To further verify the control effect of the controller when the guidance system operates away from the equilibrium point for an extended period, the simplified linear guidance system model is replaced with the complete nonlinear guidance system model. The output voltage from the chopper to the guidance electromagnet in the actual circumstance cannot be negative. For this reason, an amplitude limiting module is added to the electromagnet input voltage in the simulation model. The simulation is also carried out with sudden changes in the guidance gap and disturbance forces. The results are shown in Fig. 4.5, the gap between two electromagnets in the left joint structure abruptly increases and the gap between two electromagnets in the right joint structure abruptly decreases.

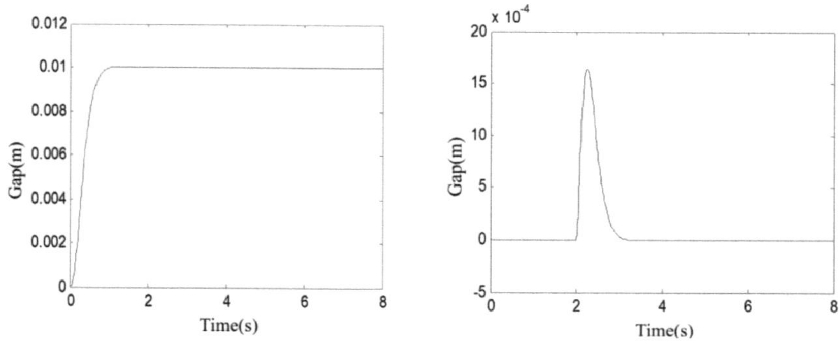

(a) Given a 10 mm step input **(b)** Given a 500 N disturbance force step input

Fig. 4.4 Response curves of the simulation for the liner guidance model

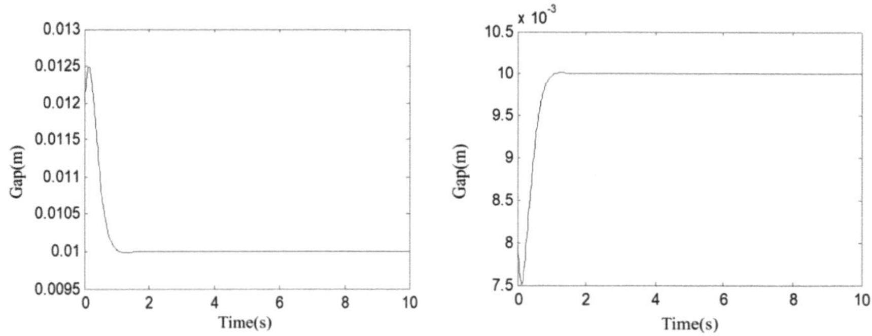

(a) Left side (b) Right side

Fig. 4.5 Gap response curves on both sides of given gap step inputs

The simulation results in Fig. 4.5 show the closed-loop guidance system regulated by this controller can adapt to changes in the gap, maintaining equal guidance gaps on both sides. However, the simulation figures depict that when a sudden change in the track causes the guidance gap on one side to increase, the guidance system will initially operate in the direction of the increasing rise in the gap, resulting in a reverse overshoot. The reason is that the electromagnetic force is proportional to the square of the current and inversely proportional to the square of the gap. When the gap suddenly changes, the current in the electromagnets on both sides is equal, and the electromagnetic force on the side with the larger gap is smaller than the electromagnetic force on the side with the smaller gap, causing an acceleration in the direction of the larger gap. While the guidance gap increases by another small distance, the controller will recognize the increased gap and then immediately increase the output voltage to increase the current of the electromagnet on one side and decrease the current on the opposite side. Gradually, the electromagnetic force on one side would be bigger than the electromagnetic force on the opposite side. And eventually the gap on both sides would be back into alignment. The simulation results of the model are consistent with the theoretical analysis, which also verifies the validity of the model. The simulation results of the guidance model with disturbances are shown in Figs. 4.6, 4.7 and 4.8.

When the system is subjected to a small static disturbance force, as shown in Fig. 4.6, the guidance gap recovers immediately after a short adjustment without overshoots or steady-state errors, which demonstrates the controller can respond rapidly. When the disturbance force is increased to 3200 N as shown in Fig. 4.7, the system gap variation has a slight overshoot, and the controller briefly gets 0 current in the right-side electromagnet during the adjustment process, with only the left electromagnet running alone. As the disturbance force continues to rise, as shown in Fig. 4.8, the performance of the guidance system starts to deteriorate noticeably. Although the guidance gap can eventually be restored to the desired gap,

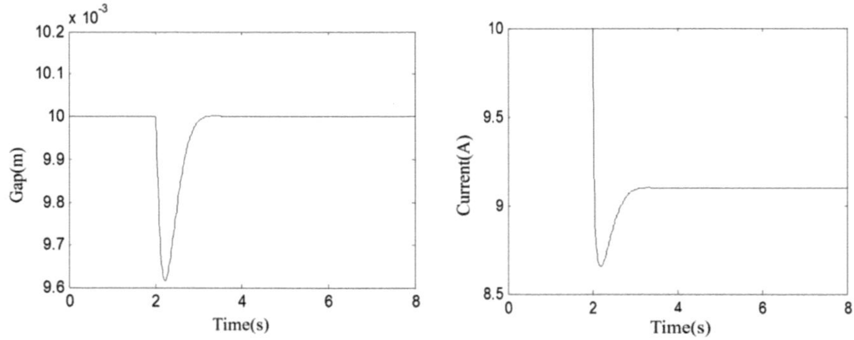

(a) Gap curve **(b)** Current curve

Fig. 4.6 Right-side gap and current curve for a 500 N disturbance force step input

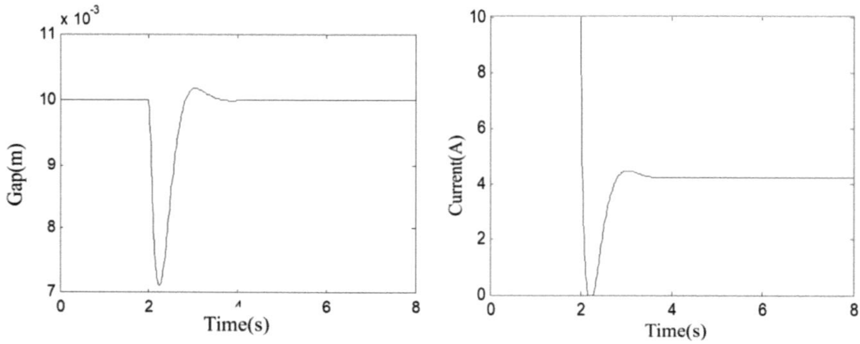

(a) Gap curve **(b)** Current curve

Fig. 4.7 Right-side gap and current curve for a 3200 N disturbance force step input

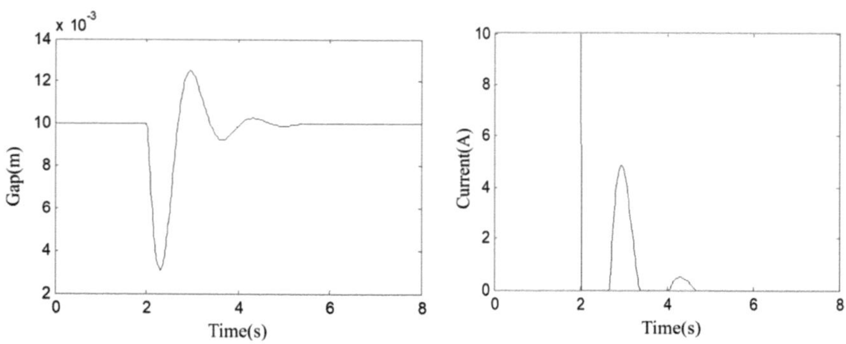

(a) Gap curve **(b)** Current curve

Fig. 4.8 Right-side gap and current curve for a 5600 N disturbance force step input there is a significant decay oscillation in the adjustment process, eventually leaving the right-side electromagnet in an inoperative state.

4.2.4 Analysis of Guidance System's Characteristics

Based on the simulation results in the preceding section, when the system is fault-free and disturbance-free, the guidance controller using LQR has a good control performance. However, when external disturbance forces are applied to the system, the controller's capacity to control the real nonlinear guidance system is significantly diminished since the current in the guidance electromagnet will diverge from the current equilibrium point prescribed in the controller design. The adaptability of the controller decreases when the electromagnet on one side of the system doesn't work. To provide a logical explanation for the causes, the analysis is made as follows in terms of the guidance model's characteristics.

When a static disturbance force of constant magnitude is continuously given to the guidance system, the controller regulates the system to equilibrium again. The guidance gap is the same when the system reaches a new equilibrium point because of the integrating element. Under this condition, the electromagnet currents on both sides are: $i_{l0} + \Delta i$ and $i_{r0} - \Delta i$, respectively. The guidance model is linearized at the new equilibrium points:

$$\begin{cases} \begin{bmatrix} \Delta \dot{z} \\ \Delta \ddot{z} \\ \Delta \dot{i} \end{bmatrix} = \begin{bmatrix} 0 & 1 & 0 \\ \frac{2k_e}{z_0^3 M_0}\left[(i_{l0} + \Delta i)^2 + (i_{r0} - \Delta i)^2\right] & 0 & -\frac{2k_e}{z_0^2 M_0}[[(i_{l0} + \Delta i) + (i_{r0} - \Delta i)]] \\ 0 & 0 & -\frac{R_e}{L_0} \end{bmatrix} \\ \qquad \begin{bmatrix} \Delta z \\ \Delta \dot{z} \\ \Delta i \end{bmatrix} + \begin{bmatrix} 0 \\ 0 \\ \frac{1}{L_0} \end{bmatrix} \Delta u \\ y = \begin{bmatrix} 1 & 0 & 0 \end{bmatrix} \begin{bmatrix} \Delta z \\ \Delta \dot{z} \\ \Delta i \end{bmatrix} \end{cases} \tag{4.19}$$

Compared with the previous linearized model in predefined equilibrium points, the input and output matrices of the system remain the same while the equilibrium point varies. Furthermore, only $\frac{2k_e}{z_0^3 M_0}\left[(i_{l0} + \Delta i)^2 + (i_{r0} - \Delta i)^2\right]$ of the system matrix is shown a very slight rise $\frac{4k_e \Delta i^2}{z_0^3 M_0}$. The effects of changes in the equilibrium point current are further counteracted by the state variable Δz with large feedback weighting in the controller's design, so that the closed-loop guidance system's characteristics almost stay the same. When the guidance system is exposed to a strong static disturbance force that causes the electromagnet current to be reduced to zero on one side, the guidance model reaches a new steady state:

$$\begin{cases} \begin{bmatrix} \Delta\dot{z} \\ \Delta\ddot{z} \\ \Delta\dot{i} \end{bmatrix} = \begin{bmatrix} 0 & 1 & 0 \\ \frac{2k_e}{z_0^3 M_0}(i_{10} + \Delta i)^2 & 0 & -\frac{2k_e}{z_0^2 M_0}(i_{10} + \Delta i) \\ 0 & 0 & -\frac{R_e}{L_0} \end{bmatrix} \begin{bmatrix} \Delta z \\ \Delta\dot{z} \\ \Delta i \end{bmatrix} + \begin{bmatrix} 0 \\ 0 \\ \frac{1}{L_0} \end{bmatrix} \Delta u \\ y = \begin{bmatrix} 1 & 0 & 0 \end{bmatrix} \begin{bmatrix} \Delta z \\ \Delta\dot{z} \\ \Delta i \end{bmatrix} \end{cases} \tag{4.20}$$

As the equilibrium point changes, $\frac{2k_e}{z_0^3 M_0}(i_{10} + \Delta i)^2$ and $\frac{2k_e}{z_0^2 M_0}(i_{10} + \Delta i)$ also change. The change in $-R_e/L_0$ indicates that the system's input characteristics change, and the system's response to the same voltage input to the controller at various equilibrium points varies. That is, the guidance system with single-side electromagnet operation is a time-varying system that varies with the current equilibrium point, and the control performance of a single controller is difficult to ensure throughout the entire operating interval. There is no difference in essence between the single electromagnet model and the double electromagnet model. One side electromagnet is inoperative when the current variation Δi gradually increases to equal the static bias current, thus there are no dramatic changes in the system characteristics. After a static disturbance force of 500 N is given to the guidance system in differential operation and single operation respectively, the right-side gap variation is illustrated in Fig. 4.9. The simulation results are consistent with the above analysis.

Theoretically, importing a larger static bias current in the guidance electromagnet will guarantee the proper operation of both sides of the guidance system, and the current variation range is smaller when the system is subjected to static disturbances of the same magnitude. The difference between the system model reaching the new equilibrium and the nominal guidance model is smaller. Therefore, the guidance

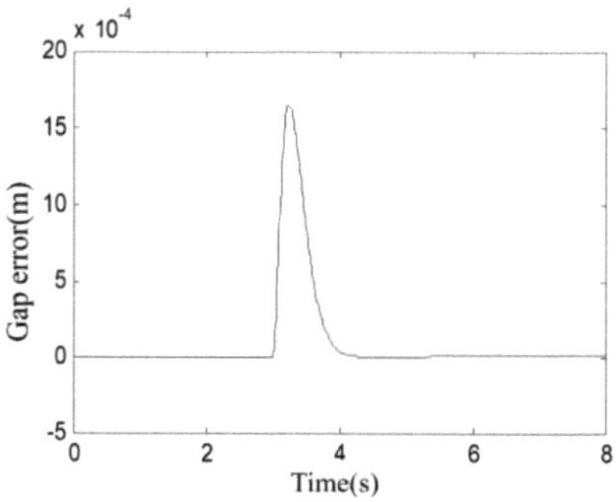

(a) Single operation

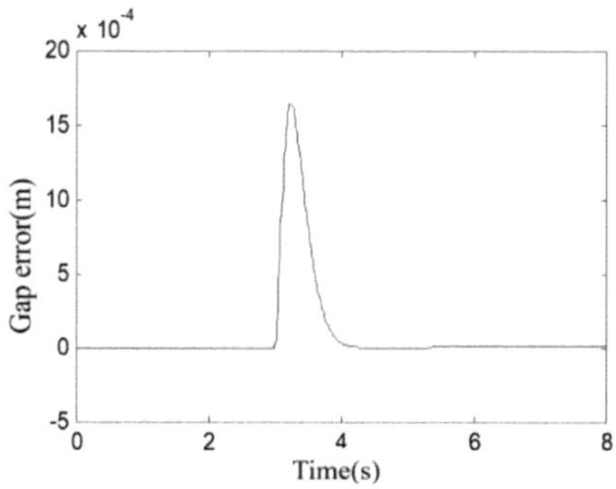

(b) Differential operation

Fig. 4.9 Curves of the right-side gap changes when 500 N static disturbance force is applied

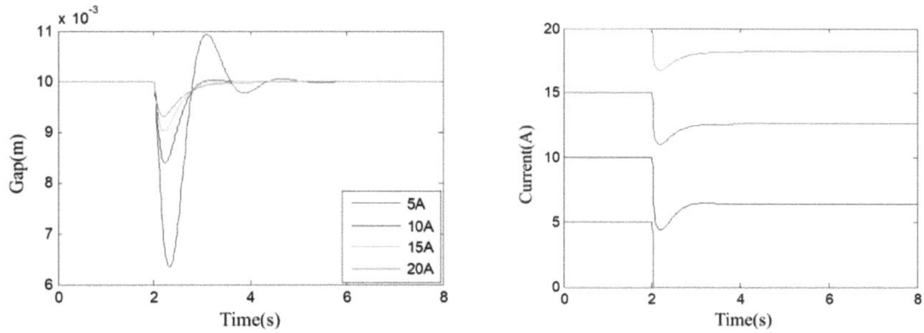

(a) Change of right-side gap (b) Change of right-side electromagnet's current

Fig. 4.10 Simulation curves of 2 kN disturbance force applied to the system with different bias currents

system can maintain reasonably good operating characteristics throughout the entire operating interval as shown in Fig. 4.10. To prevent the electromagnet coils from heating up adversely over an extended period and to reduce energy waste, the static bias current of the guidance system is set at 10 A in consideration of the actual circumstances of the train operation.

4.3 Design of Robust Controller for Guidance System

The guidance nominal controller is changed to H_2 robust control based on the preceding content to improve the resistance of the guidance system to disturbances and its capacity to adapt to uncertainties.

4.3.1 Design of Robust Controller

The standard H_2 control uses the general structure of a robust control system as shown in Fig. 4.11.

Fig. 4.11 Standard H_2 control's block diagram

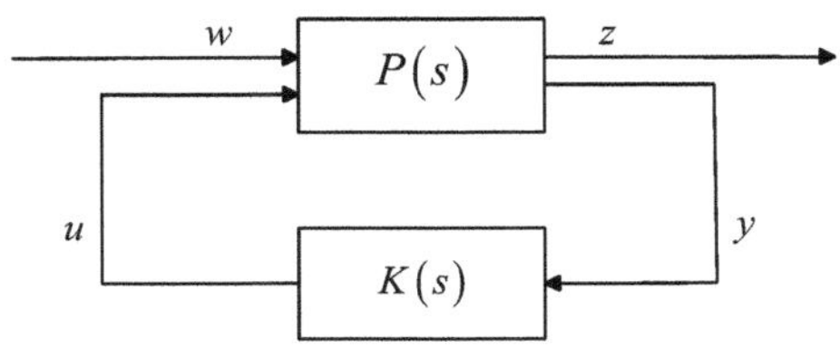

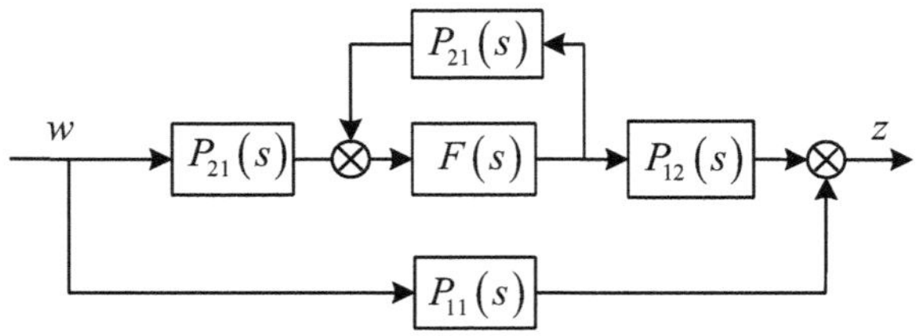

Fig. 4.12 Another description of standard H_2 control's block diagram

Where $\boldsymbol{P}(s)$ is an augmented model for the control object, $\boldsymbol{K}(s)$ is the controller, u and y are the input and output of the controller, w is disturbance input, and z is the output of the error.

The state space expression of the augmented model is shown as:

$$\begin{bmatrix} \dot{\boldsymbol{x}} \\ z \\ y \end{bmatrix} = \begin{bmatrix} \boldsymbol{P}_{11}(s) & \boldsymbol{P}_{12}(s) \\ \boldsymbol{P}_{21}(s) & \boldsymbol{P}_{22}(s) \end{bmatrix} \begin{bmatrix} \boldsymbol{x} \\ w \\ u \end{bmatrix} = \left[\begin{array}{c|cc} \boldsymbol{A} & \boldsymbol{B}_1 & \boldsymbol{B}_2 \\ \hline \boldsymbol{C}_1 & \boldsymbol{D}_{11} & \boldsymbol{D}_{12} \\ \boldsymbol{C}_2 & \boldsymbol{D}_{21} & \boldsymbol{D}_{22} \end{array}\right] \begin{bmatrix} \boldsymbol{x} \\ w \\ u \end{bmatrix} \tag{4.21}$$

The robust controller $\boldsymbol{K}(s)$ is designed to make the norm of the closed-loop transfer function $G_{zw}(s)$ from w to z less than 1. Wherein, $\boldsymbol{K}(s) = \boldsymbol{F}(s)y(s)$. The standard block diagram of the control system can be expressed in Fig. 4.12. The aim of H_2 control is to find a feedback control rate $\boldsymbol{F}(s)$ to make $\|G_{zw}(s)\|_2$ minimum.

By solving the Riccati equation for Eqs. (4.22) and (4.23), the state feedback coefficient $\boldsymbol{F}_2 = -\boldsymbol{B}_2^{\mathrm{T}} X_2$ and observer coefficient $\boldsymbol{L}_2 = -Y_2 \boldsymbol{C}_2^{\mathrm{T}}$ can be obtained according to the separation principle of controller design, respectively.

$$\boldsymbol{A}^T X_2 + X_2 \boldsymbol{A} - X_2 \boldsymbol{B}_2 \boldsymbol{B}_2^{\mathrm{T}} X_2 + \boldsymbol{C}_1^{\mathrm{T}} \boldsymbol{C}_1 = 0 \tag{4.22}$$

$$\boldsymbol{A} Y_2 + Y_2 \boldsymbol{A}^{\mathrm{T}} - Y_2 \boldsymbol{C}_2^{\mathrm{T}} \boldsymbol{C}_2 Y_2 + \boldsymbol{B}_1 \boldsymbol{B}_1^{\mathrm{T}} = 0 \tag{4.23}$$

The transfer function matrices for the controller and the observer are obtained as follows:

$$\boldsymbol{G}_{\mathrm{c}}(s) = \left[\begin{array}{c|c} \boldsymbol{A} + \boldsymbol{B}_2 \boldsymbol{F}_2 & \boldsymbol{I} \\ \hline \boldsymbol{C}_1 + \boldsymbol{D}_{12} \boldsymbol{F}_2 & 0 \end{array}\right], \ \boldsymbol{G}_{\mathrm{f}}(s) = \left[\begin{array}{c|c} \boldsymbol{A} + \boldsymbol{L}_2 \boldsymbol{C}_2 & \boldsymbol{B}_1 + \boldsymbol{L}_2 \boldsymbol{D}_{21} \\ \hline \boldsymbol{I} & 0 \end{array}\right] \tag{4.24}$$

The unique H_2 optimal controller is obtained as shown:

$$\boldsymbol{K}(s) = \left[\begin{array}{c|c} \hat{\boldsymbol{A}}_2 & -\boldsymbol{L}_2 \\ \hline \boldsymbol{F}_2 & 0 \end{array}\right] \tag{4.25}$$

where $\hat{\boldsymbol{A}}_2 = \boldsymbol{A} + \boldsymbol{B}_2 \boldsymbol{F}_2 + \boldsymbol{L}_2 \boldsymbol{C}_2$.

The simulation results and theoretical analysis in the preceding content show that the characteristics of the guidance system are significantly worsened when only one side of the electromagnet is operating than in the differential state. For this reason, in this section, the single-electromagnet operating model in Eq. (4.19) is used as the subject to design the controller for it to make the system more robust throughout the entire operating interval.

The first stage in developing H_2 controller for guidance systems is to establish the augmented model based on the theory shown in the preceding section, resulting in an increase in the model orders and a more complicated controller. Therefore, refer to the levitation system, the inner loop feedback is added to the part from the voltage to the current of the electromagnet. The transfer function between the linearized guidance electromagnet current i and the control voltage u is:

$$G_i(s) = \frac{i(s)}{u(s)} = \frac{1/R_e}{\frac{L_0}{R_e}s + 1} \tag{4.26}$$

At this time, the electromagnet voltage-current characteristic can be regarded as a first-order inertial element, which means the output current i cannot immediately track the voltage u, and have a delay time $T_i = L_0/R_e = 0.05$ s. Considering the guidance system should be a rapid tracking system and the train must be able to recover quickly when it deviates laterally. The current feedback is added into the guidance system to decrease the time constant to the millisecond level. Then, the inertial element can be approximated as a proportional element with a magnification of 1. The current loop is designed utilizing the series control as shown in Fig. 4.13.

Where $\hat{u}$ is the input voltage, the input of the current feedback is $u = k_{c1} \cdot (\hat{u} - k_{c2} \cdot i)$. The transfer function is:

$$\hat{G}_i(s) = \frac{i(s)}{\hat{u}(s)} = \frac{k_{c1}/(R_e + k_{c1}k_{c2})}{\frac{L_0}{(R_e + k_{c1}k_{c2})} \cdot s + 1} \tag{4.27}$$

The current loop parameters are set as follows: the time constant $T_i = 0.001$, proportionality is 1, $k_{c1} = 276$ and $k_{c2} = 0.99$. A step signal is given as input, and the simulation results before and after correction are shown in Fig. 4.14.

The results in Fig. 4.14 show that the corrected voltage-to-current transfer function can be reduced to a proportional element with a proportionality of 1. The guidance model can be obtained as:

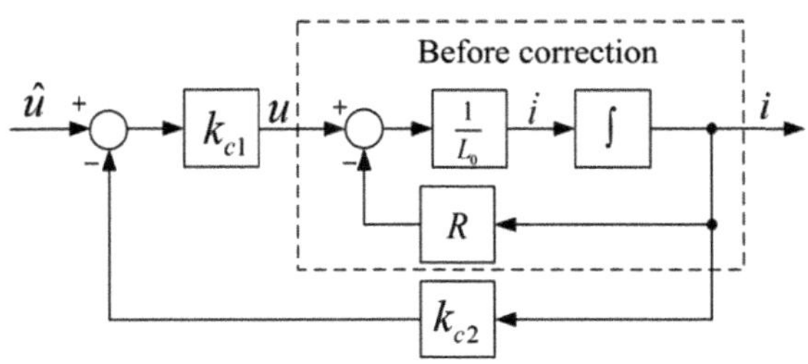

Fig. 4.13 Cascade current feedback block diagram

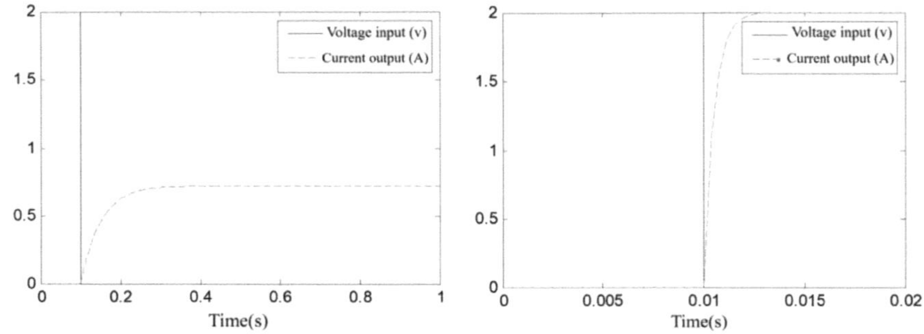

Fig. 4.14 Current response curves before (left) and after (right) correction

$$\begin{cases} \begin{bmatrix} \Delta\dot{z} \\ \Delta\ddot{z} \end{bmatrix} = \begin{bmatrix} 0 & 1 \\ \frac{2i_{10}^2 k_e}{z_0^3 M_0} & 0 \end{bmatrix} \begin{bmatrix} \Delta z \\ \Delta\dot{z} \end{bmatrix} + \begin{bmatrix} 0 \\ -\frac{2k_e i_{10}}{z_0^2 M_0} \end{bmatrix} \Delta u \\ y = \begin{bmatrix} 1 & 0 \end{bmatrix} \begin{bmatrix} \Delta z \\ \Delta\dot{z} \end{bmatrix} \end{cases} \tag{4.28}$$

The inputs and outputs are weighted while establishing the augmented model of the guidance system to increase the system's robustness, as shown in Fig. 4.15.

The selection of the weight function is very critical, which determines the optimal performance index of the system. According to experience and continuous trial and error, the final weightings are:

$$\boldsymbol{W}_1 = \begin{bmatrix} 0 & 100 \\ 0.1 & 1 \end{bmatrix}, \quad \boldsymbol{W}_2 = 1 \times 10^{-5}, \quad \boldsymbol{W}_3 = \begin{bmatrix} 1 & 0 \\ 0 & 0.001 \end{bmatrix}$$

The final augmented model of the guidance system can be obtained as follows. After substituting these realistic parameters:

Fig. 4.15 Weighted augmented model's block diagram

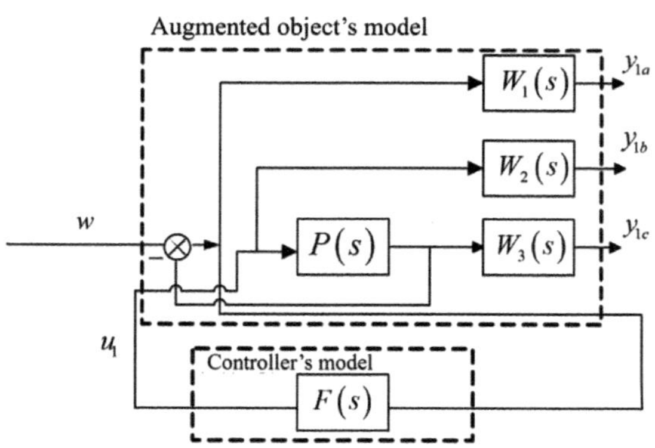

$$P(s) = \left[\begin{array}{c|c} A & B_1\ B_2 \\ \hline C_1 & D_{11}\ D_{12} \\ C_2 & D_{21}\ D_{22} \end{array}\right] = \left[\begin{array}{ccc|cc} 0 & 1 & 0 & 0 & 0 \\ 104.3 & 0 & 0 & 0 & -0.1 \\ -1 & 0 & -10 & 1 & 0 \\ \hline 0 & 0 & 1 & 0 & 0 \\ 0 & 0 & 0 & 0 & 0 \\ 0 & 0 & 0 & 0 & 0 \\ -0.001 & 0 & 0 & 1 & 0 \end{array}\right] \tag{4.29}$$

4.3.2 Simulation Analysis

Firstly, the prerequisites for the solution's existence are verified during the design of the H_2 controller.

(1) $(A\ B_2)$ is completely controllable, and $(A\ C_2)$ is completely observable.
(2) D_{12} is full column rank, and D_{21} is full row rank.
(3) $\begin{bmatrix} A - jwI & B_2 \\ C_1 & D_{12} \end{bmatrix}$ is full column rank for all the w.
(4) $\begin{bmatrix} A - jwI & B_1 \\ C_2 & D_{21} \end{bmatrix}$ is full row rank for all the w.

The augmented guidance model is verified to be solvable, and the final H_2 controller's transfer function can be obtained as follows:

$$K(s) = \frac{-1.1 \times 10^8 (s + 10.21)(s + 10.17)}{(s + 10)\left(s^2 + 448s + 1 \times 10^5\right)} \tag{4.30}$$

The controller of Eq. (4.30) is used to simulate the guidance system, and the simulation curves when a gap and a disturbance force step are given are shown in Fig. 4.16.

Compared the simulation results in Fig. 4.16 with the simulation results of the controller designed using LQR in Sect. 4.2.3, the guidance system regulated by the H_2 controller is more interference-resistant and has greater robustness.

4.4 Design and Implementation of Controller for Guidance System

Based on the simulation analysis of the guidance system, and to further verify the effectiveness of the guidance control algorithm, the physical implementation of the guidance control algorithm and the corresponding guidance control experiment are realized using the high-speed maglev single bogie experiment platform.

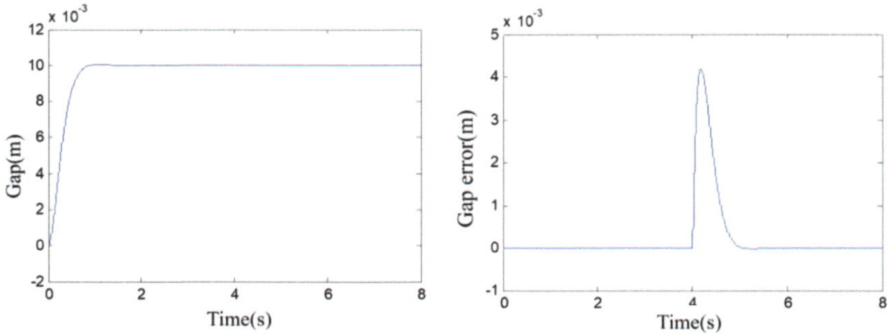

(a) Given a 10 mm step input **(b)** Given an 8.5 kN step input as a disturbance force

Fig. 4.16 Simulation of the guidance system under H_2 control

4.4.1 Test Platform for Guidance System

The high-speed maglev single bogie experiment platform, which is designed according to the actual structure, is used to complete the guidance control experiment and is shown in Fig. 4.17.

The guidance joint structure on one side is clearly shown in Fig. 4.17. The guidance function is carried out by its two guidance control points in conjunction with the other two points on the opposite side. Each guidance control point consists of a set of electromagnets, a guidance sensor, and a guidance control cubicle.

Fig. 4.17 Guidance experiment platform

The guidance control cubicle is composed of the signal pre-processing, a digital guidance controller, and the chopper management. The signal pre-processing module is mainly responsible for collecting the guidance sensor signals, using the FPGA to filter and convert the signals from the sensors to the guidance controller. The chopper is the primary component of the chopper management, where the PWM signal from the digital controller drives the chopper to produce the current that controls the guidance electromagnet. The guidance control algorithm is realized by using the digital guidance controller, namely two DSPs: the TMS28335 as the central processing unit. The two DSPs are connected through a dual-port RAM and perform the sampling, pre-processing (DSPA) and control operation (DSPB) respectively. The information from adjacent and relative points is transmitted to this point through 485 communication, and data interaction between this point and the train host is realized by CAN communication.

The software implementation of the guidance controller consists of two main parts: the main program and the interrupt. The DSPB first detects state signals after the initialization to ensure that the system is operating in a normal state. And then the data transferred from the DSPA to the dual-port RAM is read and the PWM signal corresponding to the control variable is obtained by the guidance control algorithm. The interrupt section responsible for triggering the A/D sampling module to continuously obtain the sensor data required by the guidance control algorithm.

4.4.2 Performance Test of Guidance System

The guidance control experiment is performed on the guidance experiment platform after commissioning the controller hardware and software. The guidance gap signals from the sensor to the control cubicle are read through the CAN communication protocol by using LabVIEW, and the discrete data points obtained are plotted as curves shown in Fig. 4.18.

The experiment results shown in Fig. 4.18 demonstrate that the initial guidance gaps on the left and right sides are approximately 4 mm and 14.6 mm respectively. Under the preceding designed guidance controller, the system moved towards the central line of the guideway, eventually reaching an equal guidance gap of approximately 9.2 mm on both sides.

To observe the characteristics of the guidance system with a single electromagnet working, the power supply to one side of the guidance electromagnet of the guidance experiment platform is cut off and only one side of the electromagnet is used to realize the guidance function. The guidance force in the opposite direction is provided by the component force of gravity. Given a gap target of 10 mm, and a gap disturbance signal with a period of 4 s and an amplitude of 3 mm are added to the system, the gap variation curve of the guidance platform is shown in Fig. 4.19.

The results of the experiment demonstrate the effectiveness and good control effect of the designed guidance control algorithm. Furthermore, it is entirely consistent with the analysis and simulation results in Sect. 4.2.3 that the control of the

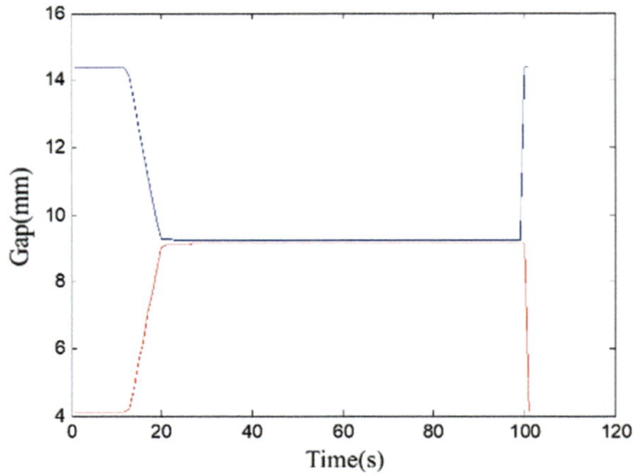

Fig. 4.18 Gap variation curves of the guidance experiment

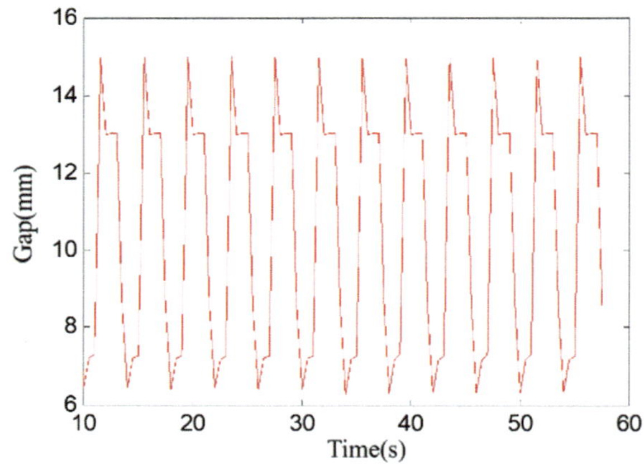

Fig. 4.19 Gap variation curve after adding disturbance with only one side electromagnet working

guidance experiment platform with only one side of the electromagnet operating is less effective than in differential operation. The step-up of the guidance experiment platform at the first start-up shown in Fig. 4.18 is a slow start-up link designed to avoid violent swaying, considering the comfort of the actual train ride.

4.5 Conclusion

Firstly, the guidance system's fundamental structure is analyzed, followed by a force analysis of the guidance electromagnet and establishment of the guidance system's fundamental mathematical model based on the joint structure. Further, a more comprehensive mathematical model taking disturbances and faults into account is built. Then, the guidance system model is simplified at the equilibrium point and

LQR approach is adopted to design the controller. The simulation analysis for the guidance system carried out under the nominal state and the disturbances showed that the LQR based controller performed well on the whole. The performance may decline when there exist external disturbances. To resolve this issue, the H_2 robust controller is proposed to remedy the deficiencies existed in LQR. The simulation results show that the control quality is significantly improved by adoption of the H_2 robust control method. Finally, the guidance control algorithm is conducted on the high-speed maglev single bogie experiment platform, and the corresponding guidance control experiments with good control quality are performed.

Bibliography

1. Ding S (2022) 600 km/h high-speed maglev transportation system. Shanghai Scientific and Technical Publishers, Shanghai
2. Hao A (2008) Research on key technologies of permanent magnet guideway high-speed maglev train suspension and guidance system. Ph.D. dissertation, National University of Defense Technology Doctor Degree, Changsha
3. Hao A, She L, Liu H et al (2007) Research of force feedback controlling on high speed maglev train guidance system. In: ICCA 2007, Guangzhou, China, pp 1969–1972
4. Hao A, Long Z, Chang W (2008a) Design of the robust controller of the guidance system in high-speed maglev train. J China Railw Soc 30(6):40–45
5. Hao A, She L, Chang W (2008b) Adaptive controller design of guidance system of EMS high speed maglev train. Control Eng China 15(2):116–119, 170
6. Huaqing Liu Compile (1995) The German maglev train-Transrapid. University of Electronic Science and Technology Press, Sichuang
7. Kim CH, Ha CW, Lim J et al (2016) Yaw motion control of electromagnetic guidance system for high-speed maglev vehicles. J Electr Eng Technol 11(5):1299–1304
8. Li Y, Chang W (1997) Guidance control research of module structure EMS maglev vehicle. J Natl Univ Defense Technol 19(3):116–120
9. Li B, Zhao C, Li X, Long Z (2020) Dynamics modeling analysis and experiment of the guidance control system of high-speed maglev train. IEEE Access 8:206207–206221
10. Liu Z, Long Z, Li X (2015) Maglev trains. Springer, Berlin
11. Long Z, Dou F, Wang Z et al (2023) Current status and prospects of high-speed maglev suspension and guidance control technology. Sci Technol Foresight 2(4):78–88
12. Wu X (2003) Maglev train. Shanghai Scientific and Technical Publishers, Shanghai
13. Zhai M, Hao A, Li X, Long Z (2018) Research on the active guidance control system in high speed maglev train. IEEE Access 7:741–752
14. Zhang K, Jiang Q (1999) Research of guide mode of electromagnetic suspension vehicles. J China Railw Soc 21(C1):28–31
15. Zhao C (2014) Research on guidance dynamics of EMS high-speed maglev train. Ph.D. dissertation, National University of Defense Technology Doctor Degree, Changsha

Chapter 5
Performance Evaluation of Levitation Control System

After the completion of the levitation control system design for high-speed maglev trains, the performance of the control system under complex working conditions needs to be further analyzed. In this chapter, the nominal performance of the levitation system is firstly analyzed and evaluated using the nominal performance index. Secondly, according to the working characteristics of the levitation system under dynamic operating conditions, the dynamic performance index and evaluation criteria suitable for the levitation system of the high-speed maglev train are proposed. Finally, the performance of the levitation system under deterministic and non-deterministic disturbances is quantitatively analyzed.

5.1 Nominal Performance of Levitation Control System

To analyze the control performance of the levitation system with the implemented nominal controller, it is essential to characterize its corresponding evaluation criteria and performance indicators. The nominal performance of the system is analyzed and evaluated in the following.

The initial levitation gap is set to 20 mm, while the desired gap is set to 12 mm. When the levitation command is sent, the reference input signal of the levitation system can be represented by:

$$\omega(t) = \begin{cases} 0.02 & t < 0 \\ \int_{0.02}^{0.012} -0.003\mathrm{d}t & t \geq 0 \end{cases} \tag{5.1}$$

When the levitation system receives the levitation command, the dynamic responses of the levitation electromagnets on the left and right sides of the joint structure are shown in Figs. 5.1 and 5.2.

Z. Long et al., *High-Speed Maglev Train's Levitation and Guidance Control*, Advances in High-speed Rail Technology, https://doi.org/10.1007/978-981-97-2309-6_5

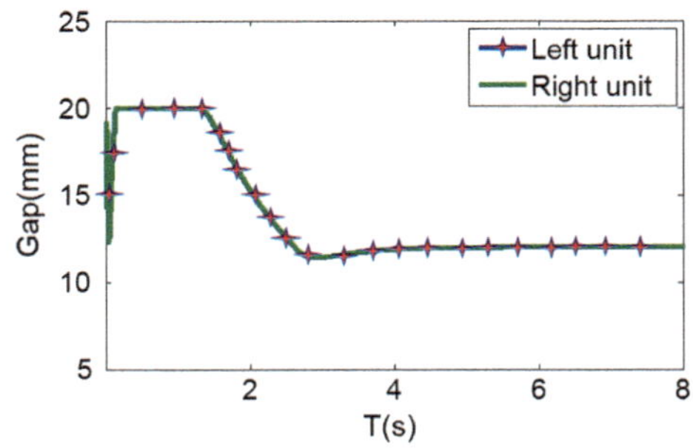

Fig. 5.1 Dynamic responses of the levitation gap

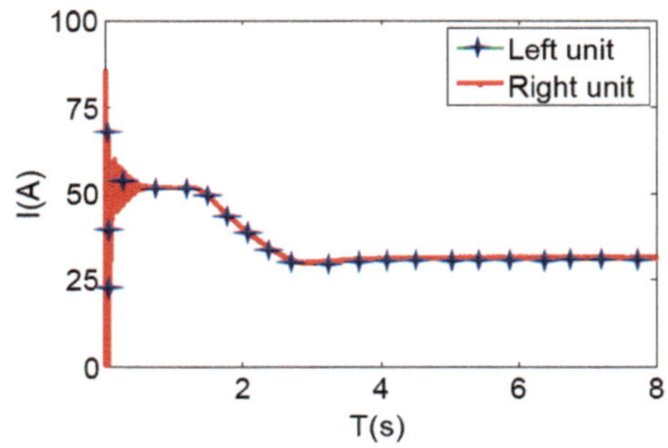

Fig. 5.2 Dynamic responses of the levitation current

The levitation electromagnets on both sides can finally be stabilized at the target gap, which indicates that the entire closed-loop levitation control system with the nominal controller is stable. In the complete dynamic process, the levitation electromagnet exhibits good dynamic characteristics.

5.1.1 Stability and Relative Stability of Levitation Control System

The stability and relative stability of the levitation control system are critical for high-speed maglev trains, which determine whether the levitation system is stable and controllable under various complicated disturbance conditions. If the closed-loop system is unstable, it may result in divergence of the levitation control system, leading to a loss of control. Even if the levitation control system itself is stable, once the levitation control system exceeds its stability threshold, it may still crash. In general, to judge the stability of the closed-loop control system, Bode diagram of the open-loop system can be utilized and analyzed. In the meantime, the gain margin and phase margin of the levitation system can be further obtained from the Bode plot, which acts as indicators for the relative stability of the system. However, the levitation system of the high-speed maglev train is normally a non-minimum phase system, and the stability and relative stability analysis method for the closed-loop system using Bode diagram might not always be applicable.

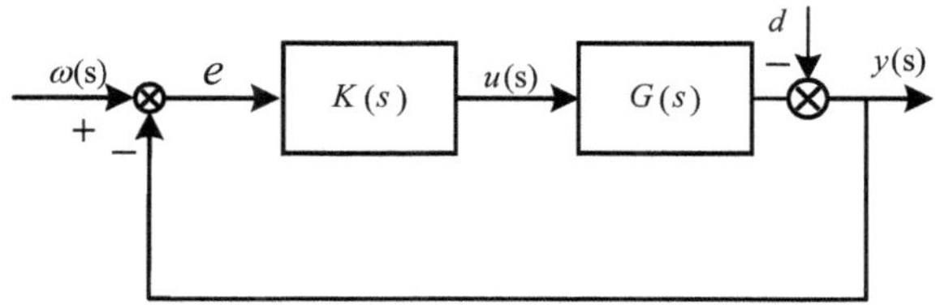

Fig. 5.3 Block diagram of a standard feedback control system

Since the left and right levitation units are symmetric, their structures are exactly the same, and so are the functions. For simplicity, now only consider the left levitation unit. Through equivalent transformation, its original closed-loop control system is simplified as a standard feedback control system as shown in Fig. 5.3. $K(s)$ represents the nominal controller, and $G(s)$ is the left levitation electromagnet model as the controlled object. $\omega(s)$ is the reference, e is the tracking error, d represents the external disturbance, $u(s)$ represents the input of the entire system, and $y(s)$ represents the system output.

The open-loop transfer function of the system is:

$$W_{\mathrm{K}}(s) = K(s)G(s) \tag{5.2}$$

To better reflect the system stability, the integrator in the nominal controller is removed when evaluating the gain and phase characteristics of the open-loop transfer function, and the Nyquist plot is sketched in Fig. 5.4. In the $W_{\mathrm{K}}(\mathrm{j}\omega)$ plane, the Nyquist curve encircles point $(-1, \mathrm{j}0)$ one time ($N = 1$) in the counterclockwise direction, while the open-loop levitation system has one right-half plane pole ($P = 1$). In this case, $Z = P - N = 0$, which means the closed-loop levitation control system is stable according to the Nyquist stability criterion.

Once the plant $G(s)$ is subject to changes or external disturbances, the corresponding closed-loop transfer function $W_{\mathrm{B}}(s)$ will also change. The state variation in the closed-loop control system is usually described quantitatively by the sensitivity function. As defined by Bode, the logarithmic derivative of the closed-loop transfer function with respect to system parameters gives the sensitivity function.

Therefore, the transfer function of the closed-loop control system shown in Fig. 5.3 can be expressed as:

$$W_{\mathrm{B}}(s) = \frac{K(s)G(s)}{1 + K(s)G(s)} \tag{5.3}$$

which yields the following sensitivity function for the closed-loop control system:

$$S(s) = \frac{\mathrm{d}\ln W_{\mathrm{B}}(s)}{\mathrm{d}\ln G(s)} = \frac{\mathrm{d}W_{\mathrm{B}}(s)/W_{\mathrm{B}}(s)}{\mathrm{d}G(s)/G(s)} = \frac{1}{1 + K(s)G(s)} \tag{5.4}$$

According to Eqs. (5.3) and (5.4), the closer the Nyquist curve to the point $(-1, \mathrm{j}0)$ in $W_{\mathrm{K}}(\mathrm{j}\omega)$ plane, the more sensitive the system is. The closer the system

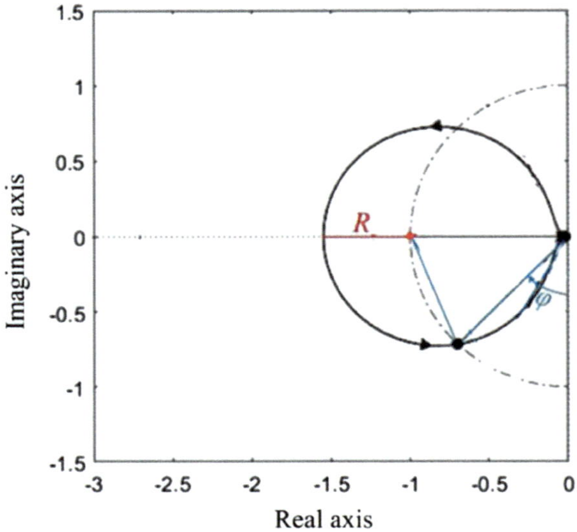

Fig. 5.4 Nyquist plot of the open-loop system

is to its stability boundary, the more liable it is to be destabilized under external disturbances. Hence, the minimum distance between the Nyquist curve and the point (− 1, j0) in $W_K(j\omega)$ plane is:

$$R_{\min} = \min|1 + K(j\omega)G(j\omega)| \tag{5.5}$$

By expressing Eq. (5.4) in the frequency domain, it is found that the denominator of Eq. (5.3) is equivalent to the distance between the Nyquist curve and the point (− 1, j0). When Eq. (5.5) holds, the sensitivity function (5.4) reaches its maximum value.

For the above non-minimum phase system, the maximum value of the sensitivity function can be used to analyze and evaluate the relative stability of the system. In other words, the maximum value of the system sensitivity function can be seen as a performance index of the relative stability. By analysis and calculation, the maximum value of the sensitivity function of the above closed-loop control system is 3, indicating that such a levitation system has a considerable stability margin and meets the requirements on the relative stability of the levitation control system for the high-speed maglev trains.

5.1.2 Steady-State Performance of Levitation Control System

Although the stability and relative stability of the levitation control system are analyzed and evaluated above, they do not represent steady-state performance. The steady-state performance reflects the control accuracy of the closed-loop levitation control system. The control target of the levitation control system should track the target position without error. Nevertheless, when the system is perturbed by external force or the reference input changes during the tracking process, the output of the system will fluctuate. As the system reaches a steady state again with the action of the controller, the deviation between the system output and reference input may arise. When the system operates in a steady state, the deviation between the output and the reference input is called the steady-state error of the system. The magnitude of steady-state error indicates the steady-state accuracy of the closed-loop control system.

To eliminate the steady-state error, integral feedback is introduced to the design of the levitation nominal controller, which converts the original system from Type 0 to Type I. According to the definition of steady-state error, the steady-state error of the levitation system is zero under the action of the nominal controller, and the steady-state performance of the levitation control system is good enough to meet the requirements of the levitation control system for the high-speed maglev train.

5.1.3 Transient Performance of Levitation Control System

Although the steady-state performance can reflect the control accuracy of the closed-loop control system, its dynamic characteristics when the system is in variation are also worthy of attention. This is because a system has to go through a transient transition process when moving from a certain steady state to a new one. In this subsection, the time domain analysis method is mainly used to evaluate the control performance of the levitation system objectively.

For example, when the given reference signal of the levitation system experiences a sudden step change with an amplitude of 13 mm at 10 s, the control system must be able to adapt to it immediately. At this point, the transient response of the levitation gap is shown in Fig. 5.5.

By analyzing the transient response given in Fig. 5.5, the transient performance of the levitation system for the high-speed maglev train under the action of the nominal controller is summarized in Table 5.1.

In conclusion, this section quantitatively analyzes and evaluates the stability and relative stability, steady-state performance and transient performance of the high-speed train levitation system. The results show that the performance of levitation control system is good to meet the control requirements.

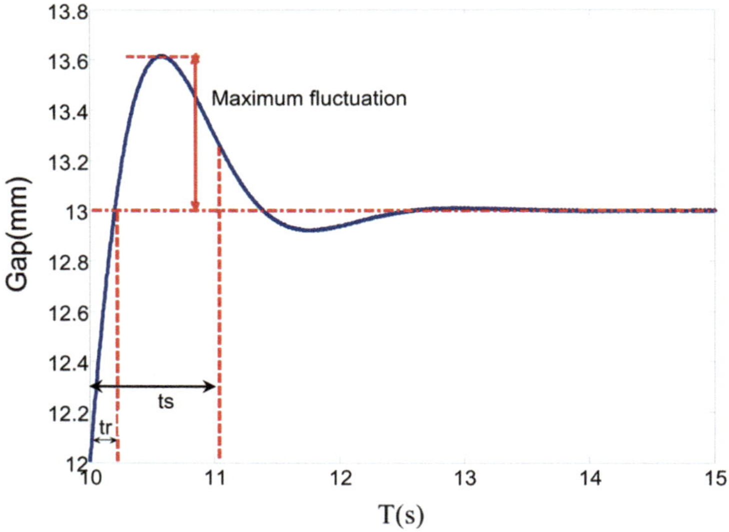

Fig. 5.5 Transient response of the levitation control system for high-speed maglev train

Table 5.1 Transient performance index of levitation control system for high-speed maglev train

Transient performance index	Maximum overshoot σ (%)	Rise time tr	Settling time ts	Cycles of oscillation μ
Value	4.7%	0.2	1.03	1

5.2 Dynamic Performance Evaluation of Levitation System

After the design of the nominal controller for the levitation system is completed, the nominal performance index has been applied to analyze and evaluate the nominal performance of the levitation system. It has been demonstrated that with the implementation of the nominal controller, each performance index is acceptable and satisfies the control requirements. However, during the operation phase of the high-speed maglev train, the levitation system is consistently in a dynamic balance state, in which renders the nominal performance index to evaluate the performance of the levitation system is deficient. Therefore, to further investigate the control performance of the levitation system during the operation of the high-speed maglev train, a dynamic performance evaluation index that applies to the levitation system is proposed in this section.

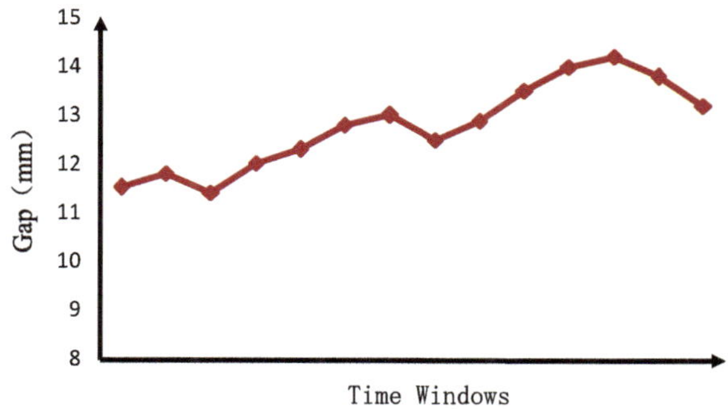

Fig. 5.6 Levitation gap fluctuation curve

5.2.1 Deficiencies in Using Nominal Performance Index

The levitation system of high-speed maglev train is a tracking control system, and the control objective is to achieve reference tracking and disturbance rejection. On one hand, the output of the system is not allowed to deviate from the reference input, which means the actual levitation gap must always be kept in the vicinity of the commanded gap. On the other hand, however, since the high-speed maglev train is subject to various known or unpredictable disturbances during its operation, the output of the levitation control system will inevitably deviate from the input due to such disturbances. Therefore, the levitation control algorithm is always regulating the system in real time so that a dynamic balance state can be maintained between the output and input of the levitation system.

Figure 5.6 displays a segment of the levitation gap fluctuation curve that is randomly truncated during the operation of the high-speed maglev train. As expected, the levitation control system is in a dynamic regulation status all the time, and the levitation gap oscillates around the reference gap of 11–14 mm.

Aware of the dynamic characteristics of the levitation system during the operation, evaluating the system performance based on the nominal performance index has its limitation. Thus, it is very necessary to propose a dynamic performance evaluation method and index that are suitable for the levitation system of the high-speed maglev train. The index should be able to objectively evaluate the control performance of the levitation system during the operation of the high-speed maglev train.

5.2.2 Performance Evaluation Under Dynamic Operating Conditions

Generally, for the curve depicted in Fig. 5.6, a smaller fluctuation range of the levitation gap indicates better control performance of the levitation system. Based on the above analysis, the fluctuation range of the levitation gap can be directly leveraged to evaluate the control performance of the levitation system for high-speed maglev trains during the operation phase. For instance, let the 12 mm be the central datum line, and set the time window to 1 s. In this period, the fluctuation range of the levitation gap is considered to be small if the levitation gap oscillates between 10 and 14 mm, which concludes that the levitation gap in the operating process falls into a high-quality range. If the levitation gap can always be kept in the high-quality range, then it indicates the good control performance of the levitation system during the operation of the high-speed maglev train.

When the levitation gap fluctuates within the interval 14–16 mm or 8–10 mm, it means that the fluctuation range of the levitation gap is enlarged during this period, and the levitation gap starts deviating from the central datum line, although such range is still considered acceptable during the operation. As long as the levitation gap can always be maintained in the high-quality range and the acceptable range during the operation of the high-speed maglev train, the control performance of the levitation system during the operation is qualified.

However, if the levitation gap is above 16 mm or below 8 mm, it indicates that the control performance of the levitation system has been seriously deteriorated during this period. With the levitation gap entering such interval during the operation of the train, the control performance of the levitation system during the operation is unqualified.

Through the above analysis and discussion, it is certain that the fluctuation interval of the levitation gap can be directly used to evaluate the control performance level of the levitation system during the operation of the high-speed maglev train. The fluctuation interval of the levitation gap and the corresponding control performance level are defined in Fig. 5.7.

However, when the fluctuation range of the levitation gap is within the range of 10–14 mm, and the levitation electromagnet vibrates violently at this time, it will also affect the stability and comfort of the high-speed maglev train. It is found that it is not enough to rely on the fluctuation range of the levitation gap as the basis for evaluating the control performance only. Therefore, to evaluate the control performance of the

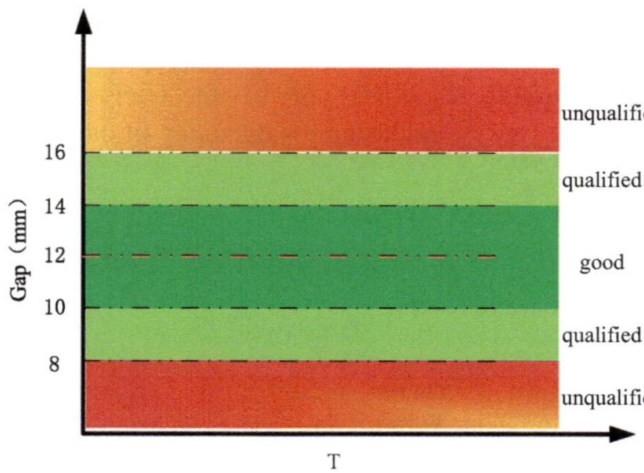

Fig. 5.7 Levitation gap fluctuation range and control performance level

levitation system comprehensively and objectively, this section proposes the sperling stability index reflecting the vibration impulse and vibration kinetic energy of the levitation electromagnet.

Some national railway cooperation organizations such as the European Union of Railways use the sperling stability index to evaluate the running quality of vehicles. The sperling stability index reflecting the vibration impulse and vibration kinetic energy is calculated and used as the basis for judging the stability of the train. However, the sperling stability index of the levitation electromagnet is different from the sperling stability index of the vehicle. It only reflects the vibration impulse and vibration kinetic energy of the levitation electromagnet. An air spring is installed between the levitation frame and the carriage of a high-speed maglev train, and the vibration and impact of the levitation electromagnet cannot directly act on the carriage. As the secondary system of the high-speed maglev train, the air spring is able to isolate the high-frequency vibration and shock well. The sperling stability index of the levitation electromagnet is used as the basis for evaluating the stability of the high-speed maglev train. Considering the different structural and functional characteristics, the calculation equation of the sperling stability index of the levitation electromagnet as shown in Eq. (5.6) is given in this section. Aiming at the dynamic characteristics of the levitation system during the operation of the high-speed maglev train, this section defines the sperling stability index of the levitation electromagnet as the dynamic performance index of the levitation system of the high-speed maglev train.

$$w = \begin{cases} 0.896\sqrt[10]{0.325a^3 f} & f \in [0.5,\ 5.9] \\ 0.896\sqrt[10]{\frac{400a^3}{f^3}} & f \in [5.9,\ 20] \\ 0.896\sqrt[10]{\frac{a^3}{f}} & f \in [20,\ 30] \end{cases} \tag{5.6}$$

where a is the vibration acceleration of the levitation electromagnet, and the unit is cm/s^2. f is the vibration frequency of the levitation electromagnet, and the unit is Hz.

During the actual operation of the high-speed maglev train, the vibration of the levitation electromagnet is random. Therefore, the vibration acceleration of the levitation electromagnet can be decomposed according to the frequency, and the vibration amplitude of the levitation electromagnet in each frequency range can be obtained. According to Eq. (5.6), each frequency band corresponds to a different dynamic performance index calculation formula. The dynamic performance indicators of the levitation electromagnet in the entire frequency range are:

$$W = \left(w_1^{10} + w_2^{10} + \cdots w_n^{10}\right)^{0.1} \tag{5.7}$$

According to the actual operation of the high-speed maglev train and the on-site riding experience, the following conclusions are found. If the dynamic performance index value of the levitation electromagnet is less than 2.1 for a long time, the

Table 5.2 Dynamic performance index and control performance level

Control performance level	Assessment	Dynamic performance indicators
1	Excellent	< 2.1
2	Good	2.1–2.3
3	Qualified	2.3–2.5
4	Failed	> 2.5

control performance of the levitation system is regarded as excellent. If the dynamic performance index of the levitation electromagnet is between 2.1 and 2.3 for a long time, the control performance of the levitation system is considered to be good. If the dynamic performance index value of the levitation electromagnet is between 2.3 and 2.5 for a long time, the control performance of the levitation system is considered to be qualified. If the dynamic performance index value of the levitation electromagnet is greater than 2.5 for a long time, it is considered that the control performance of the levitation system is unqualified, and the control performance of the levitation system has been seriously degraded.

As shown in Table 5.2, the dynamic performance index and control performance level of the levitation electromagnet are given.

In summary, during the operation of the high-speed maglev train, the fluctuation interval of the levitation gap and the dynamic performance index of the levitation electromagnet can comprehensively evaluate the control performance of the levitation system.

5.3 Performance of Levitation System Under Deterministic Disturbance

During the operation of the high-speed maglev train, there are many kinds of deterministic disturbances to the levitation system, but the deterministic disturbances are known, clear and obtainable. This section mainly analyzes and discusses the adverse effects of load changes and line curve changes on the performance of the levitation system.

5.3.1 Influence Analysis of Load Changes

During the operation of the high-speed maglev train, when the load changes, the control performance of the levitation system is reduced due to the impact of the load change. To analyze and study the impact of load changes on the performance of the levitation system, a nonlinear levitation system model was built to observe the dynamic response. When the load of the left levitation unit increases by 10 kN from 10 s and lasts for 5 s, the levitation gap and levitation current of the entire levitation system change as shown in Figs. 5.8 and 5.9.

The load of the left levitation unit has returned to normal since 15 s. The levitation gap and levitation current of the entire levitation system are as shown in Figs. 5.10 and 5.11.

As shown in Figs. 5.8, 5.9, 5.10 and 5.11, the levitation gap between the left and right levitation units of the high-speed maglev train levitation system is no longer maintained at the set gap, and the levitation current becomes larger. The maximum levitation gap is 13 mm, and the minimum levitation gap is 11.6 mm. At the same time, the instantaneous vibration of the levitation electromagnets on the left and right sides is strong. The instantaneous dynamic performance index of the levitation system reaches 2.7346 when the load of the left levitation unit increases by 10 kN. The change of load has an adverse effect on the performance of the levitation system, and the performance of the levitation system is degraded.

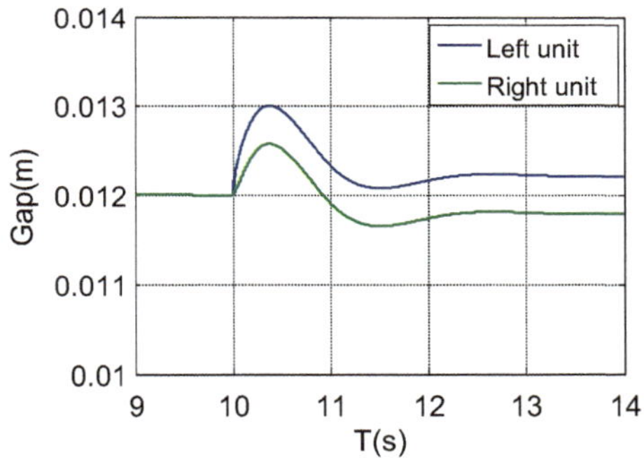

Fig. 5.8 Levitation gap curve when the load changes

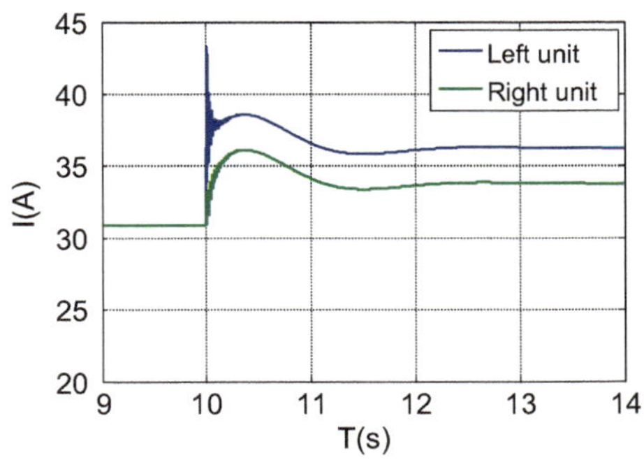

Fig. 5.9 Levitation current curve when the load changes

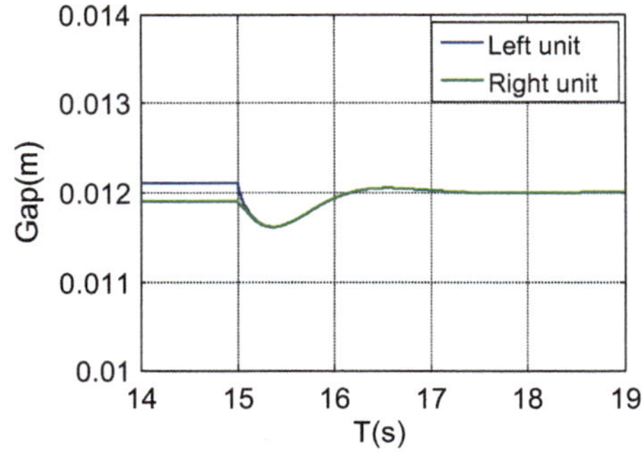

Fig. 5.10 Levitation gap curve when the load returns to normal

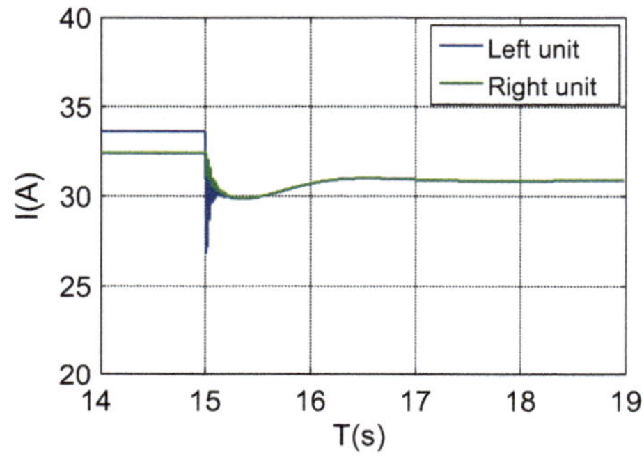

Fig. 5.11 Levitation current curve when the load returns to normal

5.3.2 *Influence Analysis of Line Curve Changes*

During the design and construction of high-speed maglev lines, due to terrain fluctuations or other space constraints, the curve of high-speed maglev train lines will change. Taking the vertical curve on the high-speed maglev line as an example, when the similar gradient value exceeds the allowable range, it is necessary to ease the vertical curve to connect. Although the appearance of the vertical curve is unfavorable for the levitation system, the vertical curve on the high-speed maglev line track is known and the parameters are clear, which is a deterministic disturbance. This section takes the vertical curve in the high-speed maglev line as an example to analyze and discuss the change of the control performance of the levitation system when the high-speed maglev train passes through the vertical curve.

As shown in Fig. 5.12, the profile of the maglev train line is composed of five parts, the strait section *AB*, the concave curve of section *BC*, the oblique curve of section *CD*, the convex curve of section *DE*, and the strait section *EF*. The radius of circular curve is r and the angle of slope is α. The displacement of the track relative to the absolute reference plane is z_G. It is assumed that the tangent velocity along the track is V and remains constant.

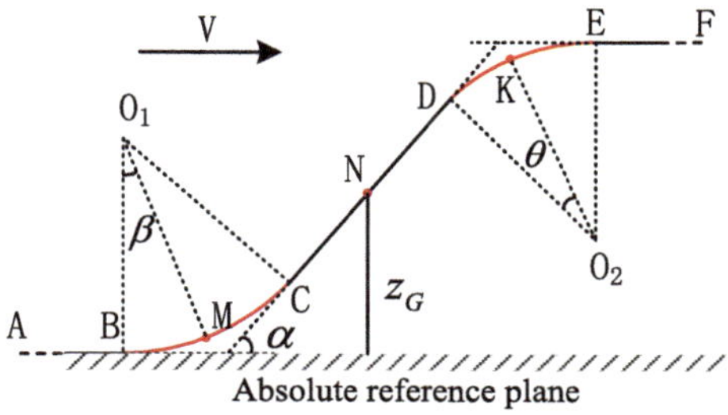

Fig. 5.12 Vertical curve section of maglev line

After the line design and construction is completed, the parameters such as the radius of the circular curve and the inclination angle of the oblique line α have been determined. The operating speed is also strictly limited and determined. Therefore, the deterministic disturbance characteristics are clear and can be described by mathematical modeling.

(1) the strait section AB:

$$z_{\mathrm{G}}(t) = 0 \tag{5.8}$$

(2) the concave curve of section BC:

$$z_{\mathrm{G}}(t) = R \cdot (1 - \cos\beta) = R \cdot 2\left(\sin\frac{\beta}{2}\right)^2 \approx R \cdot \frac{\beta^2}{2} = \frac{V^2 \cdot t_{\mathrm{BM}}^2}{2R} \tag{5.9}$$

In the equation, M is any point on the concave curve segment BC, and t_{BM} is the running time of the high-speed maglev train from point B to point M.

(3) The oblique curve of section CD:

$$z_{\mathrm{G}}(t) = R \cdot (1 - \cos\alpha) + V \cdot t_{\mathrm{CN}} \cdot \sin\alpha \tag{5.10}$$

In the equation, N is any point on the concave curve segment CD, and t_{CN} is the running time of the high-speed maglev train from point C to point N.

(4) The convex curve of section DE:

$$z_{\mathrm{G}}(t) = R \cdot (1 - \cos\alpha) + \overline{CD} \cdot \sin\alpha + R \cdot \sin\left(\frac{\pi}{2} - \alpha + \frac{Vt_{\mathrm{DK}}}{R}\right) - R \cdot \cos\alpha \tag{5.11}$$

In the equation, K is any point on the concave curve segment DE, and t_{DK} is the running time of the high-speed maglev train from point D to point K.

(5) The strait section EF:

$$z_{\mathrm{G}}(t) = R \cdot (1 - \cos\alpha) + \overline{CD} \cdot \sin\alpha \tag{5.12}$$

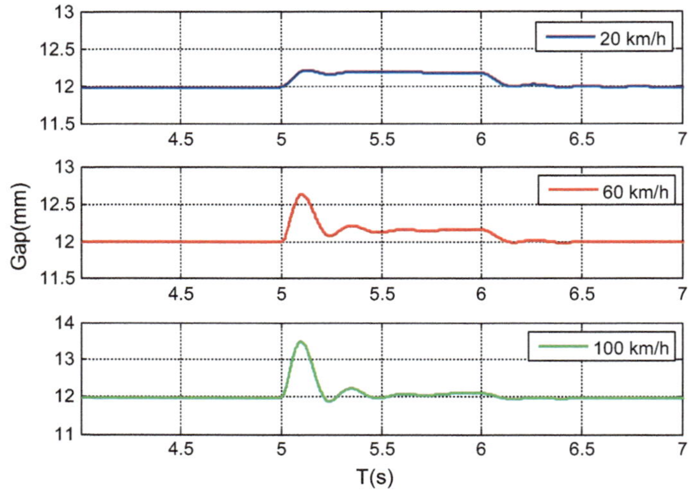

Fig. 5.13 The levitation gap curves when the high-speed maglev train enters the concave curve

When the high-speed maglev train passes through the vertical curve section shown in Fig. 5.12, the vertical curve will be added to the levitation control system as a deterministic disturbance. Taking the left levitation unit as an example, observe the dynamic response of the levitation system when it passes through the vertical curve.

When the high-speed maglev train passes through the vertical curve section shown in Fig. 5.12 at different speeds, the dynamic response of the levitation system is also different. Since the speed of the high-speed maglev train passing through the vertical curve section is strictly limited and determined, it is usually lower than 100 km/h. The levitation gap curves of the left levitation unit at the set speeds of 20, 60, and 100 km/h are shown in Fig. 5.13.

When the high-speed maglev train exits the concave curve section at the set speeds of 20 km/h, 60 km/h, and 100 km/h, respectively, the levitation gap of the levitation electromagnet will fluctuate. The levitation gap curves of the left levitation unit are shown in Fig. 5.14.

When the high-speed maglev train enters the convex curve section at the set speed of 20, 60, and 100 km/h, the levitation gap of the levitation electromagnet will fluctuate. The levitation gap curves of the left levitation unit are shown in Fig. 5.15.

When the high-speed maglev train exits the convex curve section at the set speed of 20, 60, and 100 km/h, the levitation gap of the levitation electromagnet will fluctuate. The levitation gap curves of the left levitation unit are shown in Fig. 5.16.

When the high-speed maglev train passes through the vertical curve section at the set speed, the levitation gap of the levitation electromagnet will fluctuate, and the performance of the levitation system will be reduced. The fluctuation range of the levitation gap just reflects the tracking performance of the levitation control system.

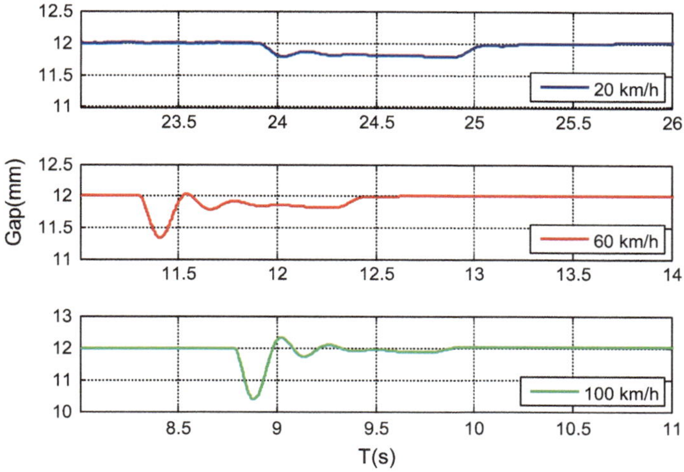

Fig. 5.14 Levitation gap curves when the high-speed maglev train exits the concave curve

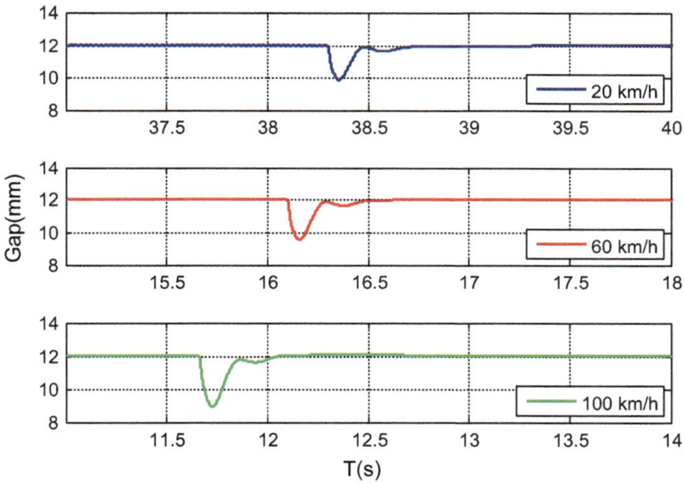

Fig. 5.15 Levitation gap curves when the high-speed maglev train enters the convex curve

As shown in Figs. 5.13, 5.14, 5.15 and 5.16, when the levitation system of the high-speed maglev train passes through the vertical curve section, the levitation gap begins to fluctuate. When the high-speed maglev train passes through the above-mentioned vertical curve section at the set speed of 20 km/h, the maximum levitation gap is 14.6 mm, and the minimum levitation gap is 9.8 mm. When the high-speed maglev train passes through the above-mentioned vertical curve section at the set speed of 60 km/h, the maximum levitation gap is 15 mm, and the minimum levitation gap

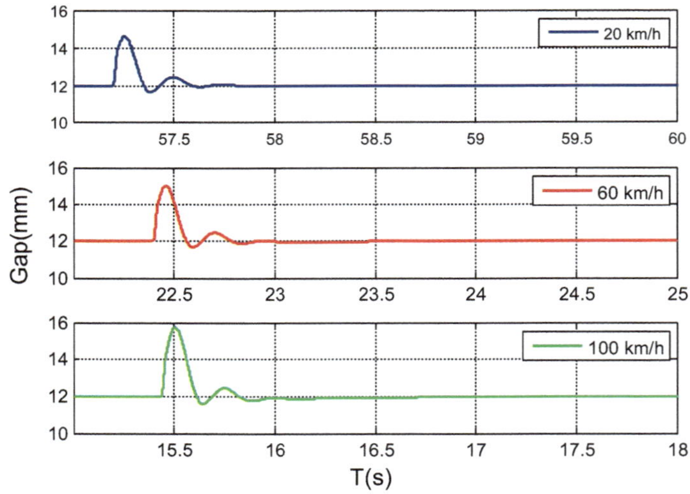

Fig. 5.16 Levitation gap curves when high-speed maglev train exits the convex curve

is 9.5 mm. When the high-speed maglev train passes through the above-mentioned vertical curve section at the set speed of 100 km/h, the maximum levitation gap is 15.7 mm, and the minimum levitation gap is 9 mm. This change in the line curve is unfavorable to the performance of the levitation system, and the performance of the levitation system is seriously degraded. However, for the actual high-speed maglev train in operation, the running speed is often limited for this special road section to reduce the adverse effect of the line curve change on the levitation system.

5.4 Performance of Levitation System Under Non-deterministic Disturbance

Considering that the high-speed maglev train runs for a long time and the external environment is complex and changeable, the non-deterministic disturbance caused by the change of the external environment will occur in an unpredictable way during the long-term operation, even if the control loop is well designed and debugged.

5.4.1 Influence Analysis of Random Track Irregularity

During the long-term operation of the high-speed maglev train, the random irregularities will appear. The track steps shown in Fig. 5.17 are a type of random track irregularity, which is particularly unfavorable for the levitation system during high-speed operation.

When the high-speed maglev train passes through the track steps of different amplitudes at a speed of 600 km/h, the dynamic characteristics of the levitation system are also different. To illustrate the influence of the track steps on the levitation system of the high-speed maglev train, the dynamic characteristics of the levitation system of the high-speed maglev train passing through the track steps with amplitudes of 1, 1.5, and 2 mm were simulated.

(1) When the amplitude of track step is 1 mm, the dynamic characteristic curves of the high-speed levitation joint structure are shown in Fig. 5.18.

As shown in Fig. 5.18, the levitation gap of levitation joint structure is always within the high-quality range. At the same time, the dynamic performance index of the levitation system is 2.3673.

(2) When the amplitude of track step is 1.5 mm, the dynamic characteristic curves of the high-speed levitation joint structure are shown in Fig. 5.19.

As shown in Fig. 5.19, the levitation gap of levitation joint structure is still in the high-quality range. The fluctuation range of the levitation gap is changed to 11.3–13.5 mm, and the fluctuation range of the levitation gap is slightly increased. At the same time, the vibration of the levitation electromagnet becomes more intense. The dynamic performance index of the levitation system increases to 2.5467. The control performance of the system has degraded.

(3) When the amplitude of track step is 2 mm, the dynamic characteristic curves of the high-speed levitation joint structure are shown in Fig. 5.20.

As shown in Fig. 5.20, the levitation gap of the levitation joint structure begins to break through the high-quality range. The fluctuation range of the levitation gap is changed to 11.1–14 mm, the fluctuation range of the levitation gap is further enlarged, and the performance of the levitation system begins to enter the acceptable range

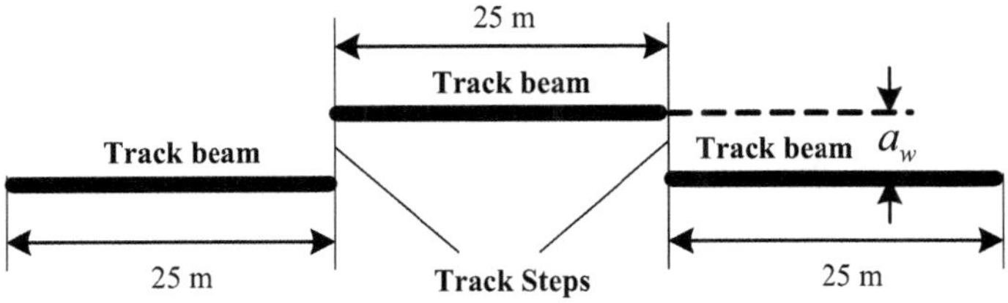

Fig. 5.17 Schematic diagram of high-speed maglev track steps

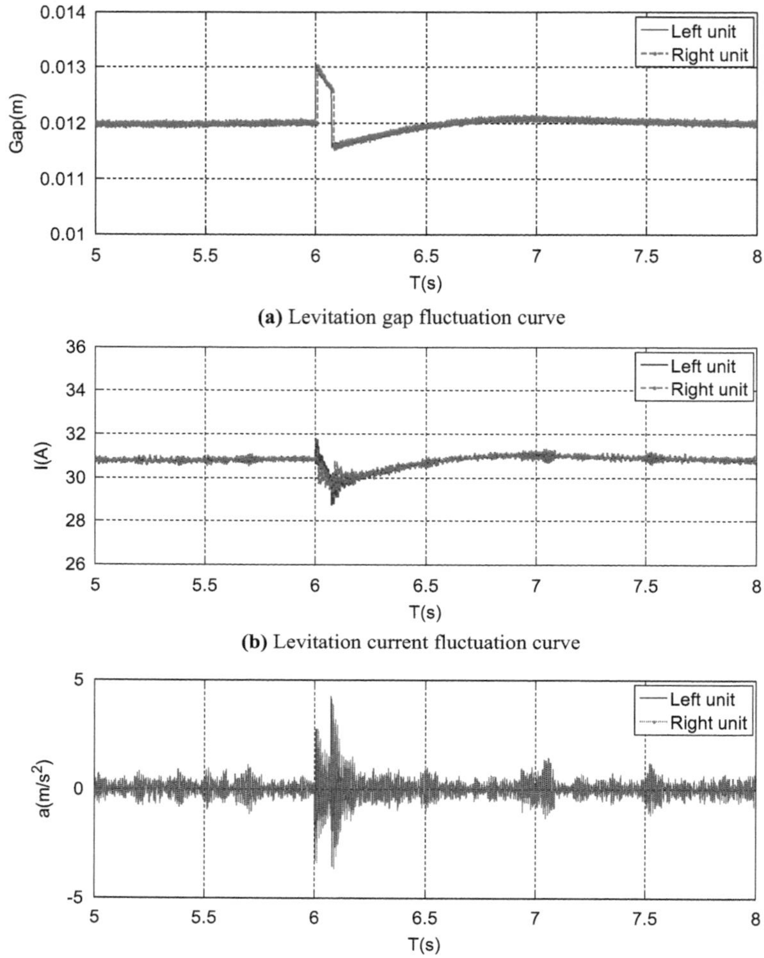

(a) Levitation gap fluctuation curve

(b) Levitation current fluctuation curve

(c) Acceleration fluctuation curve

Fig. 5.18 The dynamic characteristic curves of the levitation system when passing through a track step with an amplitude of 1 mm

from the high-quality range. The dynamic performance index of the levitation system increases to 2.7208, and the control performance of the levitation system is severely degraded. The dynamic performance index of levitation system corresponding to step amplitude is shown in Fig. 5.21.

According to the levitation gap fluctuation interval and the dynamic performance index, the control performance of the levitation system under the disturbance of the track steps of different amplitudes is evaluated in Table 5.3.

To sum up, the appearance of the track steps leads to the degradation of the control performance of the levitation system of the high-speed maglev train. The greater the

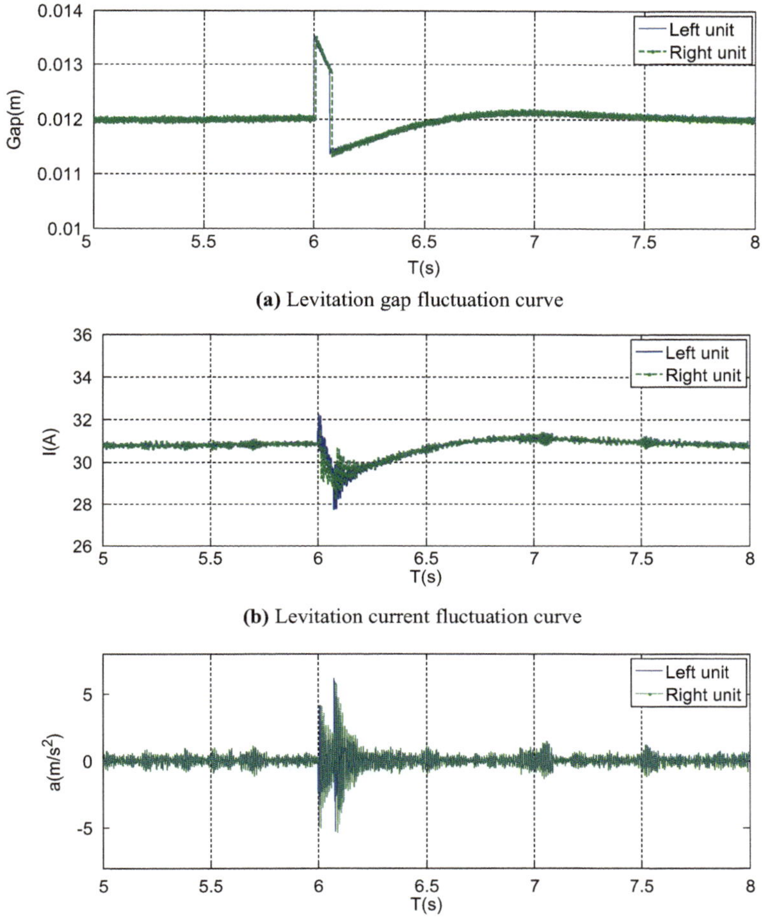

(a) Levitation gap fluctuation curve

(b) Levitation current fluctuation curve

(c) Acceleration fluctuation curve

Fig. 5.19 The dynamic characteristic curves of the levitation system when passing through a track step with an amplitude of 1.5 mm

amplitude of the track steps is, the more serious the performance degradation of the levitation system is.

At the same time, if the high-speed maglev train passes through the track steps of fixed amplitude at different speeds, the dynamic characteristics of the levitation system are also different. To illustrate the effect of speed increase on the levitation system of high-speed maglev train, the dynamic characteristics of high-speed maglev train at different speeds are studied.

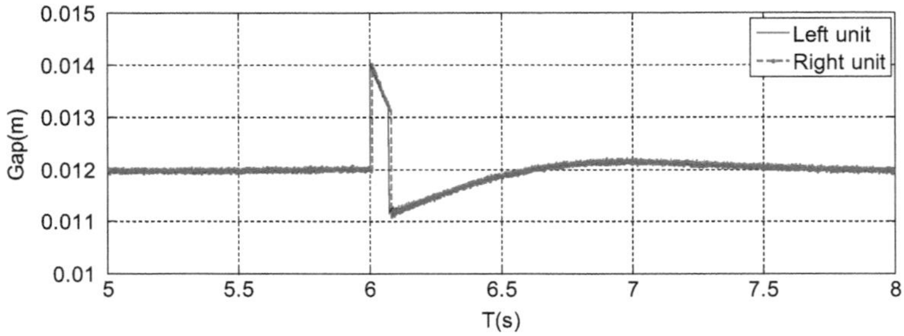

(a) Levitation gap fluctuation curve

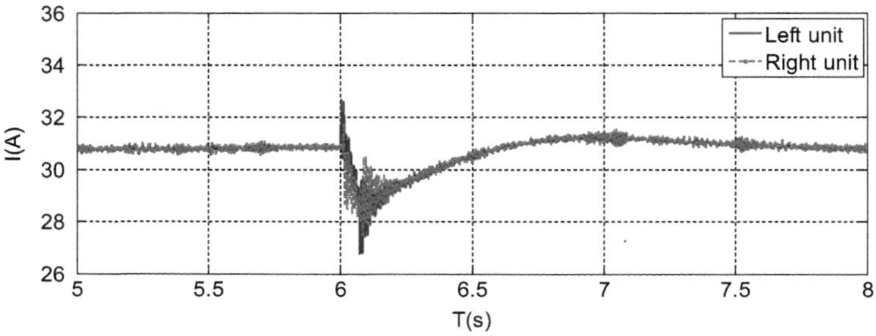

(b) Levitation current fluctuation curve

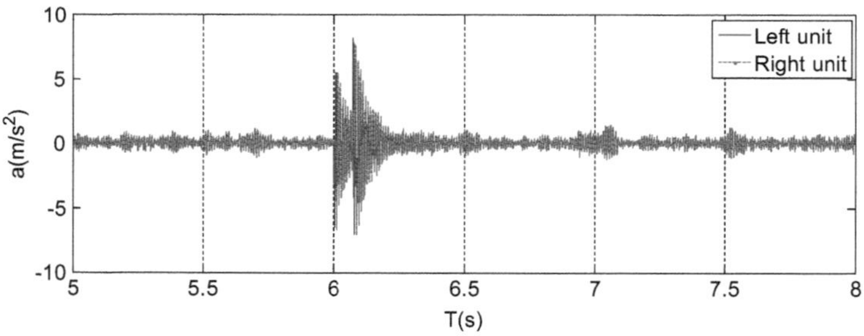

(c) Acceleration fluctuation curve

Fig. 5.20 Dynamic characteristic curves of the levitation system passing through a track step with an amplitude of 2 mm

(1) The amplitude of track step is 1 mm. When the speed is 200 km/h, the dynamic characteristic curves of the high-speed levitation joint structure are shown in Fig. 5.22.

As shown in Fig. 5.22, the levitation gap of levitation joint structure is always within the high-quality range. The maximum levitation gap becomes 13 mm, and the minimum levitation gap is 10.9 mm. At the same time, the dynamic performance index of the levitation system is 2.1039.

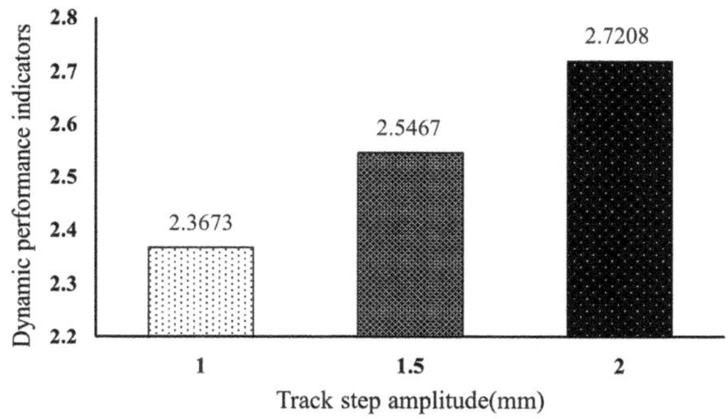

Fig. 5.21 Dynamic performance index of levitation system corresponding to step amplitude

Table 5.3 Dynamic performance of levitation system under random track irregularity disturbance

Track step amplitude (mm)	1	1.5	2
Levitation gap fluctuation range	11.5–13	11.3–13.5	11.1–14
Dynamic performance indicators	2.3673	2.5467	2.7208
Performance evaluation	Good	Acceptable	Severely degraded

(2) The amplitude of track step is 1 mm. When the speed is 400 km/h, the dynamic characteristic curves of the high-speed levitation joint structure are shown in Fig. 5.23.

As shown in Fig. 5.23, the fluctuation range of the levitation gap becomes 11.3–13 mm, and the fluctuation is within the high-quality range. At the same time, the dynamic performance index of the levitation system increases to 2.2826, which indicates that the performance of the levitation system will be affected by the speed increase.

(3) The amplitude of track step is 1 mm. When the speed is 600 km/h, the dynamic characteristic curves of the high-speed levitation joint structure are shown in Fig. 5.24.

As shown in Fig. 5.24, the fluctuation range of the levitation gap becomes 11.5–13 mm, and the fluctuation is within the high-quality range. At the same time, the dynamic performance index of the levitation system increases to 2.3595 when the speed reaches 600 km/h. The dynamic performance index of the levitation system when passing the track steps at different speeds is shown in Fig. 5.25.

To sum up, the appearance of track steps leads to the degradation of the levitation system. The greater the amplitude of the track steps is, the more serious the degradation of the control performance of the levitation system is. When the amplitude of the track step increases to 2 mm, the control performance of the levitation system has been seriously degraded, and the dynamic performance index reaches 2.7208. In addition, the running speed has little effect on the fluctuation range of the levitation gap but has a great influence on the dynamic performance indicator. When the speed increases, the dynamic performance index becomes larger.

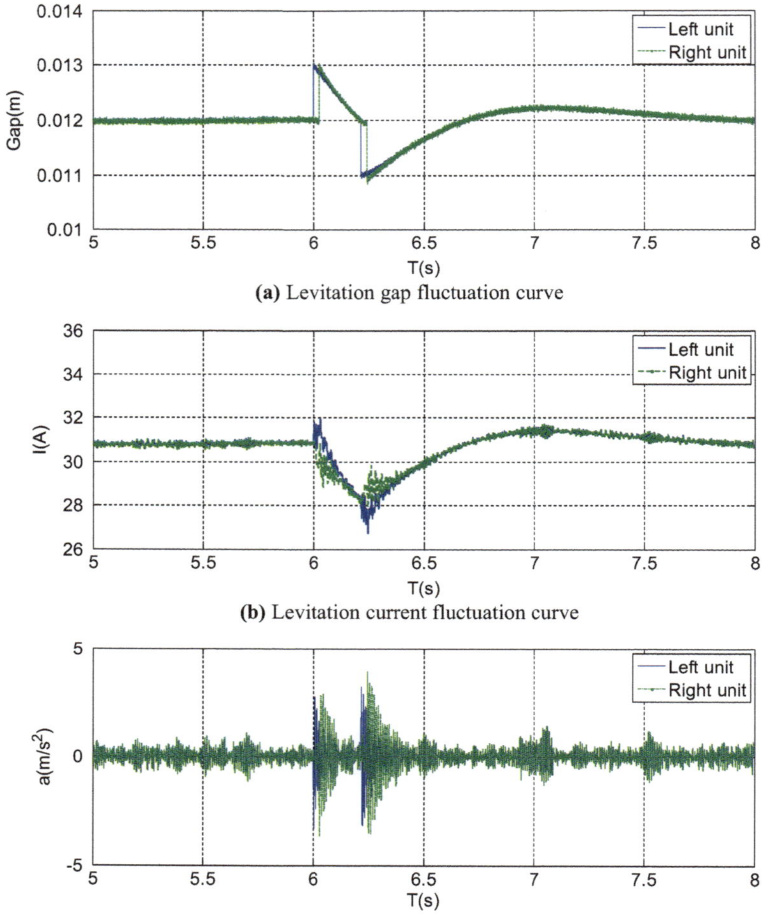

(a) Levitation gap fluctuation curve

(b) Levitation current fluctuation curve

(c) Acceleration fluctuation curve

Fig. 5.22 Dynamic characteristic curves of the levitation system when it passes through the track steps at a speed of 200 km/h

5.4.2 Influence Analysis of Torsion Angle Variation

As shown in Fig. 5.26, the high-speed maglev train usually uses an elevated track structure. The track beams are basically single-span simply supported beams, and the standard diameter of the line is 30.96 m. Ideally, the high-speed maglev track is a straight rigid body, and the deflection-span ratio can be infinitely small. However, considering the difficulty of construction and the cost, the track deflection-span ratio of high-speed maglev lines cannot be infinitely small.

To improve the comfort and stability of the high-speed maglev train, the tangential torsion angle of the beam end of the track is controlled to be very small. The schematic

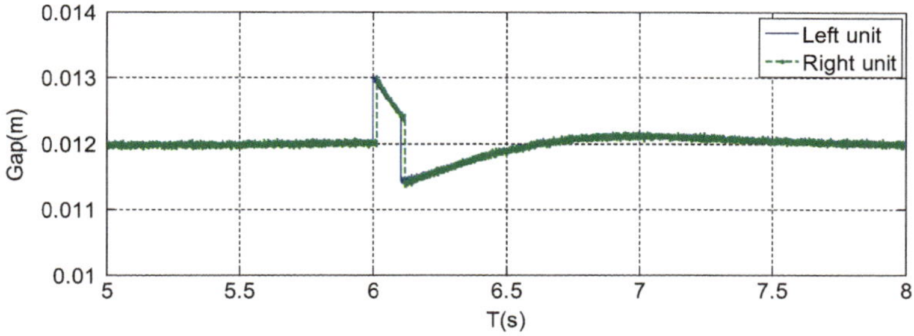

(a) Levitation gap fluctuation curve

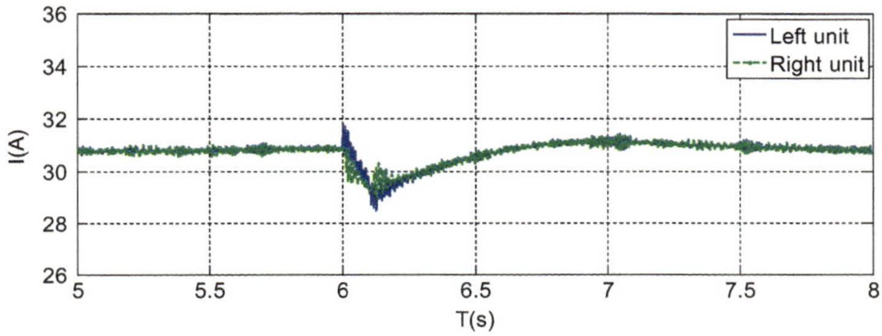

(b) Levitation current fluctuation curve

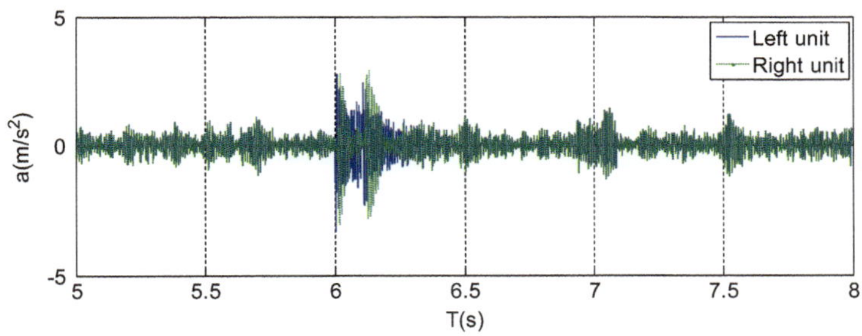

(c) Acceleration fluctuation curve

Fig. 5.23 Dynamic characteristic curves of the levitation system when passing through the track steps at a speed of 400 km/h

diagram of the tangential torsion angle of the beam end of the high-speed maglev line is shown in Fig. 5.27. Taking the Shanghai high-speed maglev line as an example, the initial design value of the beam end tangential torsion angle reaches 0.00021 rad. However, changes in the external environment such as foundation settlement and changes in ambient temperature will lead to the deformation of the track beam, and the tangential torsion angle of the beam end will also change accordingly. This change of the tangential torsion angle of the beam end is random, periodic, and changes in an unpredictable way with the change of the external environment.

The larger the tangential torsion angle of the beam end is, the larger the deflection-span ratio is. When the high-speed maglev train passes through the above-mentioned

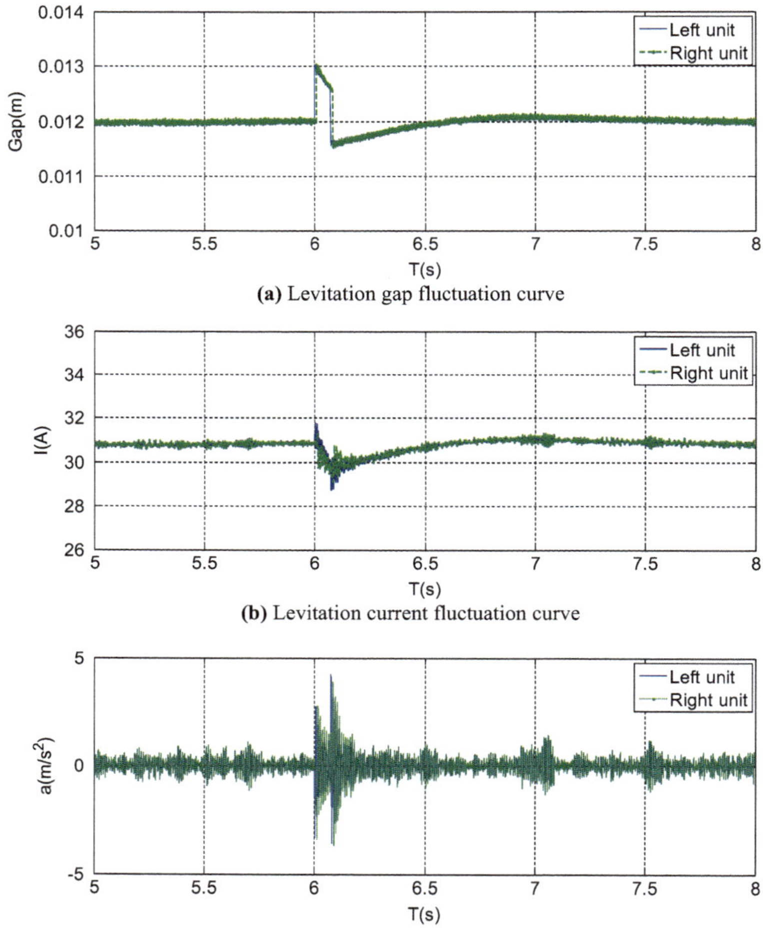

(a) Levitation gap fluctuation curve

(b) Levitation current fluctuation curve

(c) Acceleration fluctuation curve

Fig. 5.24 Dynamic characteristic curves of the levitation system when it passes through the track steps at a speed of 600 km/h

road section, the levitation control system will fluctuate in the levitation gap, and the ride comfort of the vehicle will be poor. Combined with the operation of the Shanghai high-speed maglev line, the change of the tangential torsion angle of the beam end is one of the main reasons for the deterioration of the control performance of the high-speed maglev train levitation system.

When the high-speed maglev train passes through the above-mentioned road section at a speed of 600 km/h, the levitation gap will fluctuate, and the levitation electromagnet will vibrate with the change of the track. To study the dynamic characteristics of the high-speed maglev train levitation system, the designed nominal

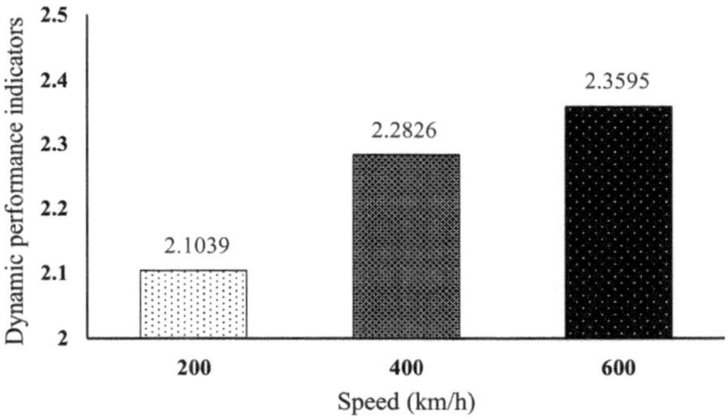

Fig. 5.25 Dynamic performance index of the levitation system when passing the track steps at different speeds

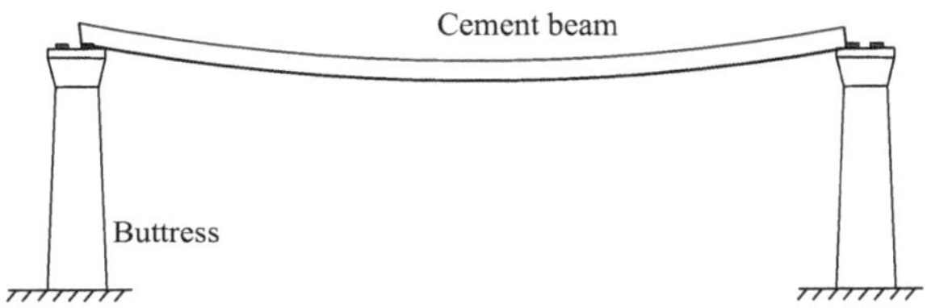

Fig. 5.26 Schematic diagram of the elevated simply supported beam of the high-speed maglev line

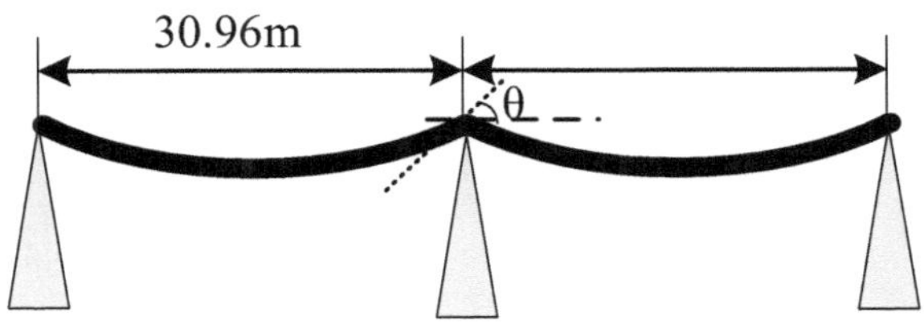

Fig. 5.27 Schematic diagram of the tangential torsion angle of the beam end of the high-speed maglev line

controller is brought into the nonlinear model. The dynamic characteristic curves of the high-speed levitation joint structure are obtained as shown in Fig. 5.28.

As shown in Fig. 5.28, the levitation gap of the levitation joint structure is always within the high-quality range. At the same time, the dynamic performance index of the levitation system is 2.0803. According to the original line track design index, when the tangential torsion angle of the beam end is 0.00021 rad, according to the standard for the classification of performance levels, the control performance of the levitation system of the high-speed maglev train is good.

To further analyze the adverse effect of the non-deterministic disturbance of the change of the beam end tangential torsion angle on the high-speed maglev train levitation system, the dynamic characteristics of the levitation system of the high-speed maglev train passing through the track beams with the tangential torsion angles of the beam ends of 0.0004 rad, 0.0006 rad, and 0.0008 rad are simulated, respectively.

(1) When the tangential torsion angle of the beam end is 0.0004 rad and the speed is 600 km/h, the dynamic characteristic curves of the high-speed levitation joint structure are shown in Fig. 5.29.

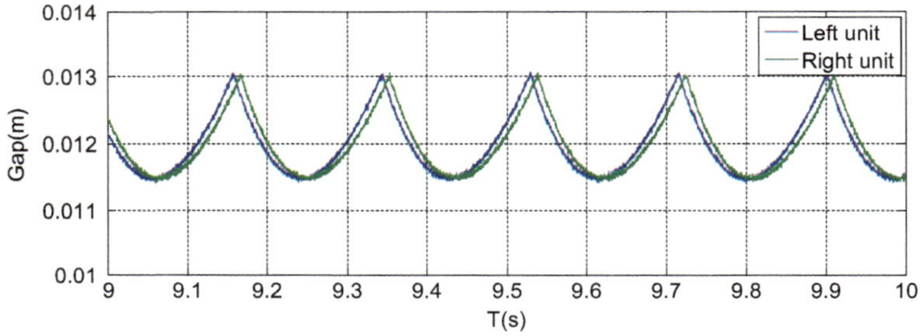

(a) Levitation gap fluctuation curve

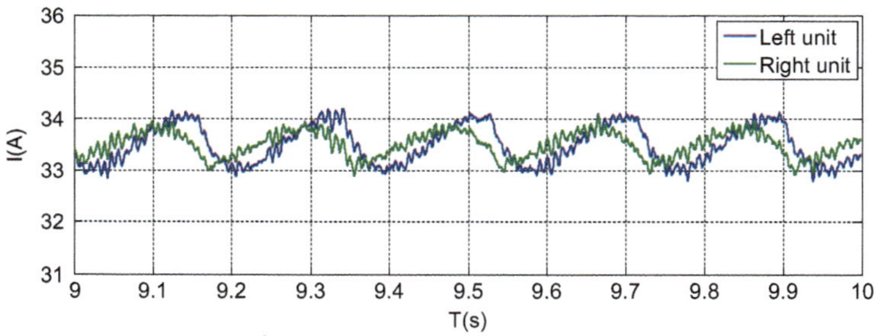

(b) Levitation current fluctuation curve

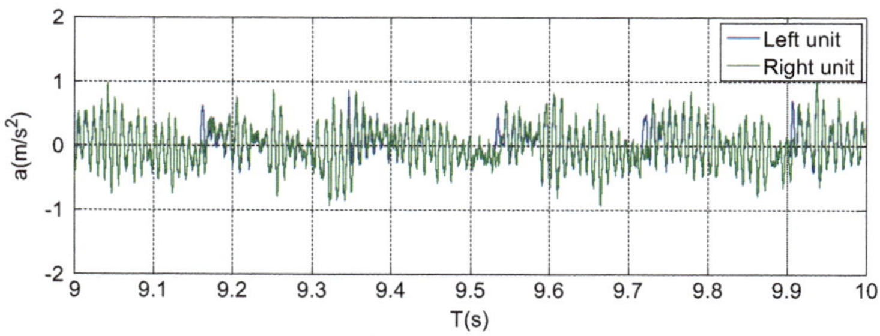

(c) Acceleration fluctuation curve

Fig. 5.28 The dynamic characteristic curves of the levitation system when the tangential torsion angle of the beam end is 0.00021 rad

As shown in Fig. 5.29, the levitation gap of levitation joint structure is still within the high-quality range. The maximum levitation gap becomes 14 mm, and the minimum levitation gap is 10.9 mm. The fluctuation range of the levitation gap becomes 10.9–14 mm, and the fluctuation range of the levitation gap becomes significantly larger. At the same time, the dynamic performance index of the levitation system increases to 2.1922. It was found that the control performance of the levitation system began to deteriorate.

(2) When the tangential torsion angle of the beam end is 0.0006 rad and the speed is 600 km/h, the dynamic characteristic curves of the high-speed levitation joint structure are shown in Fig. 5.30.

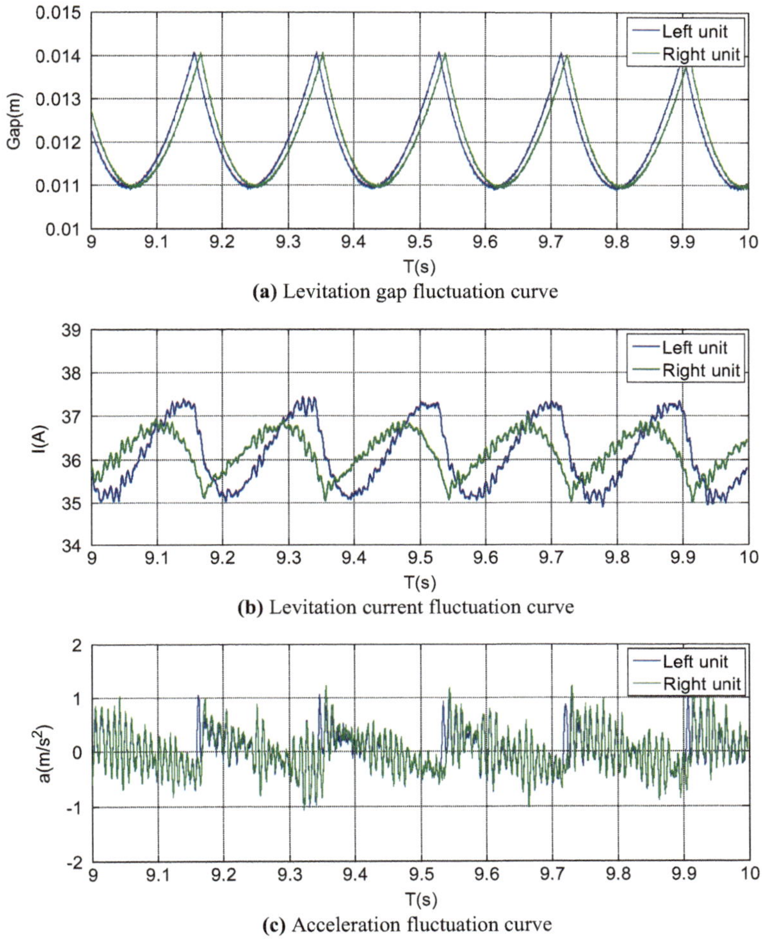

Fig. 5.29 The dynamic characteristic curves of the levitation system when the tangential torsion angle of the beam end is 0.0004 rad

As shown in Fig. 5.30, the levitation gap of levitation joint structure is no longer in the high-quality range, and the maximum levitation gap becomes 15.1 mm. The minimum levitation gap is 10.4 mm. The fluctuation range of the levitation gap becomes 10.4–15.1 mm, the fluctuation range of the levitation gap is further enlarged, and the performance of the levitation system is within the acceptable range. At the same time, the dynamic performance index of the levitation system increases to 2.3595. It can be found that the control performance of the levitation system has further deteriorated.

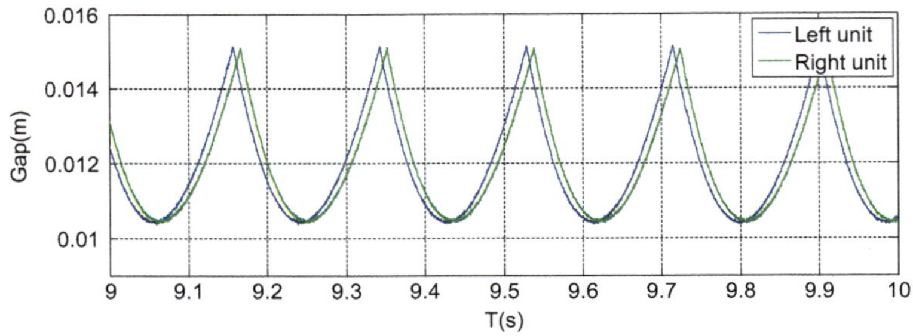

(a) Levitation gap fluctuation curve

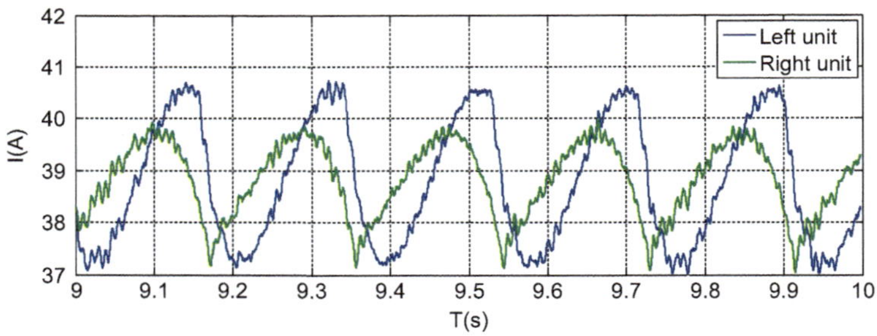

(b) Levitation current fluctuation curve

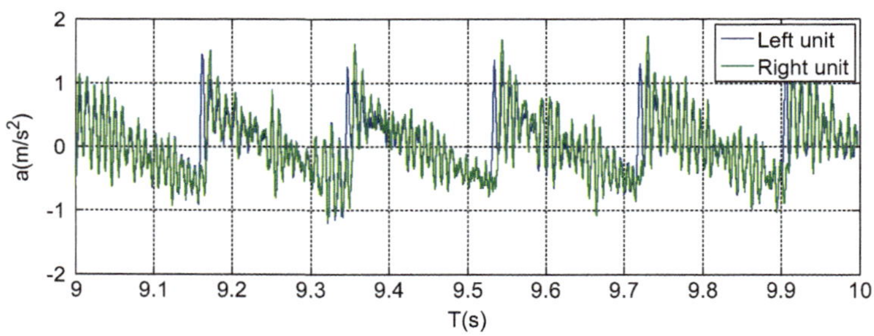

(c) Acceleration fluctuation curve

Fig. 5.30 The dynamic characteristic curves of the levitation system when the tangential torsion angle of the beam end is 0.0006 rad

(3) When the tangential torsion angle of the beam end is 0.0008 rad and the speed is 600 km/h, the dynamic characteristic curves of the high-speed levitation joint structure are shown in Fig. 5.31.

As shown in Fig. 5.31, the levitation gap of levitation joint structure begins to break through the acceptable range, and the maximum levitation gap becomes 16.1 mm. The minimum levitation gap is 9.8 mm. The fluctuation range of the levitation gap is changed to 9.8–16.1 mm, the fluctuation range of the levitation gap is further enlarged, and the performance of the levitation system begins to enter the serious degradation range from the acceptable range. At the same time, the dynamic performance index of the levitation system increases to 2.5229. It can be found that the control performance of the levitation system has been severely degraded.

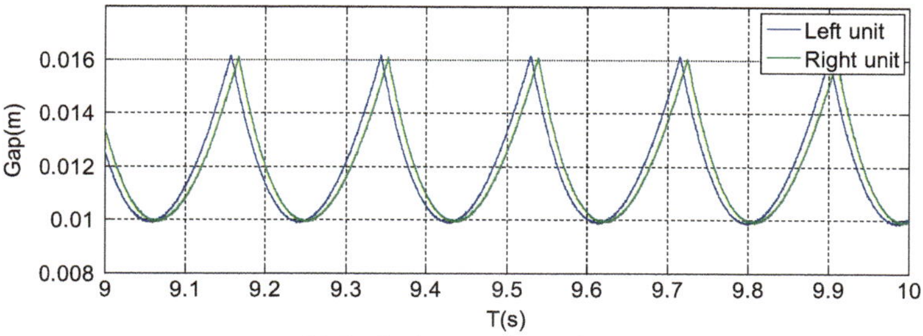

(a) Levitation gap fluctuation curve

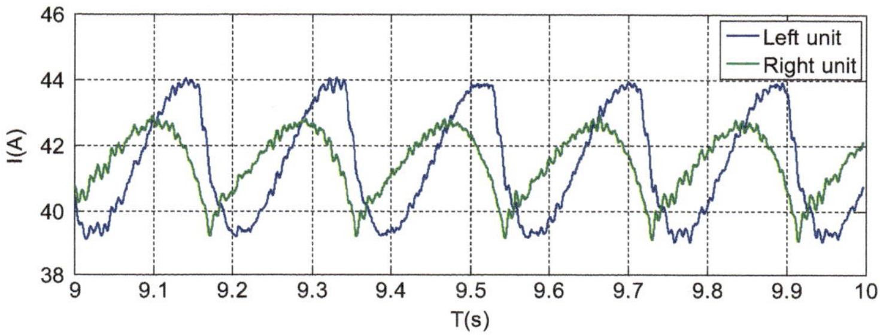

(b) Levitation current fluctuation curve

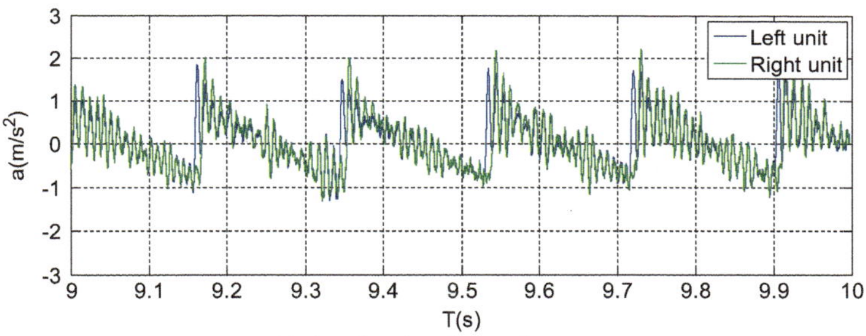

(c) Acceleration fluctuation curve

Fig. 5.31 The dynamic characteristic curves of the levitation system when the tangential torsion angle of the beam end is 0.0008 rad

As shown in Fig. 5.32, when the tangential torsion angle of the beam end of the high-speed maglev line track increases from 0.00021 to 0.0008 rad, the dynamic performance index of the high-speed maglev train levitation system also increases. The more violent it is, the less stable the levitation electromagnet becomes.

When the tangential torsion angle of the beam end increases from 0.00021 to 0.0008 rad, the fluctuation range of the levitation gap and the dynamic performance indicators also become larger. According to the classification standard of the control performance, the control performance of the levitation system under different beam end tangential torsion angles is shown in Table 5.4.

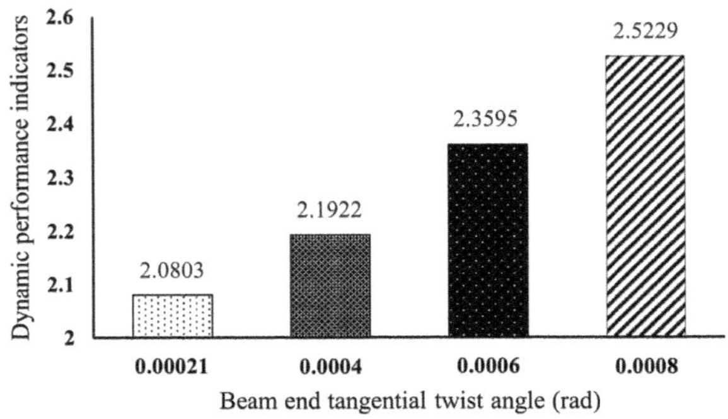

Fig. 5.32 Dynamic performance index of the levitation system corresponding to the change of the tangential torsion angle of the beam end

Table 5.4 Dynamic performance of levitation system under the disturbance of beam end tangential torsion angle

Beam end tangential twist angle (rad)	0.0002	0.0004	0.0006	0.0008
Levitation gap fluctuation range	11.4–13	10.9–14	10.4–15.1	9.8–16.1
Dynamic performance indicators	2.0803	2.1922	2.3595	2.5229
Performance evaluation	Excellent	Good	Acceptable	Severely degraded

In addition, when the high-speed maglev train passes through the track beam at different speeds, the dynamic characteristics of the levitation system are also different. To illustrate the impact of speed increase on the levitation system of high-speed maglev train, the dynamic characteristics of the levitation systems in the low-speed section, medium-speed section and high-speed section are studied. The tangential torsion angle of the beam end is 0.0006 rad.

(1) When the speed is 200 km/h, the dynamic characteristic curves of the high-speed levitation joint structure are shown in Fig. 5.33.

As shown in Fig. 5.33, the maximum levitation gap is 15.1 mm, and the minimum levitation gap is 10.1 mm. The fluctuation range of the levitation gap is 10.1–15.1 mm, and the fluctuation range of the levitation gap shows that the performance of the levitation system is within an acceptable range. At the same time, the dynamic performance index of the levitation system is 1.9872.

(2) When the speed is 400 km/h, the dynamic characteristic curves of the high-speed levitation joint structure are shown in Fig. 5.34.

As shown in Fig. 5.34, the maximum levitation gap is 15.1 mm, and the minimum levitation gap is 10.3 mm. The fluctuation range of the levitation gap becomes 10.3–15.1 mm, and the performance of the levitation system is within the acceptable range. At the same time, the dynamic performance index of the levitation system increases to 2.1728.

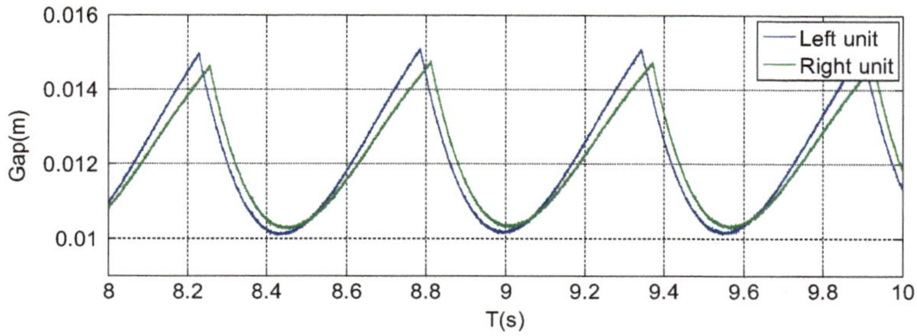

(a) Levitation gap fluctuation curve

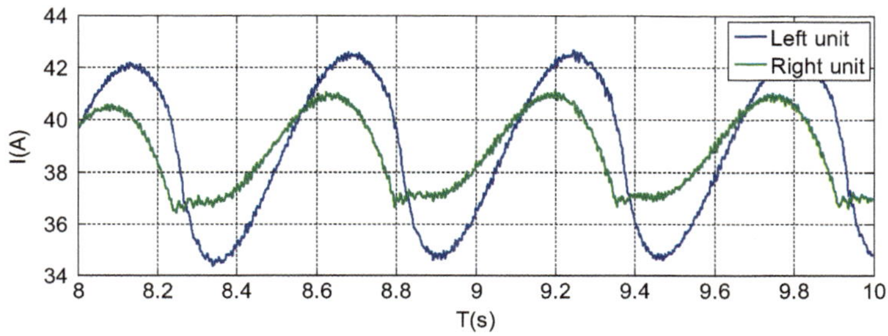

(b) Levitation current fluctuation curve

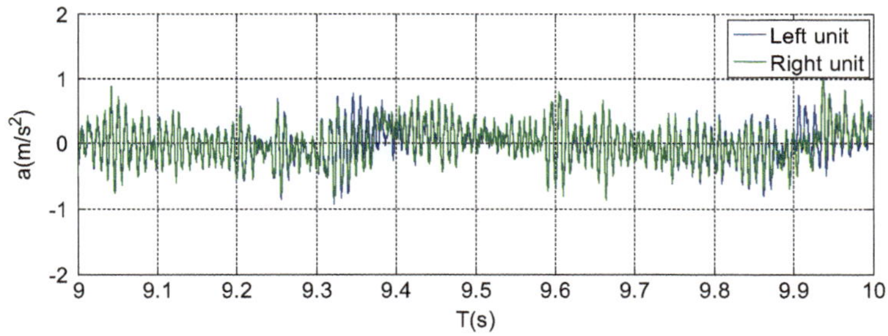

(c) Acceleration fluctuation curve

Fig. 5.33 Dynamic characteristic curves of the levitation system passing through the tangential torsion angle of the beam end at a speed of 200 km/h

(3) When the speed is 600 km/h, the dynamic characteristic curves of the high-speed levitation joint structure are shown in Fig. 5.35.

As shown in Fig. 5.35, the maximum levitation gap is 15.1 mm, and the minimum levitation gap is 10.4 mm. The fluctuation range of the levitation gap becomes 10.4–15.1 mm, and the performance of the levitation system is within the acceptable range. At the same time, the dynamic performance index of the levitation system increases to 2.3595. It can be found that the control performance of the levitation system has further deteriorated.

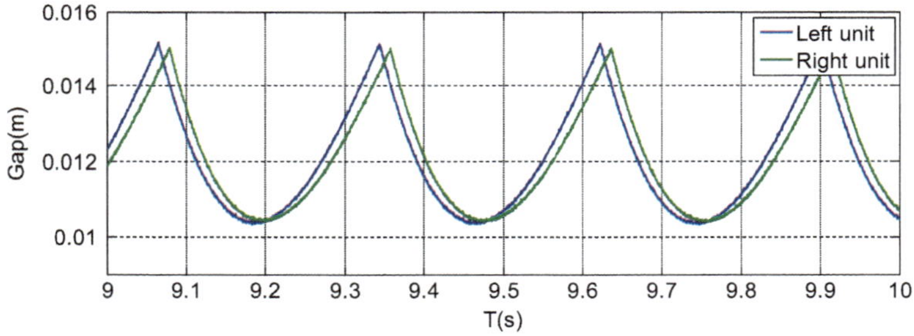

(a) Levitation gap fluctuation curve

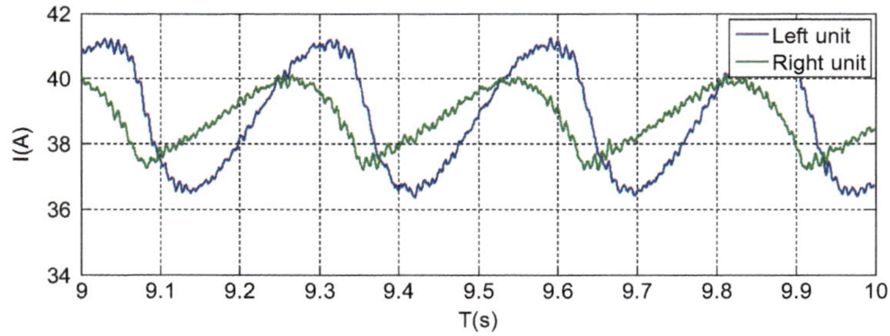

(b) Levitation current fluctuation curve

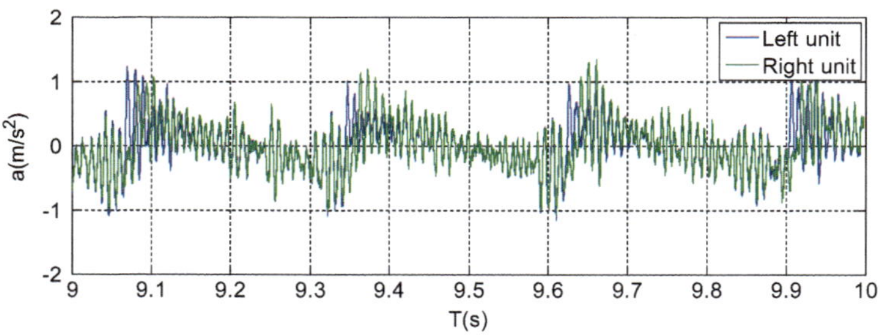

(c) Acceleration fluctuation curve

Fig. 5.34 Dynamic characteristic curves of the levitation system passing through the tangential torsion angle of the beam end at a speed of 400 km/h

When the speed of the high-speed maglev train increases from 200 to 600 km/h, the dynamic performance index of the high-speed maglev train levitation system also increases, as shown in Fig. 5.36.

To sum up, the increase of the tangential torsion angle of the beam end leads to the degradation of the control performance of the levitation system of the high-speed maglev train. The specific performance is that with the increase of the tangential torsion angle of the beam end, the fluctuation range of the levitation gap becomes larger and larger, and the dynamic performance index becomes larger and larger. When the tangential torsion angle of the beam end increases to 0.0008 rad, the control performance of the levitation system has been seriously degraded, which

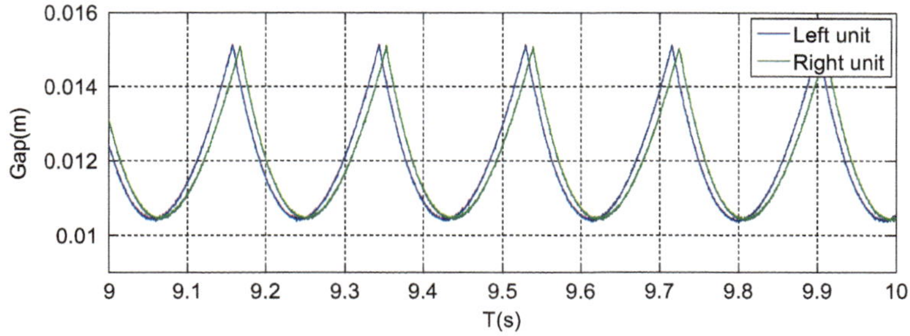

(a) Levitation gap fluctuation curve

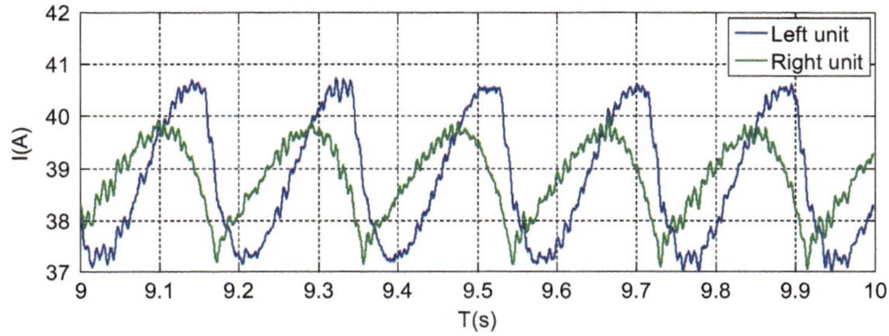

(b) Levitation current fluctuation curve

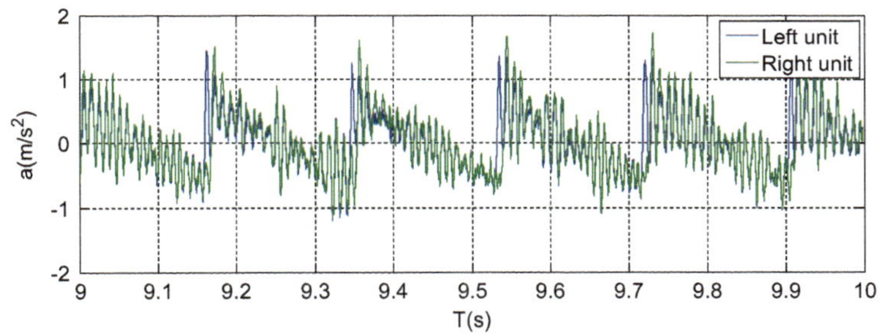

(c) Acceleration fluctuation curve

Fig. 5.35 Dynamic characteristic curves of the levitation system passing through the tangential torsion angle of the beam end at a speed of 600 km/h

seriously endangers the stability and comfort of the high-speed maglev train. At the same time, the running speed of the train has little effect on the fluctuation range of the levitation gap, but has a great influence on the dynamic performance indicators. As the speed increases, the dynamic performance index becomes larger.

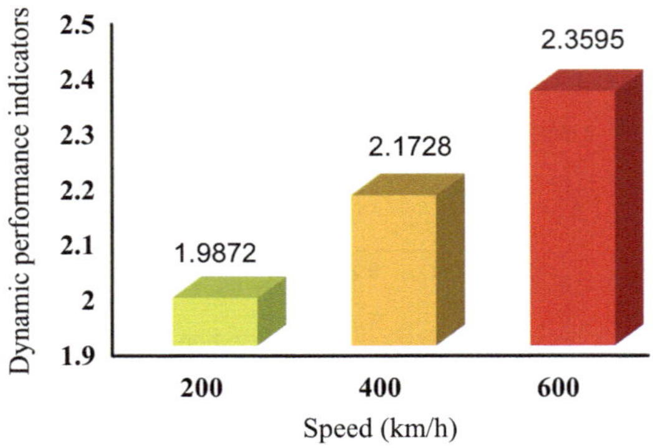

Fig. 5.36 Dynamic performance index of levitation system when different speeds pass through the beam end tangential torsion angle

5.5 Conclusion

In this chapter, the nominal performance of the levitation system by using the nominal performance index is firstly analyzed and evaluated. Secondly, according to the working characteristics of the levitation system under dynamic operating conditions, the dynamic performance index and evaluation criteria suitable for the levitation system of the high-speed maglev train are proposed. Finally, the performance of the levitation system under deterministic and non-deterministic disturbances is quantitatively analyzed.

Bibliography

1. Chen H, Long Z, Chang W (2006) Dynamic analysis of maglev train running on the ramp. Control Eng China 13:1–3
2. Ding S (2022) 600 km/h high-speed maglev transportation system. Shanghai Scientific and Technical Publishers, Shanghai
3. Janic M, Milan J (2018) Multicriteria evaluation of the high speed rail, transrapid maglev and hyperloop systems. Transp Syst Technol 4(4):5–31
4. Long Z, Hao A, Chang W (2003) Suspension controller design of maglev train considering the rail track periodical irregularity. J Natl Univ Defense Technol 25(2):84–89
5. Long Z, Dou F, Wang Z et al (2023) Current status and prospects of high-speed maglev suspension and guidance control technology. Sci Technol Foresight 2(4):78–88
6. Shi X, Long Z (2009) Nonlinear vibration analysis of the maglev guideway-vehicle coupling control system. J China Railw Soc 31(4):38–42
7. Wang Z, Long Z, Li X (2019) Track irregularity disturbance rejection for maglev train based on online optimization of PnP control architecture. IEEE Access 7:12610–12619
8. Wu X (2003) Maglev train. Shanghai Scientific and Technical Publishers, Shanghai
9. Yamamoto T, Kinoshita T, Ohnishi Y et al (2016) Design and experimental evaluation of a performance-driven PID controller. J Robot Mechatron 28(5):616–624
10. Zhai M (2019) Research on performance optimization of magnetic levitation system in high speed maglev train. Ph.D. dissertation, National University of Defense Technology Doctor Degree, Changsha

11. Zhai M, Li X, Long Z (2019a) A New strategy for improving the tracking performance of magnetic levitation system in maglev train. Symmetry 11(8):2073–8994
12. Zhai M, Long Z, Li X (2019b) Calculation and evaluation of load performance of magnetic levitation system in medium-low speed maglev train. Int J Appl Electromagn Mech 61(4):519–536
13. Zhai M, Long Z, Li X (2019c) Fault-tolerant control of magnetic levitation system based on state observer in high speed maglev train. IEEE Access 7:31624–31633
14. Zhai M, Long Z, Li X et al (2023) Control and optimization design of magnetic levitation system of high-speed maglev train based on Youla parameterization. J Tongji Univ (Nat Sci) 51(3):341–350

Chapter 6
Performance Evaluation of Guidance Control System

After the completion of the guidance control system design for high-speed maglev trains, the performance of the control system under complex working conditions needs to be further analyzed and evaluated. To scientifically and accurately analyze and evaluate the performance of the high-speed maglev train passing through the plane curve line, the active guidance ability of the high-speed maglev train is evaluated and verified. Firstly, a quantifiable evaluation index is proposed based on a fuzzy comprehensive evaluation method by using the state information of guidance gap, current, and acceleration. Secondly, under different curve radii, running speeds and payloads, the evaluation index proposed in this chapter is used to analyze and evaluate the active guidance ability of the train. Finally, the effectiveness of the evaluation results is verified by the actual line test. The results indicate that the active guidance ability of the high-speed maglev train is closely related not only to the ability of the guidance system itself but also to the line radius, running speed and other factors.

6.1 Guidance System Model Considering Curve Changes of the Line

To pass the curve smoothly and resist lateral interference, the high-speed maglev train must rely on active guidance system. The guidance system is one of the important subsystems of high-speed maglev train. The left and right guidance electromagnets are connected by the levitation frame, and the lateral forces produced by the guidance electromagnet are mainly transmitted by levitation frame. There is only attraction between the guidance electromagnets and the guideway. The value of this attraction changes with the guidance current and guidance gap, but the direction of this attraction is unidirectional. The direction of attraction between the left guidance electromagnets and the guideway is toward right, while the direction of attraction

Z. Long et al., *High-Speed Maglev Train's Levitation and Guidance Control*, Advances in High-speed Rail Technology, https://doi.org/10.1007/978-981-97-2309-6_6

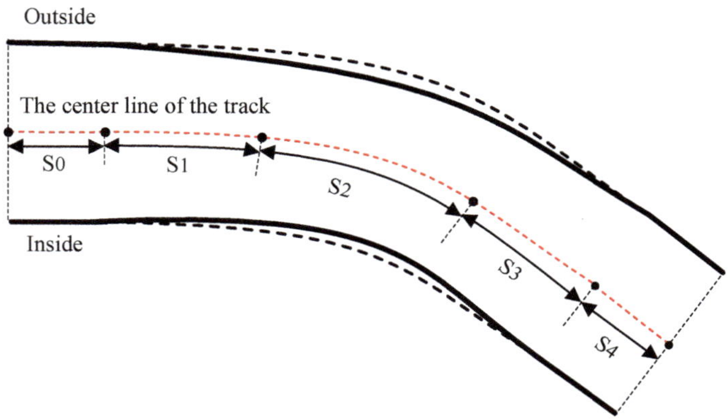

Fig. 6.1 Plane curve of high-speed maglev line

between the right guidance electromagnets and the guideway is toward left. Guidance controllers adjust the value of the left and right attraction, respectively, to ensure that the whole levitation frame is able to locate directly above the guideway centerline and doesn't move left or right.

There is no contact and friction between high-speed maglev train and guideway, the route selection is more flexible. The high-speed maglev train has small turning radius, large slope, complex line condition and high actual running speed. The plane of high-speed maglev line is generally composed of straight line, circular curve and relaxation curve. The two lines need to be connected by a circular curve. The radius of the circular curve limits the safe passing speed of high-speed maglev trains. To ensure the smooth running of the high-speed maglev train, the connection between the straight line and the circular curve is required (Fig. 6.1).

Whether the high-speed maglev train can safely pass the plane curve line is not only related to the ability of the guidance system itself, but also closely related to the radius of the line, the speed of the train and other factors. Therefore, it is necessary to establish a guidance system model considering the curve changes of the line, analyze and study the performance of the active guidance system under different curve radii, running speeds, and payloads, and test the guidance ability of the guidance system under different line conditions.

To facilitate the modeling, the movement of the left and right directions are only considered, while the yaw and roll of the guidance electromagnet are ignored. The stiffness of guidance electromagnet is very high and the deformation is neglected. Therefore, the equivalent action-point of guidance force is in the center of mass. The levitation beam, corbel, and levitation device are equivalent to spring damping systems.

The levitation frame has 4 sets of guidance unit in high-speed maglev train. When the current is switched on, the electromagnetic force is generated by the module of guidance electromagnet composed of two guidance electromagnets. The guidance module working diagram is illustrated in Fig. 6.2.

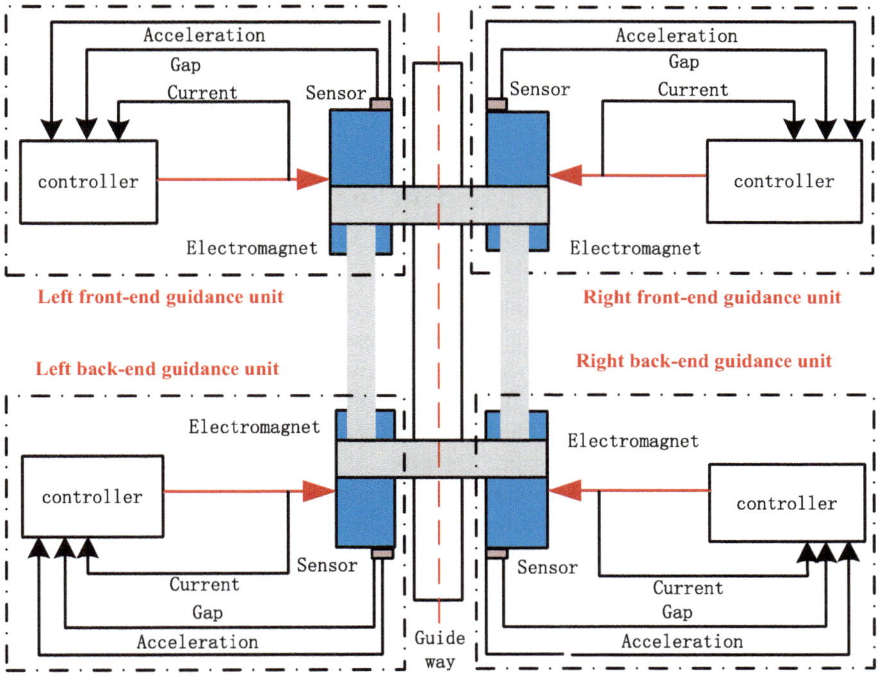

Fig. 6.2 Guidance module working diagram

The dynamic equation of the high-speed maglev train guidance system considering the curve change of the line is as follows:

$$\begin{cases} m\Delta\ddot{z} = \sum F_y + m\kappa V^2 \cos\theta \\ I_z\ddot{\alpha} = \sum M_z - I_z\dot{\kappa} V^2 \cos\theta \end{cases}, \tag{6.1}$$

where m is the system mass, I_z is the moment of inertia, V is the train speed, θ is the cross slope angle, κ is the curvature of the guide way and α is the yaw angle. ΣF_y is the guidance system of lateral force, including the electromagnetic force, interference force and gravity. ΣM_z is the external moment of force, including the moment of electromagnetic force and interference force.

$$\begin{cases} \sum F_y = (F_{\mathrm{rf}} - F_{\mathrm{lf}} + F_{\mathrm{rb}} - F_{\mathrm{lb}}) + F_{\mathrm{d}} - mg\sin\theta \\ \sum M_z = (F_{\mathrm{lf}} + F_{\mathrm{rb}}) \cdot r_2 - (F_{\mathrm{lb}} + F_{\mathrm{rf}}) \cdot r_1 + M_{\mathrm{d}} \end{cases} \tag{6.2}$$

In the equation, F_{lf}, F_{lb}, F_{rf}, F_{rb}, respectively, represent left front-end guidance electromagnetic force, left back-end guidance electromagnetic force, right front-end guidance electromagnetic force, and right back-end guidance electromagnetic force. F_{d} is transverse interference force, M_{d} is the interference moment, r_1 and r_2 are the relative moment arms on both sides of the system. According to the above equations, a mathematical model of the active guidance system of high-speed maglev train can be established considering curve change of the line.

6.2 Guidance Evaluation Based on Fuzzy Comprehensive Method

Combined with the working principle and engineering practice of the active guidance system of the high-speed maglev train, the performance of the active guidance system can be evaluated from the three dimensions of safety, stability and comfort. During the operation of the guidance system, the measurement data of the guidance gap, current and acceleration obtained by the sensor can accurately reflect the working state and performance of the guidance system. Therefore, this section uses the measurement data, establishes the evaluation index and evaluation set, constructs the weighting matrix and the membership function based on the fuzzy comprehensive evaluation method, and finally establishes the evaluation standard suitable for the high-speed maglev train.

6.2.1 *Evaluation Indicators and Evaluation Sets*

The safety of the high-speed maglev train passing through the curve is directly related to the minimum value of the guidance gap. The dispersion degree of the guidance gap and guidance current in the running process of the train reflects the stability of the guidance system. The comfort of the train passing through the curve correlates with the degree of variation in guidance acceleration. Therefore, the minimum value of the guidance gap, the dispersion degree of the guidance gap, the dispersion degree of the guidance current, and the dispersion degree of the guidance acceleration are selected as the evaluation indexes of the guidance system.

The factor set of the guidance system is $U = \{$the minimum value of the guidance gap, the discrete degree of the guidance gap, the discrete degree of the guidance current, the discrete degree of the guidance acceleration$\}$.

To evaluate the guidance ability of a trainer clearly, a corresponding evaluation set needs to be established. The evaluation set established in this section is $B = \{$excellent, good, fair, poor$\}$.

Table 6.1 Grading result

Parameter	C1	C2	C3	C4
C1	1	$9^{1/9}$	$9^{4/9}$	$9^{2/3}$
C2	$9^{-1/9}$	1	$9^{2/9}$	$9^{4/9}$
C3	$9^{-4/9}$	$9^{-2/9}$	1	$9^{2/9}$
C4	$9^{-2/3}$	$9^{-4/9}$	$9^{-2/9}$	1

6.2.2 Build a Weight Vector

Parameters C1, C2, C3, and C4 represent the minimum value of the guidance gap, the discrete degree of the guidance gap, the discrete degree of the guidance current, and the discrete degree of the guidance acceleration, respectively. Based on the exponential scaling method proposed by Hou Yueheng, the corresponding scoring results are obtained in Table 6.1.

After normalization, the weight vector is obtained as $\boldsymbol{d}$. After the consistency check, the weight vector is expressed as follows:

$$\boldsymbol{d} = \begin{bmatrix} 0.4247 & 0.2944 & 0.1700 & 0.1109 \end{bmatrix}^{\mathrm{T}}$$

6.2.3 Build a Membership Function

The membership function is constructed by the fuzzy distribution method and the trapezoidal distribution is selected. The calculation is simple and the distribution is reasonable. It is suitable for the evaluation model of the active guidance system of the high-speed maglev train.

Figure 6.3 is a graph of the membership function of the trapezoidal distribution. The x-axis represents a certain parameter value to be judged, the y-axis represents the membership degree of a certain parameter value, and the membership degree is between 0 and 1.

It is assumed that y_1 represents the excellent membership function in the evaluation set, y_2 represents the good membership function, y_3 represents the general membership function, and y_4 represents the poor membership function. The membership function for the evaluation set of "guidance gap minimum value" can be expressed by the following equations.

$$y_1 = \begin{cases} 1 & 0 \le x \le 0.5 \\ \dfrac{0.6 - x}{0.6 - 0.5} & 0.5 < x < 0.6 \\ 0 & x \ge 0.6 \end{cases}$$

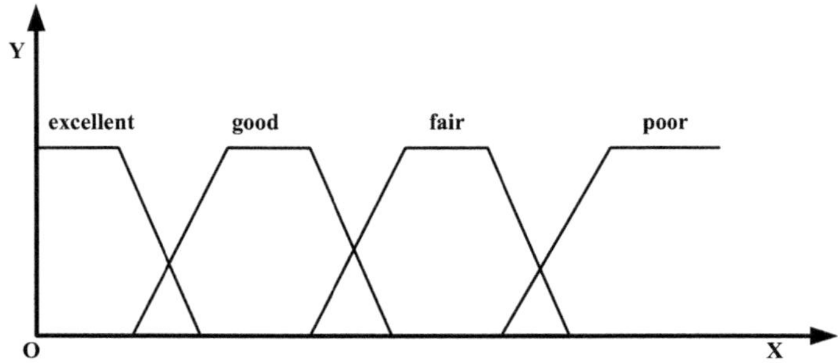

Fig. 6.3 Membership function graph of the trapezoidal distribution

$$y_2 = \begin{cases} 0 & x \leq 0.5 \text{ or } x \geq 0.8 \\ \dfrac{x - 0.5}{0.6 - 0.5} & 0.5 < x < 0.6 \\ 1 & 0.6 \leq x \leq 0.7 \\ \dfrac{0.8 - x}{0.8 - 0.7} & 0.7 < x < 0.8 \end{cases}$$

$$y_3 = \begin{cases} 0 & x \leq 0.7 \text{ or } x \geq 1 \\ \dfrac{x - 0.7}{0.8 - 0.7} & 0.7 < x < 0.8 \\ 1 & 0.8 \leq x \leq 0.9 \\ \dfrac{0.9 - x}{0.9 - 0.8} & 0.9 < x < 1 \end{cases}$$

$$y_4 = \begin{cases} 0 & x \leq 0.9 \\ \dfrac{x - 0.9}{1 - 0.9} & 0.9 < x < 1 \\ 1 & x \geq 1 \end{cases}$$

The membership function of the evaluation set of "discrete degree of guidance gap" can be expressed by the following equations.

$$y_1 = \begin{cases} 1 & 0 \leq x \leq 0.009 \\ \dfrac{0.9 - x}{0.9 - 0.009} & 0.009 < x < 0.01 \\ 0 & x \geq 0.01 \end{cases}$$

$$y_2 = \begin{cases} 0 & x \leq 0.009 \text{ or } x \geq 0.016 \\ \dfrac{x - 0.009}{0.9 - 0.009} & 0.009 < x < 0.01 \\ 1 & 0.01 \leq x \leq 0.015 \\ \dfrac{0.016 - x}{0.016 - 0.015} & 0.015 < x < 0.016 \end{cases}$$

$$y_3 = \begin{cases} 0 & x \le 0.015 \text{ or } x \ge 0.022 \\ \dfrac{x - 0.015}{0.016 - 0.015} & 0.015 < x < 0.016 \\ 1 & 0.016 \le x \le 0.021 \\ \dfrac{0.022 - x}{0.022 - 0.021} & 0.021 < x < 0.022 \end{cases}$$

$$y_4 = \begin{cases} 0 & x \le 0.021 \\ \dfrac{x - 0.021}{0.022 - 0.021} & 0.021 < x < 0.022 \\ 1 & x \ge 0.022 \end{cases}$$

The membership functions for the evaluation sets of "guidance current dispersion degree" and "guidance acceleration dispersion degree" can be represented by the following equations.

$$y_1 = \begin{cases} 1 & 0 \le x \le 1 \\ \dfrac{1.1 - x}{1.1 - 1} & 1 < x < 1.1 \\ 0 & x \ge 1.1 \end{cases}$$

$$y_2 = \begin{cases} 0 & x \le 1 \text{ or } x \ge 1.7 \\ \dfrac{x - 1}{1.1 - 1} & 1 < x < 1.1 \\ 1 & 1.1 \le x \le 1.6 \\ \dfrac{1.7 - x}{1.7 - 1.6} & 1.6 < x < 1.7 \end{cases}$$

$$y_3 = \begin{cases} 0 & x \le 1.6 \text{ or } x \ge 2.3 \\ \dfrac{x - 1.6}{1.7 - 1.6} & 1.6 < x < 1.7 \\ 1 & 1.7 \le x \le 2.2 \\ \dfrac{2.3 - x}{2.3 - 2.2} & 2.2 < x < 2.3 \end{cases}$$

$$y_4 = \begin{cases} 0 & x \le 2.2 \\ \dfrac{x - 2.2}{2.3 - 2.2} & 2.2 < x < 2.3 \\ 1 & x \ge 2.3 \end{cases}$$

To sum up, the membership functions of the evaluation sets of "guidance gap minimum value", "guidance gap dispersion degree", "guidance current dispersion degree" and "guidance acceleration dispersion degree" are completed. The membership degree matrix can be calculated by using the measurement data, and the evaluation result vector can be obtained by multiplying the membership degree matrix by the weighting vector w.

6.3 Analysis of the Guidance Ability Under Plane Curve Condition

Whether the high-speed maglev train can safely pass the plane curve line is not only related to the ability of the guidance system itself, but also closely related to the radius of the line, the speed of the train and other factors. Therefore, it is necessary to analyze and study the performance of the active guidance system under different curve radii, running speeds and payloads. In order to analyze and evaluate the performance of the guidance system, this section takes advantage of the guidance gap, current, acceleration, and other state information obtained during the operation of the guidance system to analysis and evaluate.

6.3.1 Performance of Guidance System at Different Curve Radii

To analyze and evaluate the performance of the guidance system under different curve radii, operating speeds, and payloads, the guidance system model considering line curve changes is established.

When the high-speed maglev train passes through the plane curve line at 60 km/h, and the radius of the curve is 300, 400, and 600 m, the guidance gap and current are shown in Fig. 6.4.

The guidance controller uses the gap detected by the sensor and adopts the differential control strategy to adjust the current in the guidance electromagnet to ensure that the train is always on the centerline of the guideway.

When the high-speed maglev train passes through the plane curve line with the radius of 300, 400, and 600 m, the evaluation result vectors are as follows:

$$\begin{aligned}\mathbf{w}_{r=300} &= [000.1109\ 0.8891]\\ \mathbf{w}_{r=400} &= [000.9653\ 0.0347]\\ \mathbf{w}_{r=600} &= [00.7191\ 0.28090]\end{aligned}$$

According to the principle of maximum membership degree, the corresponding evaluation result can be obtained by checking the comment corresponding to the maximum value in the evaluation result vector in evaluation set B. When the train passes through plane curve line with a radius of 300 m, the maximum value of the evaluation result vector is 0.8891, and the corresponding evaluation set comment is "poor". When the train passes through plane curve line with radius of 400 m, the maximum value of the vector is 0.9653, and the corresponding evaluation set comment is "general". When the train passes through a plane curve line with a radius of 600 m, the maximum value of the evaluation result vector is 0.7191, and the corresponding evaluation set comment is "good".

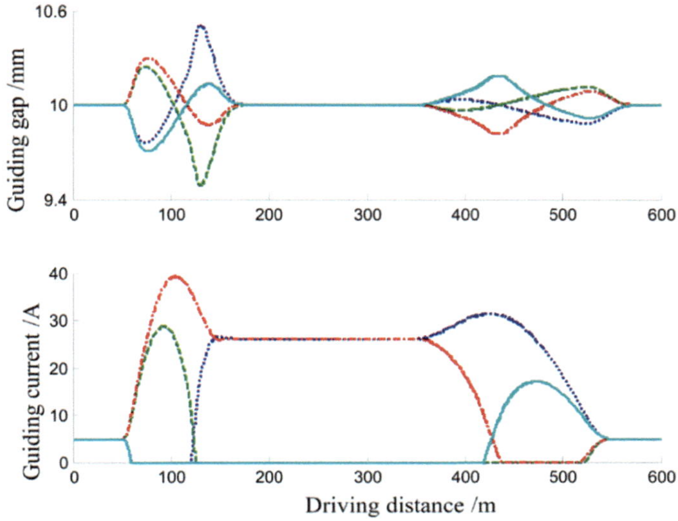

(a) The radius of the curve is 300 m

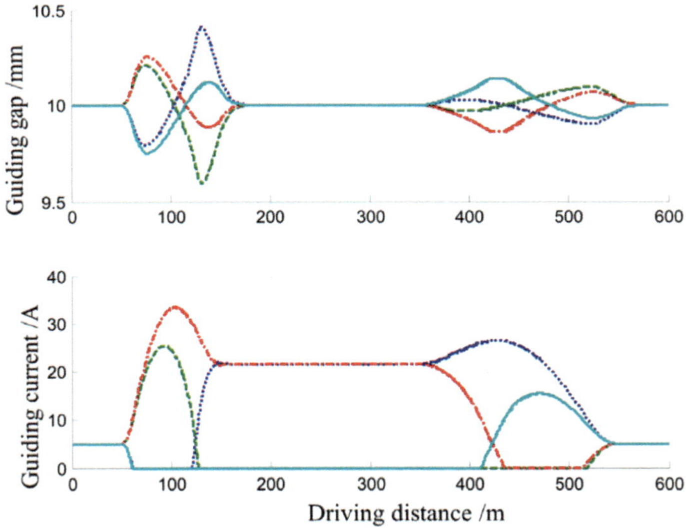

(b) The radius of the curve is 400 m

Fig. 6.4 Guidance gaps and currents with different curve radius

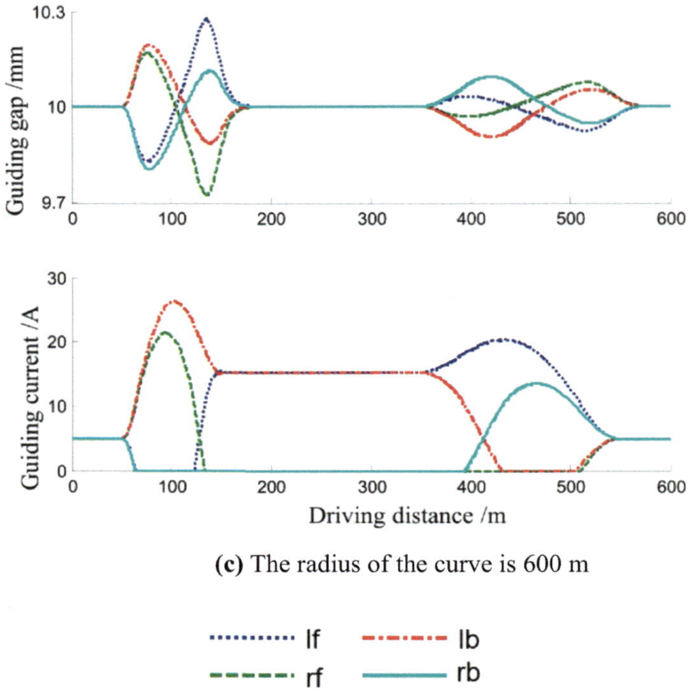

Fig. 6.4 (continued)

The larger the radius of the line curve is, the better the performance evaluation result of the active guidance system of the high-speed maglev train is. The results show that increasing the curve radius of the plane line can improve the guidance ability of the high-speed maglev train.

6.3.2 *Performance of Guidance System at Different Speeds*

The load condition is standard, and the radius of the line curve is set to 400 m. When the running speed is 60, 90, and 150 km/h, the guidance gap and current are shown in Fig. 6.5.

When the train passes through the plane curve line at different speeds, the change range of the guidance gap and the required guidance force are different, resulting in different changes of the guidance current.

When the high-speed maglev train passes through the curve at 60 km/h, 90 km/h, and 150 km/h under the conditions of standard load and set line curve radius, respectively, the evaluation result vectors are:

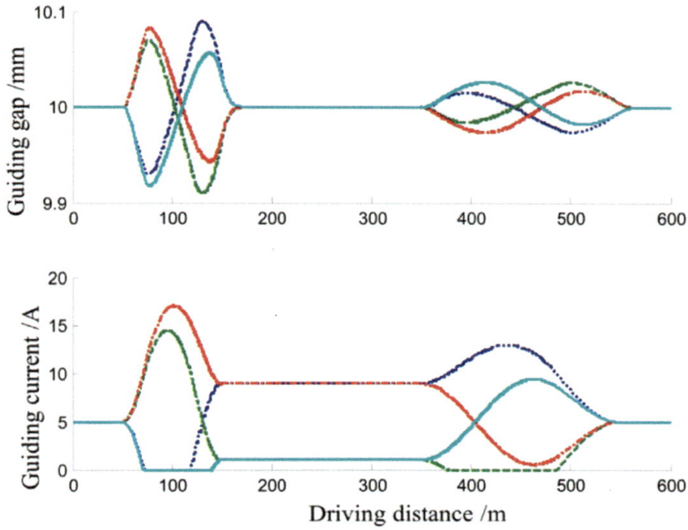

(a) 60 km/h

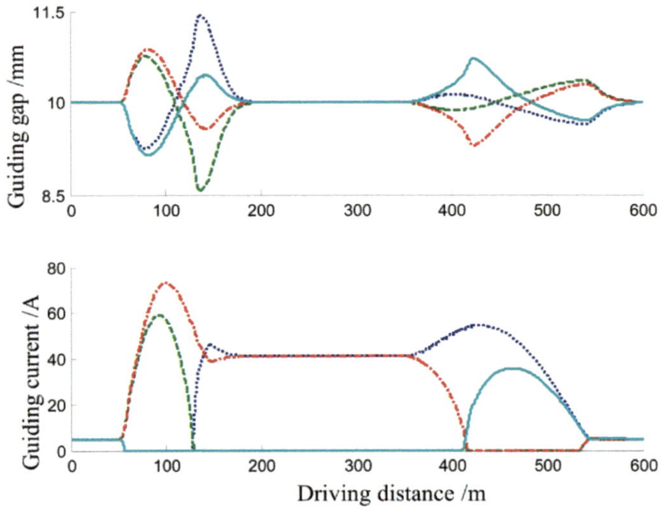

(b) 90 km/h

Fig. 6.5 Guidance gaps and currents with different speeds

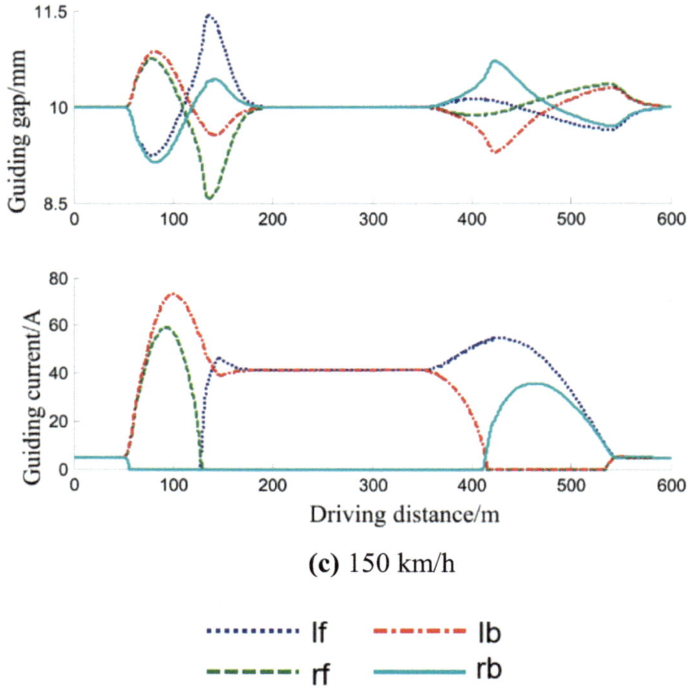

Fig. 6.5 (continued)

$$\mathbf{w}_{V=60} = [0.8891\ 0.110900]$$
$$\mathbf{w}_{V=90} = [0.0255\ 0.39920.2809\ 0.2944]$$
$$\mathbf{w}_{V=150} = [000.1109\ 0.8891]$$

When the running speed is 60 < 60 km/h, the maximum value of the evaluation result vector is 0.8891, and the corresponding evaluation set comment is “excellent”. When the running speed is 90 km/h, the maximum value of the evaluation result vector is 0.3992, the corresponding evaluation set comment is “good”. When the running speed is 150 km/h, the maximum value of the evaluation result vector is 0.8891, and the corresponding evaluation set comment is “poor”.

The results show that under the conditions of standard load and set line curve radius, increasing the running speed of the train when passing the curve line will directly weaken the guidance ability of the high-speed maglev train and damage the safety of the train.

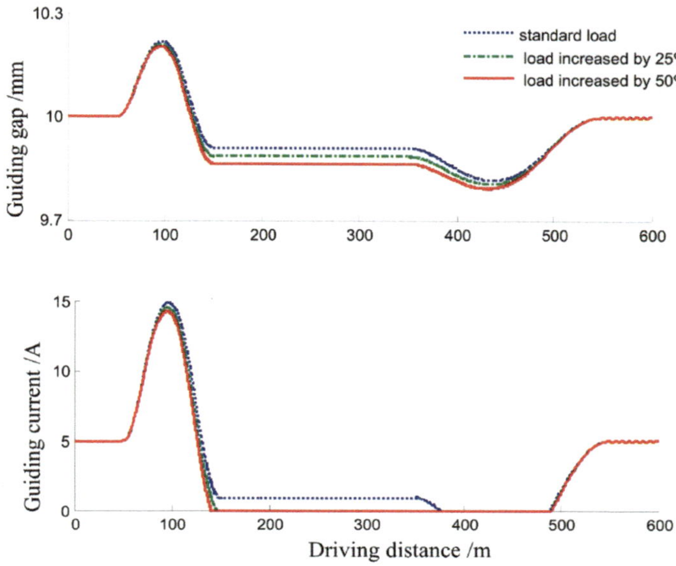

Fig. 6.6 Guidance gaps and currents with different loads

6.3.3 *Performance of Guidance System with Different Loads*

The radius of the line curve is set to 400 m, and the speed is set to 60 km/h. The guidance gap and current curves of the high-speed maglev train under the conditions of standard load, 25% load increase, and 50% load increase are shown in Fig. 6.6.

When the train passes through the plane curve line with different loads, the current of the right guide unit should decrease rapidly, while the current of the left guide unit should increase rapidly.

Under the conditions of the set running speed and line curve radius, the evaluation result vectors of the standard load, the load increased by 25% and the load increased by 50% are as follows:

$$\begin{aligned} \mathbf{w}_{M=1} &= [0.6602\ 0.27590.0639\ 0] \\ \mathbf{w}_{M=1.25} &= [0.5425\ 0.37410.0834\ 0] \\ \mathbf{w}_{M=1.5} &= [0.4247\ 0.46820.1071\ 0] \end{aligned}$$

Under the standard load condition, the maximum value of the evaluation result vector is 0.6602, and the corresponding evaluation set comment is "excellent". Under the load increase of 25%, the maximum value of the evaluation result vector is 0.5425, and the corresponding evaluation set comment is "excellent". Under the load increases by 50%, the maximum value of the evaluation result vector is 0.4682, and the corresponding evaluation set comment is "good".

With the increase of the load, the performance evaluation results of the active guidance system of the high-speed maglev train begin to deteriorate. The results show that increasing the load when the train passes the curve line will directly weaken the guidance ability of the high-speed maglev train.

6.4 Performance of Guidance System in the Actual Line

To scientifically and accurately analyze and evaluate the performance of a high-speed maglev train passing through the plane curve line, the active guidance ability of the high-speed maglev train is evaluated and verified in the 1.5 km high-speed maglev test line of Tongji University in Shanghai. The high-speed maglev test line was completed and put into use in April 2006. This test line was independently designed, developed and integrated the high-speed maglev traffic test line system. It is the only comprehensive research and development test platform for high-speed maglev traffic system in China, which can provide basic test conditions for the research and development of high-speed maglev traffic technology. The test speed of the train can reach 100 km/h.

The guidance unit of the high-speed maglev train provides active guidance ability. To further verify the guidance ability of the train under the condition of the actual line, the right front-end guidance unit of a levitation frame is selected to record the actual running data, such as the guidance gap and guidance current.

Under standard load conditions, the high-speed maglev train passes through a section with a curve radius of 400 m at a speed of 35 km/h under the action of the guidance system. The guidance gap and guidance current are shown in Fig. 6.7.

Based on the guidance ability evaluation method of the fuzzy comprehensive evaluation method, the actual running data such as guidance gap, current and acceleration are used to analyze, and evaluate the guidance ability of the high-speed maglev train concerning the evaluation criteria established above. The evaluation result vector $\mathbf{w}_t$ is as follows:

$$\mathbf{w}_t = [0.5947\ 0.110900.2944]$$

According to the principle of maximum membership, the maximum value of the evaluation result vector $\mathbf{w}_t$ of the actual line test is 0.5947, and the corresponding evaluation set comments are “excellent”.

6.5 Conclusion

In this chapter, a quantifiable evaluation index is proposed based on a fuzzy comprehensive evaluation method by using the state information of guide gap, current and acceleration in high-speed maglev train operation. Secondly, under different curve

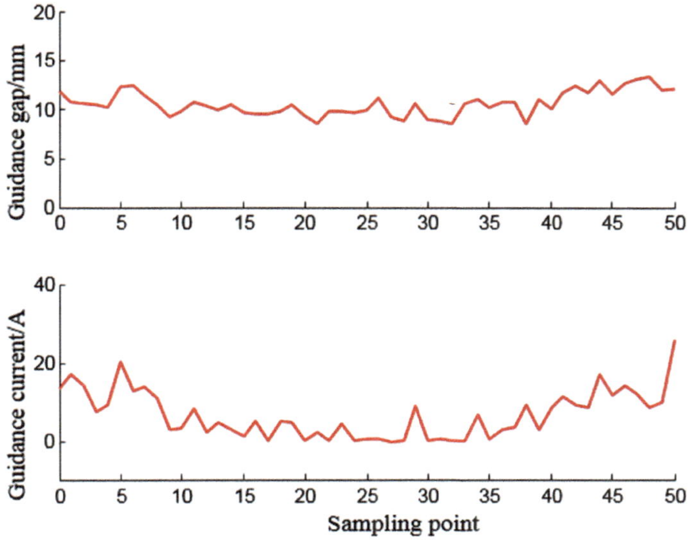

Fig. 6.7 Guidance gap and current curve in test

radii, running speed and payload, the evaluation index proposed in this chapter is used to analyze and evaluate the active guidance ability of the train. Finally, the effectiveness of the evaluation results is verified by the actual line test. The results indicate that the active guidance ability of the high-speed maglev train is closely related not only to the ability of the guidance system itself but also to the line radii, running speeds, and other factors.

Bibliography

1. Ding S (2022) 600 km/h high-speed maglev transportation system. Shanghai Scientific and Technical Publishers, Shanghai
2. Hao A, Long Z, Chang W (2008) Guidance controller design of high speed maglev train considering random irregularity of guideway. J Syst Simul 20(3):6231–6237
3. Hao A, Long Z, Chang W (2008) Design of the robust controller of the guidance system in high-speed maglev train. J China Railway Soc 30(6):40–45
4. Hao A, She L, Chang W (2008) Adaptive controller design of guidance system of EMS high speed maglev train. Control Eng China 15(2):116–119, 170
5. He (2016) Dynamic modeling of the medium speed maglev train and research on the guidance ability. Ph.D. dissertation, National University of Defense Technology Doctor Degree, Changsha
6. Hong X, Wu J, Chao C et al (2023) Guidance irregularity detection system of track for high-speed maglev transit. J Railway Sci Eng 20(7):2718–2727
7. Hou Y, Shen D (1995) Index number scale and comparison with other scales. Syst Eng Theory Practice 15(10):43–49

8. Liu S, Long C, Chen G et al (2010) Analysis of new guide electromagnet for high-speed maglev train. Electric Drive Locomot 1:49–51
9. Wang R, Cheng H, Li Y (2011) Research on turning ability of high-speed maglev train. Electr Drive for Locomot 2011(1):40–42, 46
10. Wu X (2003) Maglev train. Shanghai Scientific and Technical Publishers, Shanghai
11. Wu Y (2006) Guidance control system and simulation research of EMS type high-speed maglev vehicle. J Heilongjiang Inst Sci Technol 16(4):244–247
12. Yan L (2008) Development and application of the maglev transportation system. IEEE Trans Appl Supercond 18(2):92–99
13. Zhai M, Hao A, Li X, Long Z (2018) Research on the active guidance control system in high speed maglev train. IEEE Access 7:741–752
14. Zhai M, Zhu P, Li X et al (2022) Evaluation and verification for active guidance ability of EMS maglev train. J Southwest Jiaotong Univ 57(3):514–519
15. Zhao C (2014) Research on guidance dynamics of EMS high-speed maglev train. Ph.D. dissertation, National University of Defense Technology Doctor Degree, Changsha

Chapter 7
Control Technology of High-Speed PEM Hybrid Levitation System

In 2003, with the successful operation of Shanghai high-speed maglev line, the high-speed levitation technology as a new transportation system was gradually recognized by the industry. In the process of localization of traditional EMS-type high-speed maglev trains, the National University of Defense Technology proposed a high-speed maglev scheme with PEM hybrid levitation system. Meanwhile, with the support of the Ministry of Science and Technology and the National Maglev Center, the research and development work on the test platforms has lasted for 15 years. The test platforms include the single maglev bogie, dual maglev bogie, single-section formation high-speed maglev test train and two-section formation high-speed maglev test train. The electromagnet is the core component of the maglev train, which interacts with the track to provide electromagnetic force for the maglev train. The performance of the electromagnet is one of the important factors affecting the levitation performance of the maglev train, and it is also a key technology and difficult point of the maglev train levitation control system. For the demand of the PEM hybrid high-speed maglev train, the design technology of the electromagnet is analyzed and discussed in detail. By analyzing the model of PEM hybrid levitation system, the research of a zero-power controller based on state feedback is designed and carried out. Finally, the feasibility of the control scheme is verified through the test on the PEM hybrid levitation dual maglev bogie.

7.1 Analysis of PEM Hybrid Levitation Technology

7.1.1 Problem Formulation

The studies have shown that the value of the levitation current of traditional EMS-type high-speed maglev technology is quite big to achieve the designed load-carrying capacity. This leads the critical value of the current density of the electromagnet coil

Z. Long et al., *High-Speed Maglev Train's Levitation and Guidance Control*, Advances in High-speed Rail Technology, https://doi.org/10.1007/978-981-97-2309-6_7

to reach its rapid heating at the ultimate load. Constrained by the current density of the electromagnet coils, the levitation system of the traditional EMS-type maglev train can be further enhanced from the following four areas.

(1) The Current Density of the Electromagnet Coil Restricts the Load Capacity

Given the structural parameters of the levitation magnet, the load capacity is mainly determined by the magnetic density of the levitation field and is also related to the levitation gap and current. The experimental results show that the magnetic density of a single levitation electromagnet does not saturate at a load of 5 t. The load capacity can increase by simply increasing the magnetic density. However, when the levitation gap is 10 mm and the load is 5 t, the temperature of the aluminum coil of the electromagnet rises rapidly. It is impossible to increase the magnetic density by increasing the current. Instead, it is necessary to consider the cooling of the coil by the air. Therefore, for electromagnets, although there is still room for improvement of magnetic density, it is impossible to improve the loading capacity by simply increasing the current due to the limitation of current density.

(2) The Heat Problem of the Electromagnet Coil

To avoid the phenomena of the high temperature of the electromagnet coils during the operation of the traditional EMS-type high-speed maglev trains, the cooling of the electromagnets mainly utilizes the walking air. This cooling effect is effective when the train is running at speeds above 100 km/h. This results in the traditional EMS-type high-speed maglev trains cannot attain a long-running below 100 km/h.

(3) The Influence of the High-Temperature Environment on the Performance of the Levitation Sensors

The sensors are installed on the electromagnet of the maglev train to measure the levitation gap, etc. Their performances directly affect the levitation of the maglev train. In the TR08 maglev train, the sensor is in direct contact with the potting layer of the coil. If the surface temperature of the coil is high, a long-running task within the high-temperature environment will affect the performances of the sensors, and further degrade the levitation performance and the reliability of the maglev train.

(4) The Power Consumption of the Levitation System

The levitation force of the traditional EMS-type maglev train is provided by the electromagnet. The levitation function requires a certain current to maintain, for example, the required current is 25.7 A, to levitate the weight of 4 t. This results in the high levitation power consumption of the maglev train and the large capacity and volume of the vehicle power supply equipment. In practice, the vehicle power converter and battery of TR08 occupy a large space.

To overcome the above issues, permanent magnets are mixed into the electromagnets to form a PEM hybrid levitation system. This will decrease the levitation current without reducing the load capacity and solve the heating problem of coil. The advantage of the PEM hybrid levitation technology is that the permanent magnets provide the main static levitation force. The coils are mainly utilized to achieve stable control and the corresponding current can be small enough as per needs. The specific features are as follows:

(1) By adding the permanent magnets, the coil current density can be greatly reduced. The heating problem is fundamentally solved and the working environment of the sensor is improved.
(2) The capacity of the levitation electromagnet can be further improved. Under the same loading requirements, the coil current is no longer saturated, and there is still room to rise by increasing the current to improve the loading capacity.
(3) The levitation power consumption will be significantly reduced, which shows the advantages of maglev transit as a green and environmentally friendly rail transit.

7.1.2 Technical Analysis

Based on the introduction of the current research status of PEM hybrid levitation technology, the electromagnet design technology and PEM hybrid levitation control technology are two main focuses to be analyzed.

(1) Status of Technology Research

Researchers in America, Canada, and China have started innovative research on PEM hybrid levitation technology while conducting the promotion and application of electromagnetic levitation systems.

In the 1980s, David L. Atherton proposed a PEM hybrid levitation system scheme with a specific magnet structure, and highlighting potential control difficulties due to the inclusion of permanent magnets. David L. Atherton's research pioneered the technology of PEM hybrid maglev trains, but its progress was not smooth due to the performance of permanent magnets being difficult to meet the requirements of engineering applications and the high cost at that time. Then, there were many improvements in the manufacturing technology and processes of the permanent magnet, such as the increase of coercivity and magnetic energy product, the improvement of stability, and especially the significant decrease of price. Therefore, the PEM hybrid levitation technology has gradually attracted the attention of various countries, and a series of principle experimental platforms and corresponding levitation control methods have emerged one after another. In the late 1980s, Japan carried out research on the hybrid PEM levitation technology. In 1989, Japan designed a permanent magnet levitation device with a self-weight of 8 kg, realizing the goal of zero-power levitation under a load of 1.5 kg. The magnet structure of the device is shown in Fig. 7.1.

In 2002, MagneMotion of the United States developed the M3 maglev train, which is 1.8 m long and can run reciprocally on a 6 m long guideway, achieving zero-power stable levitation from 0.7 to 1.3 t. The actual levitation energy consumption is only 200 W/t. This type of train adopts the electromagnet structure scheme of PEM, as shown in Fig. 7.2.

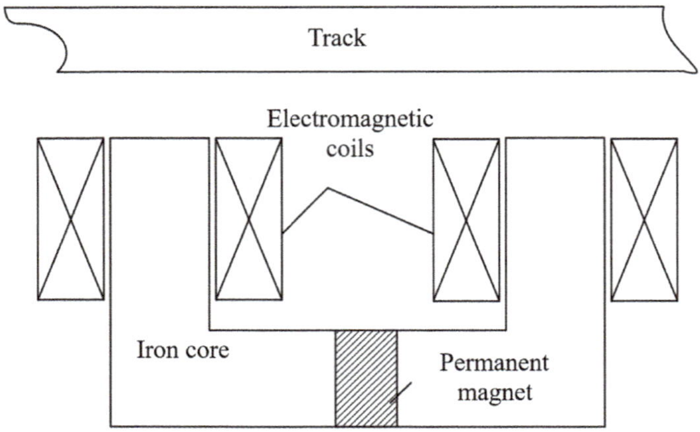

Fig. 7.1 The magnet structure designed by Japan in 1989

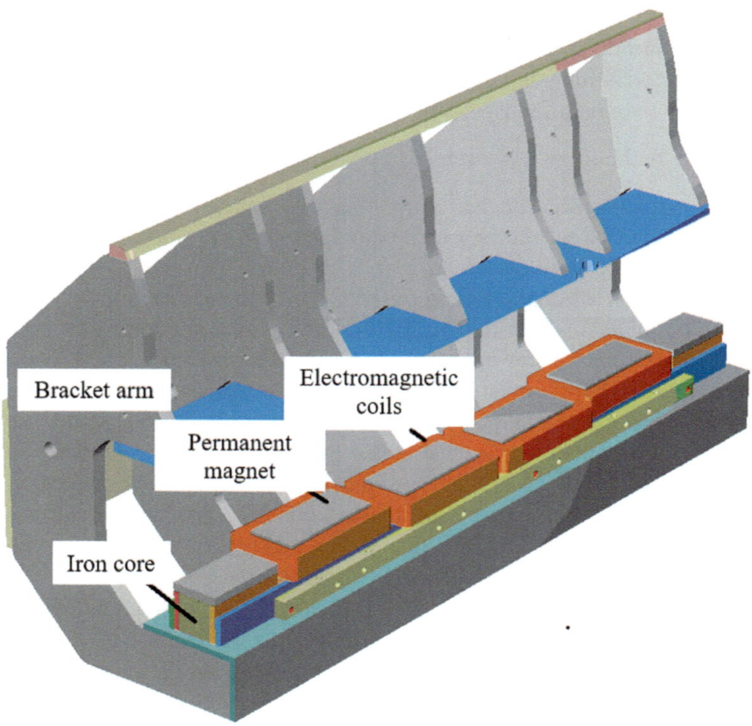

Fig. 7.2 The magnet structure of U.S. M3 maglev train

The research on PEM hybrid maglev train technology in China began in the early 1990s, with major research units such as the National University of Defense Technology and the Institute of Electrical Engineering of the Chinese Academy of Sciences. In 1996, based on the analysis of the mechanism of the PEM hybrid levitation, the National University of Defense Technology proposed a design scheme for a maglev train system with the PEM hybrid levitation and studied the zero-power control strategy. In 2002, the Institute of Electrical Engineering of the Chinese Academy of Sciences developed an experimental device for the PEM hybrid levitation with four points of levitation, which weighs 78 kg and is provided with levitation force by four hybrid magnets. A mechanical limit device is used to prevent permanent magnets from being absorbed. The Southwest Jiaotong University designed the PEM hybrid levitation system, the levitation system with the superconducting coil and the normal-conducting coil, and the high-temperature superconducting EMS levitation system. The stability control of these three systems was completed by using the same set of experimental equipment, and the steady-state current of the PEM hybrid levitation system was basically zero. In 2005, the National University of Defense Technology designed a verification device for the principle of the E-type magnet of the PEM hybrid levitation. The magnet weighs about 8kg and achieves stable levitation in the range of 8 to 16 kg. In 2007, in response to the problem of electromagnet heating up during static levitation and low-speed levitation of the traditional EMS-type maglev trains, the National University of Defense Technology developed a full-size hybrid magnet structure and constructed a single maglev bogie system for the high-speed PEM hybrid levitation. Then, the National University of Defense Technology achieved stable levitation under a load of about 10.5 t, with only 1/4 levitation power consumption of the traditional EMS-type single bogie system. With the support of the Science and Technology Support Plan of the national 11th Five-Year Plan and the national 12th Five-Year Plan, the National University of Defense Technology has taken the lead in developing the PEM hybrid levitation system of the high-speed dual maglev bogie, the high-speed maglev test train, and the low-speed maglev train. One maglev train transformed into PEM hybrid levitation has been successfully developed in December 2012 and has been tested for long-term trial operation in Tangshan urban maglev test line. By December 2021, the PEM hybrid levitation maglev train has been running for 50,000 km in total, without any safety accidents.

(2) Analysis of Key Technology

The key technologies of the PEM hybrid levitation maglev trains mainly include the design of magnet structure and the levitation control strategy. To achieve the practical and engineering goals of the PEM hybrid levitation maglev train, it is necessary to carry out an in-depth study of the above key technologies.

Design of the Magnet Structure. The electromagnet is the actuator of the PEM hybrid levitation maglev train. Its structure not only determines the levitation power consumption of the system but also directly affects the levitation performance and difficulty of controlling the system. The unreasonable magnet structure tends to lead to a larger levitation power consumption of the system, because the introduction of permanent magnets makes the equivalent levitation gap in the magnetic circuit increase. The equivalent magnetic potential required by the system increases accordingly under the premise of equal magnetic force. Also, the magnetic force is insensitive to the change of the coil current. When the magnitude of the changes of the electromagnetic force is lower than the magnitude of the changes of the magnetic force due to the gap change, the stable levitation of the system cannot even be guaranteed.

Technology of Levitation Control. The levitation control technology is the guarantee of achieving stable levitation of the PEM hybrid levitation system, and its performance directly affects the stability, safety, and ride comfort of the system. The uncontrollability of the permanent magnet increases the nonlinearity and complexity of the system model. The demagnetization phenomenon of permanent magnets increases the uncertainty of the model and the magnitude of parameter perturbation. Due to the existence of uncontrollability and demagnetization phenomenon, the control technology of the PEM hybrid levitation system has higher requirements. Accordingly, the following issues need to be considered for the levitation controller design.

① The effect of the permanent magnets on the characteristics of the levitation system.
② The effect of the current integration signal on the system in zero-power control.
③ How to ensure the stability and dynamic performance of the system when the load varies over a wide range.

In summary, compared with the simple electromagnetic levitation system, the unique technology of the PEM hybrid levitation system is reflected as follows: the design strategy of the hybrid electromagnet, including the selection of permanent magnets, the structural design and parameter calculation of the electromagnet; the control technology of the PEM hybrid levitation system.

In this chapter, facing the needs of the real-life engineering, the design technology and control technology of the PEM hybrid levitation are studied.

7.2 Design Requirements for PEM

The primary purpose of adopting PEM is to reduce the levitation power consumption. Compared with the traditional EMS-type electromagnet, the current required to provide the same levitation force can be reduced. Secondly, the fewer permanent magnets are used, the better, which can not only effectively play the role of permanent magnets, but also help to reduce the self-weight of the PEM. In addition, the levitation force should be adjusted rapidly with the change of current to ensure the stability of the levitation system. The basic design requirements of the PEM are introduced from these three aspects.

7.2.1 Energy Conservation

The effect of adopting the permanent magnets on the levitation force is mainly in two aspects. Firstly, the magnetic permeability of the permanent magnet is comparable to that of air, and the magnetic resistance of the magnetic circuit will increase, leading to the decrease of the levitation force. Secondly, the permanent magnet provides the magnetic potential to the outside. So, the magnetic potential of the magnetic circuit will increase, leading to the increase of the levitation force. Since the above two aspects have opposite trends, quantitative analysis is required.

The U-type PEM is used as an example for analysis. Depending on the installation position of the permanent magnet, there are two basic structures of the U-type PEM. One structure is to install the permanent magnet on the upper surface of the core (Fig. 7.3a), where the pole area of the permanent magnet and the core are equal. Another structure is to install the permanent magnet in the middle of the yoke (Fig. 7.3b), where the pole area of the permanent magnet and the core are not necessarily equal. By comparison, the structure in Fig. 7.3b is more practical, and this structure will be analyzed as an example.

Assuming a uniform distribution of the magnetic field and neglecting the effect of the magnetic leakage flux, the levitation force of the U-type PEM is as follows:

$$F_m = \frac{\mu_0 (Ni + H_c \delta_m)^2}{S\left(2z/S + \delta_m/(\mu_m S_m)\right)^2}, \tag{7.1}$$

where μ_0 is the vacuum magnetic permeability, N is the number of coil turns, i is the coil current, H_c is the coercivity of the permanent magnet, δ_m is the thickness of the permanent magnet, S is the pole area of the iron core, z is the levitation gap, μ_m is the relative magnetic permeability of the permanent magnet, S_m is the pole area of the permanent magnet.

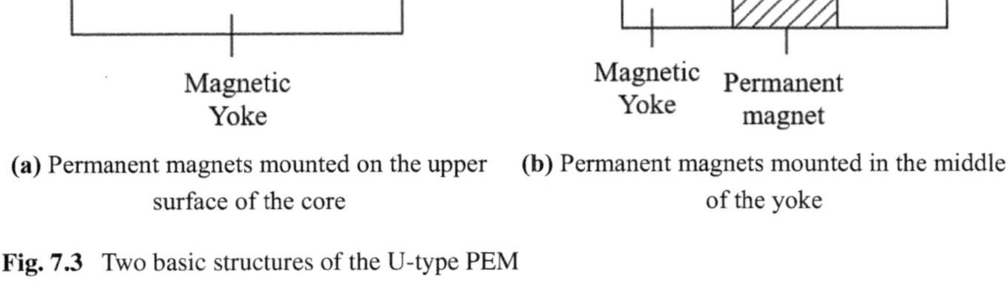

(a) Permanent magnets mounted on the upper surface of the core **(b)** Permanent magnets mounted in the middle of the yoke

Fig. 7.3 Two basic structures of the U-type PEM

According to Eq. (7.1), the formula for calculating the current can be obtained as follows:

$$i = \frac{1}{N}\sqrt{\frac{F_m S}{\mu_0}}\left(\frac{2z}{S} + \frac{\delta_m}{\mu_m S_m}\right) - \frac{H_c \delta_m}{N} \tag{7.2}$$

Let $\partial i/(\partial \delta_m) < 0$, the sufficient condition for the current to decrease after the adopting the permanent magnet is as follows:

$$H_c > \frac{1}{\mu_m S_m}\sqrt{\frac{F_m S}{\mu_0}} \tag{7.3}$$

Considering the remanence of the permanent magnet can be expressed as $B_r = \mu_0 \mu_m H_c$, Eq. (7.3) could be expressed to:

$$B_r S_m > \sqrt{F_m S \mu_0}, \tag{7.4}$$

where $B_r S_m$ is the magnetic flux inside the permanent magnet, $\sqrt{F_m S \mu_0}$ is the magnetic flux corresponding to a levitation force of F_m. Increasing the permanent magnet can reduce the current only if the magnetic flux inside the permanent magnet is bigger than the magnetic flux required to maintain levitation.

The PEM applied to maglev trains usually needs to provide a large levitation force. To ensure that Eq. (7.4) is valid, both permanent magnet materials with high remanence can be selected, and the pole area of the permanent magnet can be increased. When the improvement room of the remanence is not large, it is more effective to increase the pole area.

7.2.2 Enhance the Utilization of Permanent Magnets

Different working points of permanent magnets correspond to different magnetic densities and magnetic field intensities. The working point of permanent magnets can reflect the working state of permanent magnets, and it can be any point on the demagnetization curve. The working point of permanent magnets is set to fully utilize the role of permanent magnets and use the less permanent magnets in the control process to achieve the goal.

Take the N45 NdFeB permanent magnet as an example, its demagnetization curve is approximated as a line segment, as shown in Fig. 7.4. The working point can be any point M on the demagnetization curve, where B_m is the magnetic density of the working point, also called magnetic induction and $-H_m$ is the magnetic field intensity of the working point. B_m and $-H_m$ correspond to each other and satisfy the following relationship.

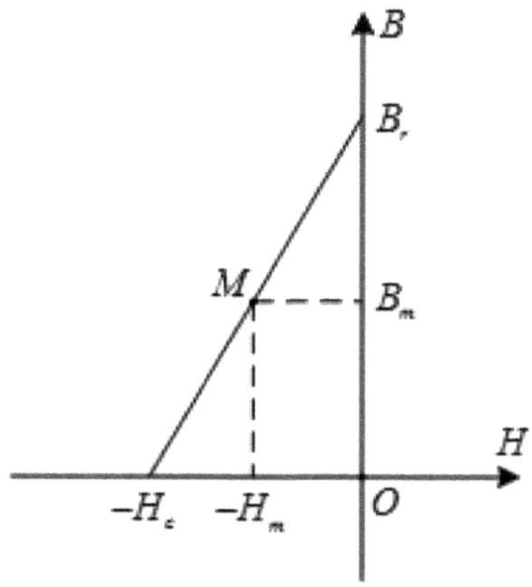

Fig. 7.4 The demagnetization curve of the N45 NdFeB

$$-H_m = -H_c + B_m / (\mu_0 \mu_m) \tag{7.5}$$

The magnetic energy product of the working point is the absolute value of the product of the magnetic density and the magnetic field intensity of the working point. Among all the working points of permanent magnets, the working point with the largest magnetic energy product is called the maximum magnetic energy product point. From Fig. 7.4, the maximum magnetic energy product point of N45 NdFeB is located at the midpoint of the demagnetization curve.

(1) Setting Method of the Working Point

The working point of the permanent magnet is affected by the environment. For the PEM shown in Fig. 7.3b, it is known from the basic knowledge of the magnetic circuit that the magnetic density and the magnetic field intensity of the working point of the permanent magnet satisfy the following relationship.

$$B_m S_m \cdot \frac{2z}{\mu_0 S} = Ni + H_m \delta_m z \tag{7.6}$$

From Eq. (7.5) and Eq. (7.6), the magnetic density of the working point is as follows:

$$B_m = \frac{\mu_0 (Ni + H_c \delta_m)}{2z S_m / S + \delta_m / \mu_m} \tag{7.7}$$

Through the above analysis, the magnetic density of the working point of the permanent magnet is not only related to the material but also related to the parameters of the magnetic circuit, including the structural parameters of the permanent magnet and the core, the levitation gap, and the current, etc. For a PEM with a given load capacity requirement, the working point of the permanent magnet can be set according to the following steps.

1. The required magnetic flux ϕ is calculated according to the load-carrying capacity requirement. Assuming that the required levitation force is F_m, the required flux is $\phi = \sqrt{F_m \mu_0 S}$. Since the designed PEM is an improvement of the original EM, the form factor should be ensured to be unchanged, and the pole area S of the core should follow the original value. Then, the value of the magnetic flux can be calculated according to the levitation force.
2. The pole area S_m of the permanent magnet is selected according to the target of the working point. Since $\phi = B_m S_m$, the magnetic density of the working point is only related to the S_m. Assuming that the working point is to be set at a magnetic density of B_{ms}, then the pole area of the permanent magnet is $S_m = \phi / B_{ms} = \sqrt{F_m \mu_0 S} / B_{ms}$. Therefore, the selection of the permanent magnet pole area is crucial. Otherwise, the designed PEM cannot balance the load-carrying capacity requirement with the setting requirement of the working point.

(2) The Basis for Setting the Working Point

According to the magnetic flux continuity, it is obtained that:

$$\phi = B_\delta S = B_m S_m, \tag{7.8}$$

where B_δ is the magnetic density at the levitation gap.

According to the loop theorem of magnetic field intensity, it can be obtained:

$$Ni = 2H_\delta z - H_m \delta_m, \tag{7.9}$$

where H_δ is magnetic field intensity at the levitation gap, Ni is the magnetomotive force provided by the coil, $2H_\delta z$ is the difference of magnetic flux intensity between levitation gaps, $-H_m \delta_m$ is the difference of magnetic flux intensity of the permanent magnet, the negative values indicate that the permanent magnet provides a magnetomotive force of $H_m \delta_m$ to the outside.

Considering $B_\delta = \mu_0 H_\delta$, the magnetic energy of the levitation magnetic field can also be expressed as follows:

$$\frac{1}{2} B_\delta H_\delta V_\delta = \frac{B_\delta^2 S}{\mu_0} z = F_m z \tag{7.10}$$

With Eq. (7.9) and Eq. (7.10), the volume of the permanent magnet can be expressed as follows:

$$V_m = \frac{2F_m z - N\phi i}{B_m H_m} \tag{7.11}$$

Equation (7.11) shows that, under the premise that the levitation force, levitation gap, current and electromagnetic structure parameters generated by PEM remain unchanged, the volume of required permanent magnet is inversely proportional to the magnetic energy product at its working point. The larger the magnetic energy product at the working point, the smaller the volume of required permanent magnet, which means the higher the efficiency of the permanent magnet. And when the volume of the permanent magnet is constant, the larger the magnetic energy product at the working point, the smaller the current required to generate the same levitation force, and the better the effect of reducing the power consumption.

It can be concluded that the best working point of the permanent magnet is its maximum magnetic energy product point, and the working point of the permanent magnet should be set near the maximum magnetic energy product point when designing the PEM.

7.2.3 Improve the Controllable Performance of PEM

After equipping permanent magnets, the magnetic resistance of the magnetic circuit increases, and the trend of levitation force changes with the number of ampere-turns slows down. The design requirements of the PEM are analyzed below from the perspective of ensuring that the levitation force changes rapidly with the current.

(1) Description of the Controllable Performance of the Levitation Electromagnet

The levitation system is theoretically controllable, and for any initial state, there always exists a control voltage that regulates the levitation force by changing the current to make the system reach the target state in a limited time. The slower the current regulates the levitation force, the higher the control voltage required. Since the actual control voltage is always limited, if the regulation of the current on the levitation force is too weak, it will lead to an unstable system. Thus, the sensitivity of the levitation force with the number of ampere-turns is considered to describe the controllable performance of the electromagnet.

For comparison, the electromagnets of the TR08 maglev train are used as an example for calculation. The full load weight of a single electromagnet is 4.3 t, and the total number of turns of the coil in the levitation state is 86000 A. The average sensitivity of the levitation force to the number of ampere-turns of the coil is calculated to be 0.49 N/A. This calculation can be used as a reference value to judge the controllability of the levitation magnet in the following.

(2) Influence of Structural Parameters of Permanent Magnets on Controllable Performance

According to Eq. (7.1), the controllable performance of PEM is obtained by taking the partial derivative of the levitation force with respect to the number of ampere-turns.

$$J_k = \frac{\partial F_m}{\partial (Ni)} = \frac{2\mu_0(Ni + H_c\delta_m)}{S(2z/S + \delta_m/(\mu_m S_m))^2} \tag{7.12}$$

The larger the permanent magnet pole area is, the higher the controllable performance of the PEM.

The effect of the thickness of the permanent magnet on the controllable performance needs to be analyzed specifically. By taking the partial derivative of the controllable performance with respect to the thickness of the permanent magnet, there is:

$$\frac{\partial J_k}{\partial \delta_m} = \frac{2\mu_0}{\mu_m S_m} \frac{2\mu_m S_m z H_c - 2NiS - H_c \delta_m S}{\left(2z/S + \delta_m/(\mu_m S_m)\right)^3} \tag{7.13}$$

Let $\partial J_k / \partial \delta_m > 0$, from Eq. (7.13), it can be obtained that:

$$\delta_m < \frac{2\mu_m S_m z H_c - 2NiS}{H_c S} \tag{7.14}$$

Only when the thickness of the permanent magnet is less than a certain upper limit, thickening the permanent magnet can improve the controllable performance of the PEM. From Eq. (7.14), the upper limit of the thickness of the permanent magnet is proportional to the pole area. If the pole area is too small and the upper limit is less than zero, then the thickening of the permanent magnet will lead to worse controllability of the PEM.

(3) Measures to Improve the Controllable Performance of PEM

According to Eq. (7.14), it is known that increasing the pole area of the permanent magnet will improve the controllable performance of the PEM. Regarding the selection of the thickness of the permanent magnet, it is discussed as follows, the critical values for the thickness of the permanent magnets are as follows:

$$\delta_{m0} = \frac{2\mu_m S_m z H_c - 2NiS}{H_c S} \tag{7.15}$$

When $\delta_m < \delta_{m0}$, $\partial J_k / (\partial \delta_m) > 0$. When $\delta_m > \delta_{m0}$, $\partial J_k / (\partial \delta_m) < 0$. Therefore, when the thickness of the permanent magnet is δ_{m0}, it is the optimal thickness of the permanent magnet, and the rate of change of the levitation force with the number of ampere-turns of the coil is the fastest. However, this result is achievable only if $\delta_{m0} > 0$ holds, when the pole area of the permanent magnet should satisfy the following conditions.

$$S_m > \frac{NiS}{\mu_m z H_c} \tag{7.16}$$

Only when the pole area of the permanent magnet exceeds a certain value, the optimal value of the thickness of the permanent magnet exists, and the controllable performance of the PEM is optimal.

In summary, to improve the controllable performance of the PEM, the method of increasing the pole area of the permanent magnet can be used. If the constraint of

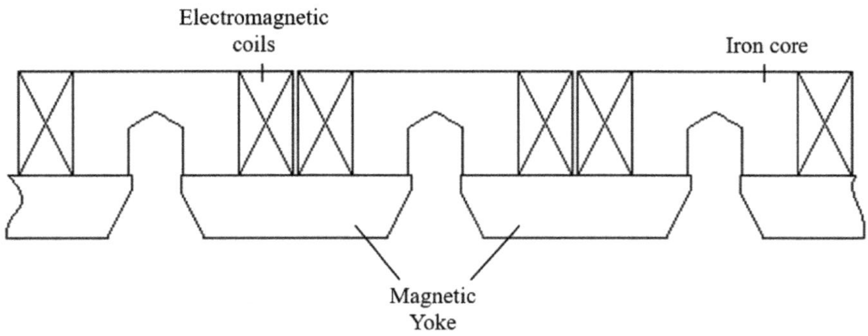

Fig. 7.5 The structure diagram of EM of TR08 Maglev train

Eq. (7.16) holds, then the thickness of the permanent magnet is the optimal thickness. If the constraint does not hold, then thickening the permanent magnet will lead to a poor controllable performance.

7.3 Structural Design of PEM

In the following, research on how to adopt permanent magnets into the existing EM, three ways are proposed according to the different installation positions of permanent magnets, and the structural design of PEM is completed by finite element calculation and comparative analysis.

7.3.1 Basic Structure of EM

Figure 7.5 shows the structure of the EM of the TR08 Maglev train. The EM mainly consists of an electromagnetic coil, iron core, and magnetic yoke. The coil is wound around each iron core to form the magnetic pole, and the two adjacent poles are connected by a magnetic yoke. A complete EM consists of 12 poles and 11 yokes.

7.3.2 Three Types of Structural Solutions for PEM

There are three schemes for adopting permanent magnets to the EM structure shown in Fig. 7.5.

(1) Scheme One

The permanent magnets are installed in the middle of the iron core and the structure is shown in Fig. 7.6. The addition of permanent magnets does not change the external dimensions of the levitation magnet. And, the electromagnetic coil and the magnetic yoke do not need to be modified, only the iron core needs to be redesigned. The addition of permanent magnets in the iron core increases the difficulty in the winding of the electromagnetic coil. And, the pole area of the permanent magnets is equal to the pole area of the iron core.

(2) Scheme Two

The permanent magnet is installed between the iron core and the magnetic yoke, and the structure of the PEM is shown in Fig. 7.7. The iron core, electromagnetic coil and magnetic yoke do not need to be modified, but the pole area of the permanent magnets is small and cannot be adjusted as needed.

(3) Scheme Three

The permanent magnet is installed in the middle of the magnetic yoke, and the structure of the PEM is shown in Fig. 7.8. The iron core and magnetic coil do not need to be modified, and the pole area of the permanent magnet can be selected flexibly according to the needs, but the structure of the magnetic yoke needs to be redesigned.

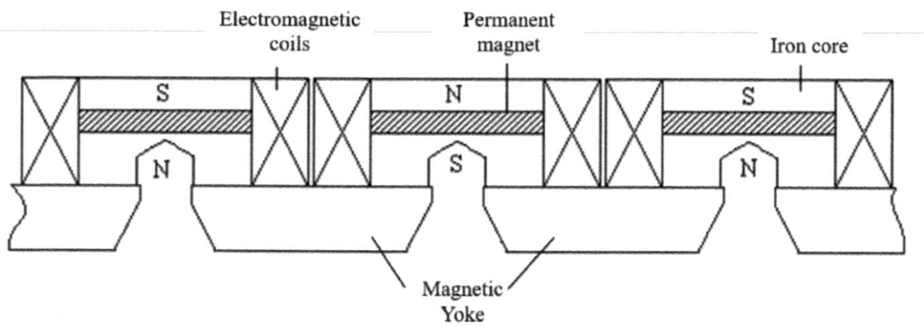

Fig. 7.6 The structure of PEM with permanent magnets installed in the middle of the iron core

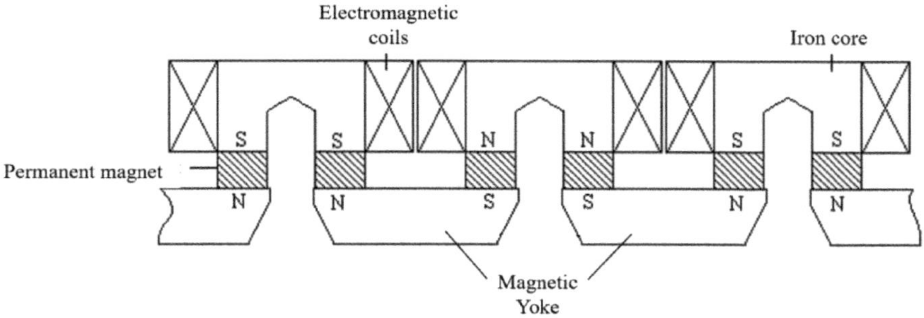

Fig. 7.7 The structure of PEM with permanent magnets installed between the iron core and the magnetic yoke

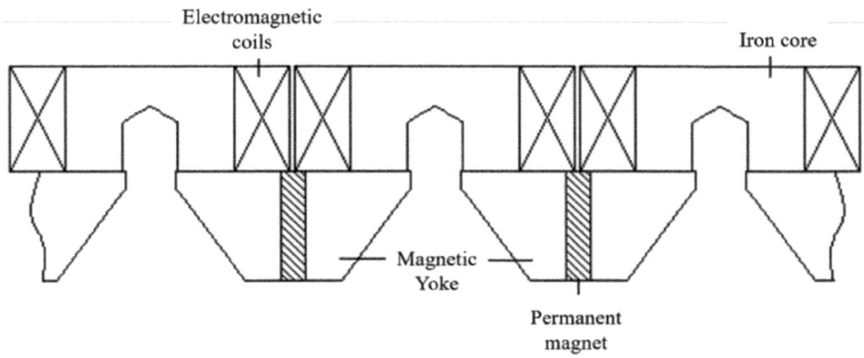

Fig. 7.8 The structure of PEM with permanent magnets installed in the middle of the magnetic yoke

7.3.3 Finite Element Calculation and Analysis

The finite element method is a numerical calculation method based on the variational principle. This method divides the whole region into many small sub-regions and solves each sub-region to finally obtain the numerical solution of the edge value problem. For the above three schemes of adopting permanent magnets, the following calculations and analysis are carried out by the finite element method.

Maxwell 3D is a tool in Ansoft to analyze the three-dimensional electromagnetic fields, which has the characteristics of powerful and accurate results. The three-dimensional electromagnetic fields can be used to accurately simulate the magnetic fields caused by DC voltage and current sources, permanent magnets and applied magnetic field excitation. And, it can calculate magnetic field intensity, magnetic flux density, levitation force, etc. Therefore, Ansoft is chosen to calculate PEM, and the relevant parameters are taken as follows.

① The permanent magnet material is N45 NdFeB with a remanent magnetization of 1.35 T. The volume of the permanent magnets is temporarily variable, but for the convenience of comparison, the volume of the permanent magnets is set equal in the three schemes.
② Considering the rated levitation gap and the effect of the guideway and pole potting layer, the levitation gap is selected as 13 mm.
③ The error of the finite element software is set to 3%.
④ For parts that do not need to be changed, such as the magnetic coil and the iron core, the corresponding values of the original EM are used.

(1) Finite Element Calculation and Analysis of Scheme 1

For the structure shown in Fig. 7.6, the thickness of the permanent magnet is initially selected as 2 mm. The levitation force of PEM and the magnetic density at the working point are calculated when the amount of ampere-turn of the coil is changed. The least squares method is used to fit the calculated results, and the analysis of the results are obtained as follows.

① The amount of ampere-turn required for the PEM to carry 4.3 t is 7,320, which is only 8.5% of the required number for the EM. This indicates that the current required to provide the same levitation force is significantly reduced by equipping permanent magnets. When the number of ampere-turn is 86,000, the loading capacity of the PEM is 7.4 t. This indicates that the loading capacity of the PEM is higher than that of the EM at the same current after adding permanent magnets.

② The magnetic density at the working point is 0.91 T when the load is 4.3 t. Combined with the demagnetization curve of N45 NdFeB, the magnetic energy product of the working point is calculated to be 89% of the maximum magnetic energy product, which indicates the utilization rate of the permanent magnets is high.

③ The average sensitivity of the levitation force to the amount of ampere-turn of the coil is 0.396 N/A within the variation range of the amount of ampere-turn selected for the calculation, indicating that the controllable performance of the PEM is well.

In general, for the PEM structure with permanent magnets installed in the middle of the iron core, it can meet the design requirements presented in Sect. 7.2.

(2) Finite Element Calculation and Analysis of Scheme 2

For the structure shown in Fig. 7.7, the thickness of the permanent magnet is selected to be 30 mm to keep the volume of the permanent magnet constant. When the amount of ampere-turn of the coil is changed, the levitation force of the PEM and the magnetic density at the working point are calculated. The least squares method is used to fit the calculated results, and the analysis of the results are obtained as follows.

① When the total number of ampere-turn is 86,000, the load capacity of PEM is calculated to be 4.32 t. This indicates that after adding permanent magnets, the levitation force of PEM does not increase under the same current. The current required to provide the same levitation force is basically equal. And adding permanent magnets cannot reduce the effect of levitation power consumption.

② When the load is 4.32 t, the magnetic density of the working point of the permanent magnet is 1.33 T. Combined with the demagnetization curve, the magnetic energy product of the working point is only 5.8% of the maximum magnetic energy product. This indicates that the working point of the permanent magnet is high, and the utilization rate is poor. The reason for the high working point is that the pole area of the permanent magnet is small, and the magnetic density of the working point is too large to provide the required magnetic flux for levitation.

③ The average sensitivity of the levitation force to the number of ampere-turn of the coil is 0.3 N/A within the variation range of the number of ampere-turn selected for the calculation, which is slightly lower compared to scheme 1. This indicates that the controllable performance of the PEM becomes worse when the pole area of the permanent magnets is reduced.

Overall, for the PEM structure with the permanent magnet installed between the iron core and the magnetic yoke, the PEM cannot meet the design requirements proposed in Sect. 7.2, and this scheme is not suitable to be selected as the PEM structure.

(3) Finite Element Calculation and Analysis of Scheme 3

For the structure shown in Fig. 7.8, compared with scheme one, the thickness of the permanent magnet is increased by a factor of 1, while the pole area is reduced to 1/2 of the original size. When the amount of ampere-turn of the coil is changed, the levitation force of the PEM and the magnetic density at the working point are calculated. The least squares method is used to fit the calculated results, and the analysis of the results are obtained as follows.

① When the load is 4.3 t, the amount of ampere-turn of the coil is 53,000, accounting for 61.6% of the EM ampere-turns. This indicates that adding permanent magnets has an energy-saving effect, and the current required to provide the same levitation force is reduced. When the amount of ampere-turn of coil is 86,000, the PEM could load 6.3 t, indicating that the levitation force generated by the PEM with the same current is larger than that generated by the EM.

② At a load of 4.3 t, the magnetic density at the working point is about 1.16 T. The magnetic energy product accounts for 48% of the maximum magnetic energy product, and the efficiency of the permanent magnet is relatively low.

③ The average sensitivity of the levitation force to the amount of ampere-turn of the coil is 0.46 N/A within the range selected for the calculation, indicating that the controllable performance of the PEM is relatively good.

In general, when the permanent magnet is installed in the middle of the magnetic yoke, the levitation power consumption of the PEM is reduced and its controllable performance is better, but the magnetic density at the working point is slightly higher. However, by increasing the pole area of the permanent magnet, the magnetic density of the working point can be reduced, so that the performance of the permanent magnet is near the maximum magnetic energy product point. The PEM obtained with such a structure can meet the design requirements presented in Sect. 7.2.

After comprehensive consideration, the magnet structure in scheme three is selected as the PEM structure.

7.4 Modeling and Analysis of PEM Hybrid Levitation System

In this section, the mathematical model of the PEM hybrid levitation system is developed considering magnetic leakage, and the open-loop characteristics and controllability of the system are analyzed.

7.4.1 System Model

In the PEM structure, the coil current is firstly generated by the voltage at both ends of the electromagnetic coil, which in turn generates the electromagnetic field. The electromagnetic field and the permanent magnetic field interact with the guideway to generate the magnetic force. Then, the change of the magnetic force changes the mechanical motion state of the system. Therefore, the system modeling needs to consider the mechanical motion characteristics of the magnet, the voltage-current relationship, and the current-magnetic force relationship.

(1) Kinetic Equation

In the vertical direction, from Newton's second law, the kinetic equation of the magnet can be obtained as follows:

$$M\ddot{z} = Mg - F = Mg - \mu_0 S_z \left(\frac{Ni}{2z + P} + \frac{H_c \delta_m}{2z + Q} \right)^2, \tag{7.17}$$

where $P = \delta_m S_z / (\varepsilon_e \mu_r S_m)$, $Q = \delta_m S_z / (\varepsilon_m \mu_r S_m)$, S_m is the cross-sectional area of the permanent magnet, S_z is the cross-sectional area of the gap, ε_m is the magnetic flux utilization factor of the permanent magnetic field, ε_e is the magnetic flux utilization factor of the electromagnetic field.

(2) Voltage Balance Equation

From the magnetic flux utilization factor ε_e, the magnetic flux Φ_t at the pole iron core is as follows:

$$\Phi_t = \frac{\mu_0 S_z N i}{\varepsilon_e (2z + P)} + \frac{\mu_0 S_z H_c \delta_m}{\varepsilon_e (2z + Q)} \tag{7.18}$$

Thus, the induced voltage (back electromotive force) of the coil due to the change of the magnetic field is

$$e = N\frac{\mathrm{d}\Phi_{\mathrm{t}}}{\mathrm{d}t} \tag{7.19}$$

Substituting Eq. (7.18) into Eq. (7.19) and simplifying to get:

$$e = \frac{\mu_0 N^2 S_z}{\varepsilon_{\mathrm{e}}(2z+P)}\frac{\mathrm{d}i}{\mathrm{d}t} - \frac{2\mu_0 N S_z}{\varepsilon_{\mathrm{e}}}\left(\frac{H_{\mathrm{c}}\delta_m}{(2z+Q)^2} + \frac{Ni}{(2z+P)^2}\right)\frac{\mathrm{d}z}{\mathrm{d}t} \tag{7.20}$$

Therefore, the voltage equation of the electromagnetic coil winding circuit satisfies:

$$u = Ri + e = Ri + \frac{C_1\dot{i}}{(2z+P)} + \left(\frac{2C_2}{(2z+Q)^2} - \frac{2C_1 i}{(2z+P)^2}\right)\dot{z}, \tag{7.21}$$

where $C_1 = \frac{\mu_0 N^2 S_z}{\varepsilon_{\mathrm{e}}} > 0$; $C_2 = -\frac{\mu_0 N S_z H_{\mathrm{c}}\delta_{\mathrm{m}}}{\varepsilon_{\mathrm{e}}} < 0$.

(3) The Boundary Conditions

Under the zero-power steady-state condition, the boundary condition of the system can be obtained as:

$$Mg = \mu_0 S_z\left(\frac{H_{\mathrm{c}}\delta_m}{2z_0+Q}\right)^2 \tag{7.22}$$

Based on the above analysis, the mathematical model of the permanent magnet single-point levitation system is completely determined by the following equations.

$$\begin{cases} \ddot{z} = g - \dfrac{\varepsilon_{\mathrm{e}}C_1}{M}\left(\dfrac{i}{(2z+P)} - \dfrac{C_2}{C_1(2z+Q)}\right)^2 \\ u = Ri + e = Ri + \dfrac{C_1\dot{i}}{(2z+P)} + \left(\dfrac{2C_2}{(2z+Q)^2} - \dfrac{2C_1 i}{(2z+P)^2}\right)\dot{z} \\ Mg = \mu_0 S_z\left(\dfrac{H_{\mathrm{c}}\delta_m}{2z_0+Q}\right)^2 \end{cases} \tag{7.23}$$

The state variable is selected as $\boldsymbol{x} = \left[x_1\ x_2\ x_3\right]^{\mathrm{T}} = \left[z\ \dot{z}\ i\right]^{\mathrm{T}}$, then the nonlinear model of the system can be expressed as follows:

$$\begin{cases} \dot{\boldsymbol{x}} = f(\boldsymbol{x}) + g(\boldsymbol{x})u \\ y = x_1 \end{cases}, \tag{7.24}$$

where

$$f(\boldsymbol{x}) = \begin{bmatrix} x_2 \\ -\frac{\varepsilon_e C_1}{M}\left(\frac{x_3}{2x_1+P} - \frac{C_2}{C_1(2x_1+Q)}\right)^2 + g \\ \frac{2x_2x_3}{2x_1+P} - \frac{2C_2(2x_1+P)x_2}{C_1(2x_1+Q)^2} - \frac{R(2x_1+P)x_3}{C_1} \end{bmatrix};$$

$$g(\boldsymbol{x}) = \begin{bmatrix} 0 \\ 0 \\ \frac{2x_1+P}{C_1} \end{bmatrix}.$$

Obviously, the single-point permanent magnet system considering the leakage condition is a highly nonlinear system, and the leakage and permanent magnets make the relationship between magnetic force and gap more complex, and the nonlinear characteristics of the system more prominent.

7.4.2 Characteristic Analysis of the Model

(1) Stability Analysis

For the single-point system of Eq. (7.24), let $\dot{\boldsymbol{x}} = 0$, solve for its singularities as shown in Eq. (7.25) and Eq. (7.26).

$$\begin{cases} x_1 = \frac{1}{4}\left[-\left(P + Q - \frac{u_0/R - C_2/C_1}{\sqrt{Mg/(\varepsilon_e C_1)}}\right) - \right. \\ \left.\left((P-Q)^2 - 2(P-Q)\cdot\frac{u_0/R + C_2/C_1}{\sqrt{Mg/(\varepsilon_e C_1)}} + \left(\frac{u_0/R - C_2/C_1}{\sqrt{Mg/(\varepsilon_e C_1)}}\right)^2\right)^{1/2}\right] \\ x_2 = 0 \\ x_3 = u_0/R \end{cases} \tag{7.25}$$

$$\begin{cases} x_1 = \dfrac{1}{4}\left[-\left(P + Q - \dfrac{u_0/R - C_2/C_1}{\sqrt{Mg/(\varepsilon_e C_1)}}\right) + \right. \\ \left.\left((P-Q)^2 - 2(P-Q)\cdot\dfrac{u_0/R + C_2/C_1}{\sqrt{Mg/(\varepsilon_e C_1)}} + \left(\dfrac{u_0/R - C_2/C_1}{\sqrt{Mg/(\varepsilon_e C_1)}}\right)^2\right)^{1/2}\right] \\ x_2 = 0 \\ x_3 = u_0/R \end{cases} \tag{7.26}$$

The gap value corresponding to the first set of solutions shown in Eq. (7.25) is negative, which does not conform to the actual situation, so this book only considers the second set of solutions shown in Eq. (7.26).

Assuming that the zero-power control algorithm is used, $u_0/R = 0$, the singularity is as follows:

$$\boldsymbol{x}_0 = \begin{bmatrix} \frac{-C_2}{2\sqrt{\frac{C_1 Mg}{\varepsilon_e}}} - \frac{Q}{2} & 0 & 0 \end{bmatrix} \tag{7.27}$$

According to the above equation, combined with the definition of variables C_1 and C_2, the effect of the leakage on the zero-power equilibrium point is determined by the factor Q, independent of the magnetic flux utilization factor of the electromagnetic field.

Near the zero-power equilibrium point $\boldsymbol{x}_0$, the state variable of the system can be expressed as $\boldsymbol{x} = \boldsymbol{x}_0 + \Delta\boldsymbol{x}$; the control voltage can be expressed as $u = u_0 + \Delta u$, where $\Delta\boldsymbol{x} = \begin{bmatrix} \Delta x_1 & \Delta x_2 & \Delta x_3 \end{bmatrix}^{\mathrm{T}} = \begin{bmatrix} \Delta z & \Delta\dot{z} & \Delta i \end{bmatrix}^{\mathrm{T}}$.

Linearizing the system near the zero-power equilibrium point $\boldsymbol{x}_0$, defining $\boldsymbol{w} = \begin{bmatrix} \Delta x_1 & \Delta x_2 & \Delta x_3 \end{bmatrix}^{\mathrm{T}}$, $v = \Delta u$ and the linear analytic equation can be obtained as follows:

$$\begin{cases} \dot{\boldsymbol{w}} = \boldsymbol{A}\boldsymbol{w} + \boldsymbol{B}v \\ y = \boldsymbol{C}\boldsymbol{w} \end{cases}, \tag{7.28}$$

where $\boldsymbol{A} = \begin{bmatrix} 0 & 1 & 0 \\ -\frac{4gL}{C_2} & 0 & \frac{2C_1 g}{T} \\ 0 & \frac{2TL}{C_1 C_2} & \frac{RT}{C_1 L} \end{bmatrix}$, $\boldsymbol{B} = \begin{bmatrix} 0 \\ 0 \\ -\frac{T}{LC_1} \end{bmatrix}$, $\boldsymbol{C} = \begin{bmatrix} 1 & 0 & 0 \end{bmatrix}$. The variables used meet $T = C_2 + (Q - P)\sqrt{C_1 Mg/\varepsilon_e} < 0$ and $L = \sqrt{\frac{C_1 Mg}{\varepsilon_e}}$.

The characteristic equation of the linear system Eq. (7.28) is obtained from the matrix $\boldsymbol{A}$ as follows:

$$\lambda^3 - \frac{RT\lambda^2}{LC_1} - \frac{4gRT}{C_1 C_2} = 0 \tag{7.29}$$

According to the Routh criterion, the linear system Eq. (7.28) is unstable, so the single-point system Eq. (7.24) is unstable at the equilibrium point $\boldsymbol{x}_0$.

(2) Controllability Analysis

From the above analysis, the permanent magnet single-point system is a typical third-order unstable system, the controllability of the system needs to be analyzed before designing the controller.

According to the state matrices $\boldsymbol{A}$ and $\boldsymbol{B}$ of the linear system Eq. (7.28), the controllability matrix $\boldsymbol{P}_c$ of the system is as follows:

$$\boldsymbol{P}_c = \left[\boldsymbol{B}\ \boldsymbol{AB}\ \boldsymbol{A}^2\boldsymbol{B}\right] = \begin{bmatrix} 0 & 0 & \frac{-2g}{L} \\ 0 & \frac{-2g}{L} & -\frac{2RTg}{L^2 C_1} \\ -\frac{T}{LC_1} & -\frac{RT^2}{L^2 C_1^2} & -\left(\frac{4gT}{C_1 C_2} + \frac{R^2 T^3}{L^3 C_1^3}\right) \end{bmatrix} \tag{7.30}$$

Obviously, the matrix $\boldsymbol{P}_c$ is of full rank, so the permanent magnet single-point system is fully controllable.

7.5 Design of Zero-Power Controller Based on State Feedback

After the modeling and analysis of PEM hybrid levitation system, a zero-power controller with state feedback plus current integration is designed, the stability conditions of the closed-loop system are analyzed, and its control effect is simulated.

7.5.1 *Design of State Feedback Controller*

According to the system controllability analysis, it is known that the permanent magnet single-point system can be stabilized by the state feedback control. However, the current integration term is necessary to achieve a real zero-power control, either directly or indirectly.

In the controller design of the traditional EMS-type maglev system, the state feedback control is usually used to ensure the dynamic performance of the system. And the gap integration term is used as an additional part to eliminate the gap static error. Based on the experience, the current integration term is taken as an additional part to eliminate the current static error, so the state feedback control law can be expressed as:

$$u = u_0 + k_{\mathrm{p}}(z - z_0) + k_{\mathrm{d}}\dot{z} + k_{\mathrm{c}}(i - i_0) + k_{\mathrm{i}}\int (i - i_0)\mathrm{d}t \tag{7.31}$$

where u is the control voltage, u_0 is the steady-state control voltage. k_p, k_d, k_c are the feedback coefficients for the gap, gap differential, and current states, respectively; k_i is the current integration coefficient; z_0 and i_0 are the steady-state gap value and current value, respectively.

The controller shown in Eq. (7.31) consists of two parts: the state feedback loop and the integration loop. The state feedback loop maintains the stability and dynamic performance of the system, and the integration loop achieves the goal of zero static error of the current. Therefore, the role of the integration loop can be ignored when analyzing the stability of the closed-loop system.

Near the zero-power equilibrium point ($i_0 = 0$), the state feedback of the control law obtained after the linearization process is described as follows:

$$v = \begin{bmatrix} k_p & k_d & k_c \end{bmatrix} \boldsymbol{w} \tag{7.32}$$

Substituting Eq. (7.32) into the Eq. (7.33), the characteristic matrix $\boldsymbol{A}_c$ of the closed-loop system is as follows:

$$\boldsymbol{A}_c = \begin{bmatrix} 0 & 1 & 0 \\ -\frac{4gL}{C_2} & 0 & \frac{2C_1 g}{T} \\ -\frac{k_p T}{C_1 L}\frac{2LT}{C_1 C_2} & -\frac{k_d T}{LC_1}\frac{RT}{LC_1} & -\frac{k_c T}{LC_1} \end{bmatrix} \tag{7.33}$$

Solving the characteristic equation of the matrix $\boldsymbol{A}_c$ leads to:

$$\lambda^3 - \frac{T(R-k_c)}{LC_1}\lambda^2 + \frac{2gk_d}{L}\lambda + \frac{2k_p g}{L} - \frac{4gT}{C_1 C_2}(R-k_c) = 0 \tag{7.34}$$

According to the Routh-Hurwitz criterion, to make the closed-loop system stable, the symbols in the first column of the Routh table should remain unchanged, there exists:

$$\begin{cases} -\dfrac{T(R-k_c)}{LC_1} > 0 \\ \dfrac{2gk_d}{L} + \dfrac{2k_p g C_1}{T(R-k_c)} - \dfrac{4gL}{C_2} > 0 \\ \dfrac{2k_p g}{L} - \dfrac{4gT}{C_1 C_2}(R-k_c) > 0 \end{cases} \tag{7.35}$$

Simplifying the above equation, the value range of the stable feedback coefficient of the closed-loop system is as follows:

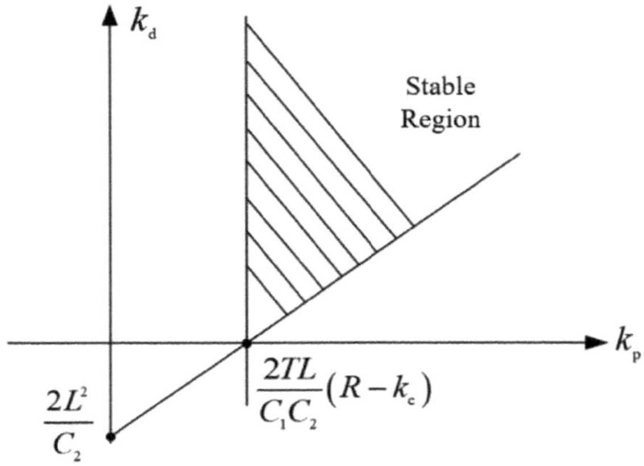

Fig. 7.9 The constraints relationship between the gap feedback coefficient and the differential feedback coefficient

$$\begin{cases} k_c < R \\ k_d > \dfrac{2L^2}{C_2} - \dfrac{k_p L C_1}{T(R-k_c)} \\ k_p > \dfrac{2TL}{C_1 C_2}(R-k_c) \end{cases} \tag{7.36}$$

From the above equation, the constraint relationship between the gap feedback coefficient and the differential feedback coefficient can be obtained as shown in Fig. 7.9.

According to the above analysis, the state feedback controller can achieve the closed-loop stability of the system near the equilibrium point. In the selection of feedback coefficients, the current feedback coefficient k_c should be less than the coil resistance R. Otherwise, no matter how to adjust the parameters k_p or k_d cannot achieve closed-loop system stability. The gap feedback coefficient k_p and the gap differential feedback coefficient k_d have a mutual constraint relationship, and the value of k_p must be larger than a certain value to meet the positive stiffness requirements of the closed-loop system. The value of k_d should at least ensure larger than zero. Otherwise, no matter how to adjust k_p cannot realize the stability of the closed-loop system.

7.5.2 Simulation Analysis of State Feedback Controller

The following is the simulation analysis of the performance of the zero-power controller based on the nonlinear simulation model of the permanent magnet single-point system. The selection of model parameters is based on the structure of the electromagnet and parameter constraints.

(1) Levitating Characteristics

The levitating method of the permanent magnet single-point system is the slow start method, and the variation of the corresponding gap and current during the levitating process are shown in Fig. 7.10.

(2) Landing Characteristics

The landing mode of the permanent magnet single-point system is slow landing, and the variation of gap and current are shown in Fig. 7.11.

(3) Variable-Load Characteristics

Under the premise of stable levitation, the system load is reduced by 10% of the rated load at 10 s and increased by 10% at 15 s. The variations of the corresponding levitation gap and current are shown in Fig. 7.12.

As can be seen from Fig. 7.12, when the load decreases, the current value decreases rapidly at first and then gradually decreases to zero by the effect of current integration.

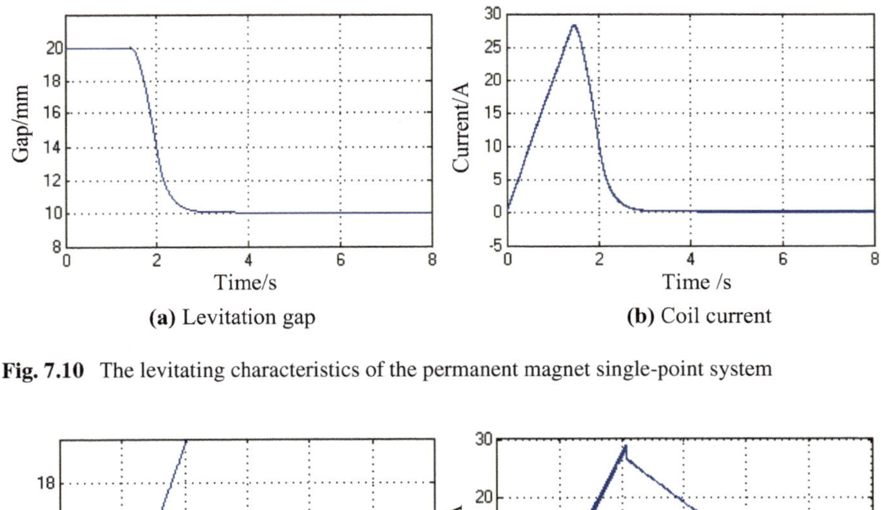

(a) Levitation gap **(b)** Coil current

Fig. 7.10 The levitating characteristics of the permanent magnet single-point system

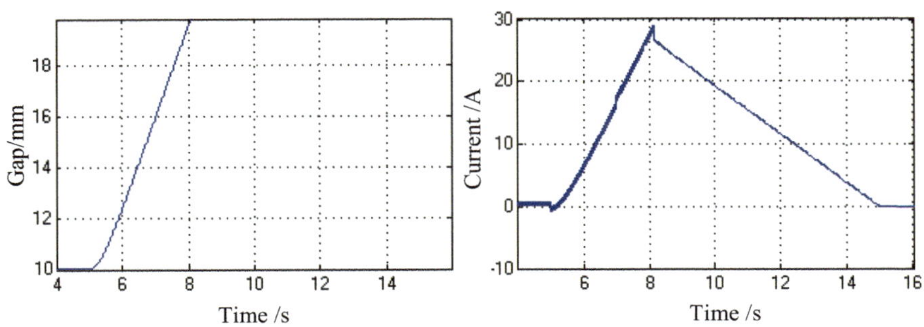

(a) Levitation gap **(b)** Coil current

Fig. 7.11 The landing characteristics of the permanent magnet single-point system

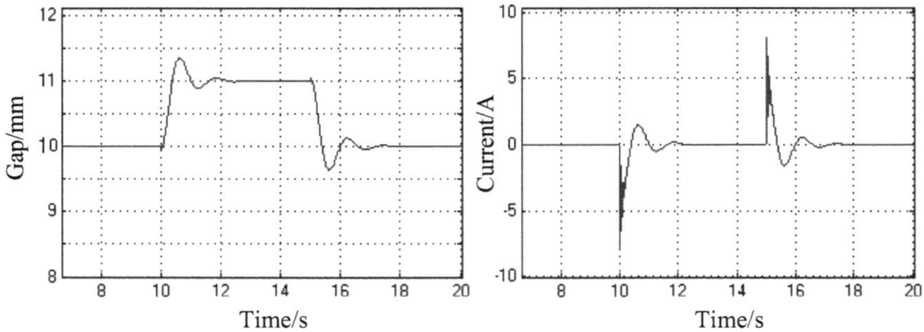

(a) Levitation gap **(b)** Coil current

Fig. 7.12 The variable-load characteristics of the permanent magnet single-point system

Fig. 7.13 The gap and current waveforms of the system under load variation

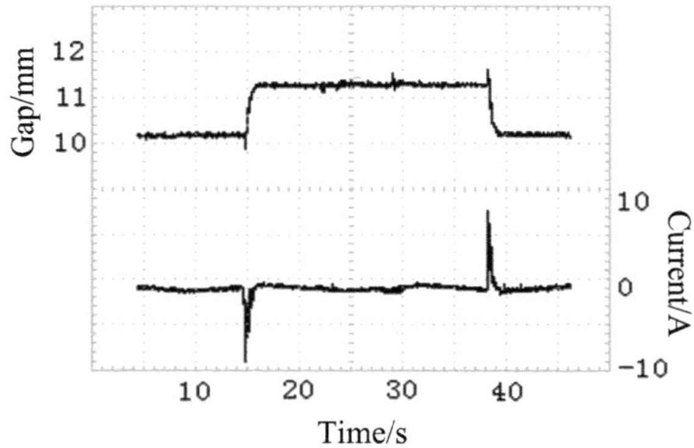

When the load increases, the current increases rapidly at first and then gradually goes to zero by the effect of current integration. The simulation result is consistent with the experimental conclusion shown in Fig. 7.13.

The above simulation experiments show that the zero-power controller based on state feedback can ensure the stability of the system in the process of levitating and landing, steady-state levitation, and automatically adjust the gap to ensure the levitation current is zero under the condition of load variation.

7.6 Experimental Study of PEM Hybrid Levitation System

The purpose of building a PEM hybrid levitation system is to conduct the experimental research and the verification of the control algorithm. The following experimental test research is carried out from two perspectives: the dual maglev bogie and the complete vehicle.

7.6.1 Experimental Test of the Dual Maglev Bogie

The experimental research based on the dual maglev bogie is aimed at solving the difficulties existing in the engineering application of the PEM hybrid levitation technology. The composition of the test platform and the functions of each part are firstly introduced, and then the experiments are carried out from two aspects of levitation control and levitation power consumption.

(1) The Test Platform of the Dual Maglev Bogie Based on the PEM Hybrid Levitation

Figure 7.14 shows the physical image of the dual maglev bogie test platform of the PEM hybrid levitation. The test platform is built on the base of the dual maglev bogie system developed during the 863 program of the 10th Five-Year Plan, using a self-developed PEM and levitation controller. The test platform is equivalent to the lower functional structure of half a maglev train, reflecting the structural characteristics and motion characteristics of the maglev train. And, the research results can be applied to the whole vehicle.

Figure 7.15 is a schematic diagram of the lower structure of the PEM hybrid dual maglev bogie experimental platform, which consists of two maglev bogies connected by two intermediate magnets.

The PEM hybrid dual maglev bogie experimental platform mainly includes a maglev bogie, PEM, levitation controllers, levitation sensors, and other auxiliary equipment, and the composition and function of each part are introduced below.

Fig. 7.14 The physical image of the dual maglev bogie test platform of the PEM hybrid levitation

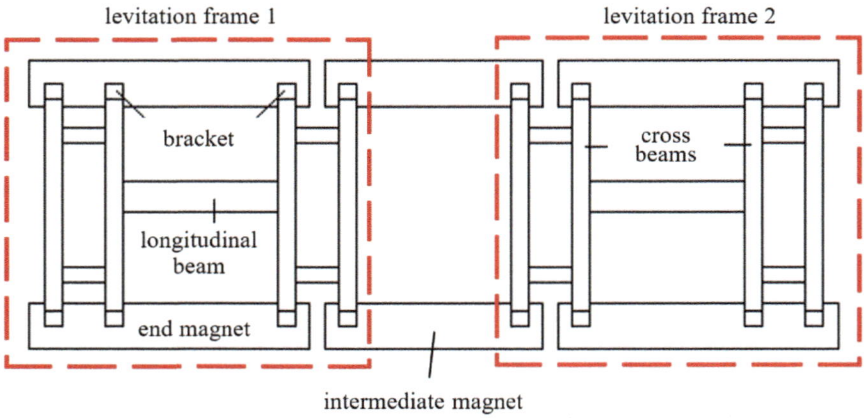

Fig. 7.15 The schematic diagram of the lower structure of the PEM hybrid dual maglev bogie experimental platform

As shown in Fig. 7.15, each maglev bogie consists of 4 cross beams, 1 longitudinal beam and 8 bracket arms, and its main function is to load magnets and transfer the levitation force, the guidance force and the propulsion force to the secondary system and carriage. Among them, 2 adjacent cross beams form the crossbeam frame to support the secondary system. The longitudinal beam connects the 2 crossbeam frames and transmits the propulsion force to the carriage through the propulsion rod. The bracket arms are used to install the levitation magnets and the guidance magnets.

A total of six PEM are installed on the experimental platform, four of which are end magnets, respectively, installed on both sides of the two maglev bogies. And the other two are intermediate magnets, which are used to connect the two maglev bogies. The length of the intermediate magnets is shorter than the length of the end magnets.

The PEM is used to provide all the levitation force required by the experimental platform. The platform has a self-weight of about 14 t and a maximum load capacity of 36 t.

The levitation controllers are used to achieve stable levitation control of the PEM. Three levitation controllers are needed for each end magnet and two levitation controllers are needed for each intermediate magnet, so there are sixteen levitation controllers in the experimental platform.

The levitation sensors are used to obtain the state information required for levitation control and transmit this information to the levitation controllers in a fixed format. Each levitation controller requires two levitation sensors, so there are thirty-two levitation sensors in the experimental platform.

Other auxiliary equipment, including secondary system, carriage, battery, debugging network, diagnostic network, air circuit, etc. In the experimental platform, the secondary system consists of an air spring, a hanger rod, a rocker, and other mechanisms, which are used to connect the maglev bogie and the carriage. The air circuit is used to adjust the distance from the carriage to the maglev bogie. The battery is installed in two groups under the bottom plate of the carriage. One set of the debugging network and one set of the diagnostic network are used to debug and diagnose the levitation controller, respectively.

(2) Levitation Control Experiment

The relevant research in this subsection carries out levitation control experiments with the help of the PEM hybrid levitation dual maglev bogie experimental platform, and conducts experiments on its steady-state levitation performance and dynamic levitation performance under the conditions of static levitation and low-speed operation of the experimental platform, respectively. Since the PEM is used on the experimental platform, any joint structure can be selected for the experiments during testing, and all experiments are done at the same location.

1. Steady-State Levitation Performance Test

In the experiment, firstly, control the experimental platform to levitate stably at a certain position in the depot, then select a levitation magnet randomly, and test the waveform of its levitation current and levitation gap with an oscilloscope, so as to verify the steady-state levitation performance of the experimental platform.

When the experimental platform is statically levitated, the levitation current and levitation gap only fluctuate in a small range near the steady-state value, where the steady-state current is 6 A, which is equal to the current setting value. The steady-state gap is 10 mm. The experimental platform has a good steady-state levitation performance and achieves the expected levitation control effect.

In this experiment, the experimental platform is also tested for its steady-state levitation performance by conducting static levitation tests at other locations inside and outside the depot. The results show that the experimental platform has good steady-state levitation performances on the steel beam inside the depot, the concrete beam outside the depot and the turnout, and there is no vehicle-track coupling resonance situation.

2. Dynamic Levitation Performance Test

Under the premise of stable levitation of the experimental platform, two electromagnets installed on the same joint structure are selected for verification. The current setting value is changed from 6 to 8 A by sending commands through the debugging network, and the waveforms of current and gap changes of one of the electromagnets are measured to verify the dynamic levitation performance of the experimental platform.

After changing the current setting value, it shows that the current suddenly decreases from 6 to 4 A and ascends to 5 A and maintains this for approximately 0.3 s, before rising further to 8 A. The gap increases from 10 mm, then decreases, holds for about 0.3 s around 11 mm, and then continues to increase to 11.5 mm. Both the gap and current have two regulation processes because the two levitation controllers in the same joint structure do not receive commands from the commissioning network at the same time.

In summary, there is a mutation of 2 A in the current during the regulation process, and the gap changes smoothly without obvious jitter, so the experimental platform has a good dynamic performance.

(3) Levitation Power Consumption Test Experiment

Due to the use of the energy-saving PEM structure, the levitation power consumption of the experimental platform is significantly reduced. However, during operation, the levitation control system will be disturbed by the track, the propulsion system and so on, resulting in fluctuations in levitation power consumption. In the following, the levitation power consumption of the experimental platform is tested and analyzed under two conditions: static levitation and cornering.

When the experimental platform is levitated statically, only the levitation control system works, it will not be affected by other systems, and reflects the power consumption required to maintain the levitation of the experimental platform. Under the static levitation of the experimental platform, the power supply current waveform of the levitation control system is tested by oscilloscope and current clamp. Since the supply voltage is kept constant at 450 V, the levitation power consumption can be calculated based on the measured supply current.

Figure 7.16a shows the current waveforms measured during the levitation of the experimental platform. As can be seen from the Fig. 7.16a, the current before levitating is about 2.5 A, corresponding to the power consumption of 1.13 kW, which is mainly consumed by the levitation controller and other equipment. At the 2.5 s, the auxiliary contactor inside the levitation controller connects, and then the support capacitor starts to charge, and the current gradually rises to 5 A. After 2 s, the support capacitor finishes charging, and the current drops to 3 A. At the 7.5 s moment, the experimental platform starts to levitate, and the current first increases to 25 A and then slowly decreases to about 7.4 A. Figure 7.16b is the current waveform measured when the experimental platform is steadily levitated. And, it can be seen from the figure that the current fluctuates around the steady-state value of 7.4 A when the experimental platform is in the static levitation state.

According to the experimental results, the average power consumption of the experimental platform in the levitation state is 3.33 kW. And, since the total weight of the experimental platform is about 18 t, the power consumption corresponding to 1 t load is calculated to be less than 200 W, which is lower than the levitation power consumption of the original EMS system of 1 kW/t. Therefore, the energy-saving PEM structure has a significant effect on reducing the levitation power consumption.

To analyze the main source of the disturbance to the levitation control system, the wavelength of the disturbance is calculated according to the bend passage speed and the period of the supply current fluctuation, and the results are shown in Table 7.1. According to the results in the Table 7.1, it is presumed that the disturbance to the levitation control system is mainly caused by the cogging structure of the long-stator track, because the length of one cogging is 86 mm, which is close to the calculated wavelength of the disturbance. Combined with the structure of the propulsion system, the normal force of the propulsion system fluctuates for one cycle after each cogging cycle, and it is the main source of the disturbance to the levitation control system.

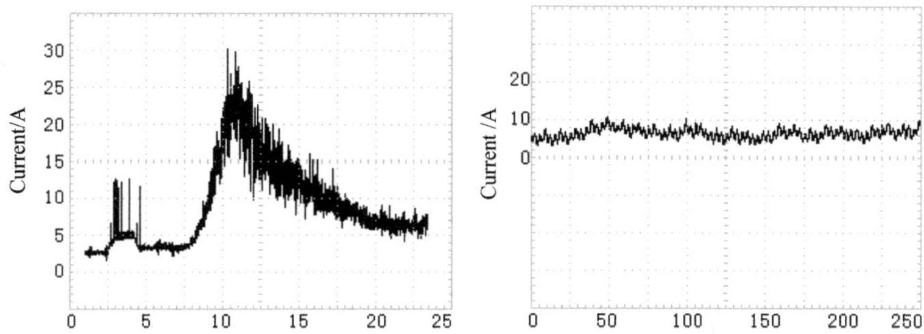

(a) Levitating process **(b)** Levitation process

Fig. 7.16 The current waveform of levitation system under static levitation condition

Table 7.1 The wavelengths of disturbances to the levitation system when passing through bends

Speed of bend passage/(km/h)	Period of fluctuation of supply current/ms	Wavelength of disturbance/mm
3.4	90	85
10	30	83.3

In summary, the levitation power consumption of the PEM hybrid dual maglev bogie system is less than 200 W/t, which is significantly lower than the power consumption of the original EMS system of 1 kW/t. The levitation control system is strongly disturbed by the propulsion system when passing through the bend, which is mainly because the normal force of the propulsion system fluctuates periodically with the long-stator track, and then the transient power consumption of the levitation system shows periodic fluctuations. And the faster the speed through the bend, the greater the frequency and amplitude of fluctuations.

(4) Operation Experiment of Dual Maglev Bogie

To test the feasibility of the engineering application of the PEM hybrid levitation technology, further operation experiments are conducted on the PEM hybrid dual maglev bogie experimental platform on the 1.5 km high-speed test line in Shanghai. With the cooperation of the levitation control system, the guidance control system, the propulsion control system and the speed measurement and positioning system, the experiments are conducted for 3 months of continuous operation for the working conditions of static levitation, trailer propulsion operation and medium power module propulsion operation of the experimental platform. The experimental results show that the dual maglev bogie experimental platform reaches the following performance.

1. The experimental platform can be statically levitated at any position on the steel beam inside the depot and the turnout outside the depot, with good static levitation performance, without the phenomenon of vehicle-track coupling resonance.

2. The longest continuous static levitation of the experimental platform is 24 h. The surface temperature of the coil winding of the PEM is roughly the same as the room temperature, and the levitation control system works normally.
3. The experimental platform is towed by the trailer along the whole line, the levitation control system works normally, and the experimental platform operates smoothly.
4. The experimental platform is towed by the ground medium power module for a total of about 400 km, the levitation control system works normally, and the experimental platform operates smoothly.

The experimental results show that the levitation power consumption of the experimental platform is significantly reduced, the load capacity is significantly improved, the heating problem of the coil winding is significantly improved, and the levitation can be stable on the steel beam and turnout. The experimental results fully verify the feasibility of the technology and lay the foundation for future engineering applications.

7.6.2 Operation Experiment of the Whole Vehicle

(1) Static Test

The comprehensive tests on the levitation system of the maglev train are divided into the following step-by-step process: the magnet performance test of the PEM structure, the static levitation and landing test of the maglev train on the steel track beam in the depot, the dynamic operation test of the maglev train along the track at different speeds, and the basic fault-tolerant control scheme test based on the joint structure. The hybrid levitation electromagnet performance test is done at the National University of Defense Technology, and the test of the PEM hybrid high-speed maglev train is done at the 1.5 km high-speed maglev test line in Shanghai. The test site can be seen in Fig. 1.7. The test results given in this section can illustrate the comprehensive performance of the maglev train levitation system from different perspectives.

A static levitation and landing test of the maglev train is conducted on the steel structure beam in the depot to verify the basic functionality of the system. The test process is to send a levitation command after the power is turned on. After the train is levitated stably at the target position, a landing command is sent. After the levitation command is sent, the current is generated in the levitation electromagnet, the levitation gap slowly decreases to the target value. There is a small fluctuation of electromagnets during steady-state levitation, which is mainly due to the light mass of the steel beam in the depot which is prone to elastic vibration. During the steady-state levitation, the range of the fluctuation of the levitation gap on the left and right sides is $\pm$ 0.28 mm and $\pm$ 0.14 mm, respectively, and the variance of the levitation gap is 0.0132 mm^2 and 0.0038 mm^2, respectively, which shows that the levitation system can maintain good stability in the steady-state. After the landing command is sent, the

Fig. 7.17 The levitation gap of left and right-side electromagnets in joint structure

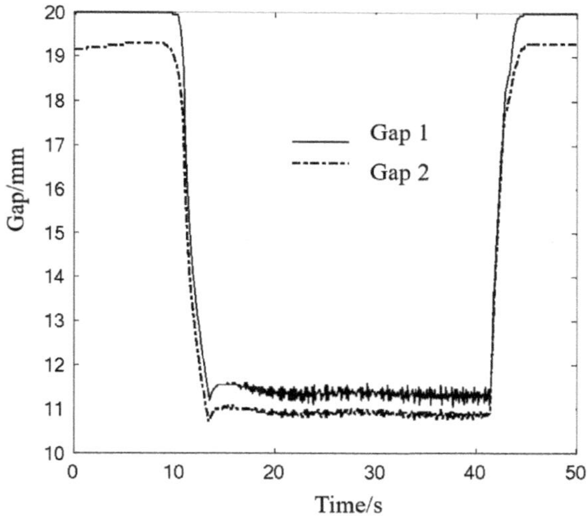

levitation electromagnets start to move downward until the support skid lands on the track, and the levitation gap value slowly increases to the initial value. The deviation of the levitation gaps between the left and right sides during the whole process is mainly caused by the drift of the sensor signal. In terms of levitation current, only the statistical results of current data are given. The maximum levitation current when levitating is less than 30 A, and the average currents of both left and right sides are 11.4 A. The maximum values of current fluctuation of both left and right sides during the steady-state levitation are ± 2.8 A and ± 1.37 A, respectively, and the variances of current fluctuation are 2.0 A^2 and 0.5 A^2, respectively (Fig. 7.17).

(2) Operation Test

Based on the static levitation landing test of the levitation system, the maglev train is towed on the cement track beam outside the depot for dynamic operation test. The levitation system can be kept stable during the train operation. The fluctuation of the levitation system is very small and negligible when the train is levitated on the concrete structure beam outside the depot. The fluctuation of the levitation gap is always limited to 3 mm, and the fluctuation of the levitation electromagnet current is within 10 A during the low-speed propulsion operation of the maglev train. After the operation test, the surface temperature of the levitation electromagnet is measured, the temperature is kept below 40 °C, and the problem of serious heating of the electromagnet is solved.

7.7 Conclusion

This chapter firstly analyzes the PEM hybrid levitation technology to obtain the requirements of PEM hybrid levitation high-speed maglev trains. On this basis, the design requirements for PEM are introduced from three aspects: the energy conservation, enhancing the utilization of permanent magnets, and improving the controllable performance of PEM. Then, based on the above design requirements and the finite element calculation results, the structural design of PEM was completed. Finally, by analyzing the model of PEM hybrid levitation system, a zero-power controller based on state feedback is designed, and the feasibility of the control scheme is verified through the test on the PEM hybrid levitation dual maglev bogie and the whole vehicle.

Bibliography

1. Chen H (2009) Research on design and control issues of permanent magnet EMS maglev train suspension magnets. Ph.D. dissertation, National University of Defense Technology Doctor Degree, Changsha
2. Chen H, Li Y, Chang W (2008) Research on levitation stiffness of hybrid suspension system. Proc CSEE 28(27):148–152
3. Chen H, Li Y, Long Z et al (2010) Optimal design and characteristic analysis of hybrid magnet in maglev system. J Syst Simul 22(4):837–840
4. Ding S (2022) 600 km/h high-speed maglev transportation system. Shanghai Scientific and Technical Publishers, Shanghai
5. Hu C (2011) Research on design and control issues of permanent magnet electromagnetic hybrid maglev train suspension magnets. Ph.D. dissertation, National University of Defense Technology Doctor Degree, Changsha
6. Hu Y, Chen B, Long Z (2023) 3D analytical modeling of the ultra-high-speed permanent magnet electrodynamic levitation system and analysis of electromagnetic force characteristics. Electr Drive Locomot 2023(6):20–30
7. Hu C, Li Y, Yan Y (2006) Design and analysis of the hybrid magnet in EMS maglev vehicles. J Nat Univ Defense Technol 28(5):94
8. Hu Y, Wang Z, Long Z (2023) Single-objective performance optimization of PM EDS system. J Traffic Transport Eng 23(6):180–192
9. Hu C, Zhang X, Li Y et al (2010) Analysis of control characteristics of electromagnetic and permanent magnet hybrid suspension system. Electric Drive Locomot 2010(2):18–21
10. Li Y, Hu C, Long Z (2005) Stability analysis and controller design of hybrid EMS maglev system. In: The 8th International Symposium on Magnetic Suspension Technology, pp 74–78
11. Li W, Jian X, Kunlun Z et al (2005) Research of hybrid suspension system made of permanent- and electromagnets. J China Railway Soc 27(3):50–54
12. Li Y, Zhang X, Hu C et al (2011) Design of zero-power hybrid magnet with permanent magnets and electromagnets in EMS maglev vehicles. Electric Drive Locomot 2011(5):30–32, 43
13. Liu Huaqing Compile (1995) The German maglev train-Transrapid. University of Electronic Science and Technology Press, Sichuang
14. Long X (2012) Research on suspension control technology of permanent magnet electromagnetic hybrid low-speed maglev trains. Ph.D. dissertation, National University of Defense Technology Doctor Degree, Changsha
15. Long Z, Li X, Cheng H (2023) Research on permanent-electromagnetic levitation technology and application. Shanghai Scientific and Technical Publishers, Shanghai

16. Long Z, Wang Z, Cheng H et al (2018) A novel design of electromagnetic levitation system for high-speed maglev train. Innov Transp Syst Technol 4(3):212–224
17. Long Z, Zhou X, Yang Q et al (1996) System study of permanent magnet maglev train. Electric Drive Locomot 3:8–11
18. Wang Z (2019) Fault diagnosis and tolerant control for high speed maglev train suspension system. Ph.D. dissertation, National University of Defense Technology Doctor Degree, Changsha
19. Wang Z, Long Z, Li X (2018) Levitation control of permanent magnet electromagnetic hybrid suspension maglev train. Proc Inst Mech Eng Part I J Syst Control Eng 232(3):315–323
20. Wang Z, Long Z, Li X (2019) Fault analysis and tolerant control for high speed PEMS maglev train end joint structure with disturbance rejection. J Electr Eng Technol 14(3):1357–1366
21. Wang Z, Long Z, Li X (2021) Fault tolerant control for joint structure in PEMS high speed maglev train. Asian J Control 23(1):486–498
22. Wang Z, Long Z, Luo J et al (2020) Data driven state monitoring of maglev system with experimental analysis. IEEE Access 8:79104–79113
23. Wu X (2003) Maglev train. Shanghai Scientific and Technical Publishers, Shanghai
24. Xu Z (2005) Research on the control strategy of controlled PM on a maglev model vehicle. Ph.D. dissertation, Institute of Electrical Engineering, Chinese Academy of Sciences, Beijing
25. Zhang K (2021) High-speed maglev railway technology. China Railway Publishing House, Beijing
26. Zhang Y, Chen H, Wu Z et al (2008) Magnet structure optimal design for hybrid maglev train. Electric Drive Locomot 2008(5):30–32

Chapter 8
Fault Diagnosis of High-Speed PEM Hybrid Levitation System

The PEMS maglev train exhibits distinct advantages in comparison to its EMS counterpart; however, the inclusion of permanent magnets also introduces new challenges. The presence of these permanent magnets gives rise to the perilous possibility of the "adsorption" of the levitation point, whereby the hybrid electromagnets become affixed to the track. As such, the levitation control system assumes paramount importance in ensuring the secure operation of the maglev train. In the event of system failure, it becomes imperative to recalibrate the levitation control algorithm to ensure the continuous safe and stable operation of the maglev train. Consequently, an in-depth exploration of fault diagnosis and fault-tolerant control technology pertaining to the levitation system of the PEM hybrid levitation train is exceedingly indispensable in augmenting the reliability and safety of the levitation system. The gap sensor, accelerometer, and current sensor are used by the levitation system to get the system's gap, acceleration, and current signals. The controller uses the processed signals to evaluate the corresponding control quantity and to control the actual system. Nonetheless, practical engineering poses challenges such as external interferences and sensor malfunctions, which render the acquired sensor signals an imperfect reflection of the true values. This chapter deals with fault diagnosis of maglev train levitation systems. Model-based, data-driven-based, and signal-based fault diagnosis methods are tailored to address the typical sensors and actuators faults encountered in maglev system.

8.1 Kalman Filter-Based Sensor Fault Diagnosis of the Levitation System

Kalman filter has a wide range of applications in engineering. The advantage of Kalman filter is that it is a powerful tool for estimating the state of a system when there is uncertainty in the measurements. It can handle noisy measurements and provide

Z. Long et al., *High-Speed Maglev Train's Levitation and Guidance Control*, Advances in High-speed Rail Technology, https://doi.org/10.1007/978-981-97-2309-6_8

accurate estimates of the system state. It is also computationally efficient and can be used in real-time applications. Additionally, the flexibility of the Kalman filter allows for easy customization to accommodate diverse system and measurement models. As Kalman filtering is a well-researched topic, this section aims to demonstrate its specific application by integrating it with a specific sensor fault.

8.1.1 Sensor Fault Detection Method Based on Kalman Filter Bank

Consider a discrete system model with sensor fault, with discrete time instance k:

$$\boldsymbol{X}(k) = \boldsymbol{A}\boldsymbol{X}(k-1) + \boldsymbol{B}\boldsymbol{U}(k-1) + \boldsymbol{G}\boldsymbol{W}(k-1) \tag{8.1}$$

$$\boldsymbol{Y}_F(k) = \left(\boldsymbol{C} + \boldsymbol{C}_f\right)\boldsymbol{X}(k) + \boldsymbol{V}(k) \tag{8.2}$$

In this context, the state vector, denoted by $\boldsymbol{X}$, has n dimensions. The input control vector, denoted by $\boldsymbol{U}$, has l dimensions. The output measurement vector, denoted by $\boldsymbol{Y}$, has m dimensions. The process noise, denoted by $\boldsymbol{W}$, is a p-dimensional vector with a covariance matrix $\boldsymbol{Q}$. The measurement noise, denoted by $\boldsymbol{V}$, is a m-dimensional vector with a covariance matrix $\boldsymbol{R}$. The state transition matrix, denoted by $\boldsymbol{A}$, the control matrix, denoted by $\boldsymbol{B}$, the coefficient matrix for the error, denoted by $\boldsymbol{G}$, and lastly, the measurement matrix, denoted by $\boldsymbol{C}$, all have proper dimensions. The notion denoted by $\boldsymbol{C}_f(k) = \left[0^{\mathrm{T}} \ldots 0^{\mathrm{T}}\ \boldsymbol{C}_{fi}(k)^{\mathrm{T}} 0^{\mathrm{T}} \ldots 0^{\mathrm{T}}\right]^{\mathrm{T}}$ symbolizes the state of sensor malfunction. In this way, the condition of the i_{th} sensor is depicted by $\boldsymbol{C}_f(k)$:

$$\begin{cases} \boldsymbol{C}_{fi}(k) = 0 \text{ sensor } i \text{ is normal} \\ \boldsymbol{C}_{fi}(k) \neq 0 \text{ sensor } i \text{ is failure} \end{cases} \tag{8.3}$$

Constructing the Kalman filter for the system model (8.1) and (8.2), the residual sequence can be obtained as:

$$\begin{aligned} \boldsymbol{r}_F(k) &= \boldsymbol{Y}_F(k) - \boldsymbol{C}(k)\boldsymbol{X}(k|k-1) \\ &= \left(\boldsymbol{C}(k) + \boldsymbol{C}_f(k)\right)\boldsymbol{X}(k) + \boldsymbol{V}(k) - \boldsymbol{C}(k)\boldsymbol{X}(k|k-1) \\ &= \boldsymbol{C}_f(k)\boldsymbol{X}(k) + \boldsymbol{Y}(k) - \boldsymbol{C}(k)\boldsymbol{X}(k|k-1) \\ &= \boldsymbol{C}_f\boldsymbol{X}(k) + \boldsymbol{r}(k) \end{aligned} \tag{8.4}$$

Since the sensor i is faulty, $\boldsymbol{C}_{fi}(k) \neq 0$, then:

$$
\begin{aligned}
\mathrm{E}[\boldsymbol{r}_F(k)] &= \mathrm{E}[\boldsymbol{C}_f\boldsymbol{X}(k)] + \mathrm{E}[\boldsymbol{r}(k)] \\
&= \mathrm{E}[\boldsymbol{C}_f\boldsymbol{X}(k)] \\
&\neq 0
\end{aligned} \tag{8.5}
$$

It can be seen from the above that when one of the sensor fails, the corresponding residual statistics will be changed. In this way, the fault can be detected, however, it is impossible to determine which sensor is faulty. Therefore, in order to locate the sensor fault, it is necessary to decouple the matrices $\boldsymbol{A}$, $\boldsymbol{C}$, and $\boldsymbol{K}(k)$ for each state variable. At this time, it can be ensured that the abnormality of the i_{th} component of the residual corresponds to the i_{th} sensor's fault.

To solve the problem that a single Kalman filter cannot locate faults, a method of fault isolation using a Kalman filter bank is proposed. The specific method is: each measurement signal corresponds to a Kalman filter, then the i_{th} Kalman filter corresponds to the i_{th} sensor fault, and the corresponding measurement matrix is $\boldsymbol{C}_i$, which is the i_{th} row of $\boldsymbol{C}$.

According to the properties of the Kalman filter, the recursive formula corresponding to the i_{th} measurement signal can be obtained as follows:

$$
\hat{\boldsymbol{X}}_i(k|k-1) = \boldsymbol{A}\hat{\boldsymbol{X}}_i(k-1|k-1) + \boldsymbol{B}u(k-1) \tag{8.6}
$$

$$
\boldsymbol{P}_i(k|k-1) = \boldsymbol{A}\boldsymbol{P}_i(k-1|k-1)\boldsymbol{A}^T + \boldsymbol{G}\boldsymbol{Q}(k-1)\boldsymbol{G}^T \tag{8.7}
$$

$$
\boldsymbol{r}_i(k) = \boldsymbol{Y}_i(k) - \boldsymbol{C}_i\hat{\boldsymbol{X}}_i(k|k-1) \tag{8.8}
$$

$$
\boldsymbol{K}_i(k) = \boldsymbol{P}_i(k|k-1)\boldsymbol{C}_i^T[\boldsymbol{C}_i\boldsymbol{P}_i(k|k-1)\boldsymbol{C}_i^T + \boldsymbol{R}_i(k)]^{-1} \tag{8.9}
$$

$$
\hat{\boldsymbol{X}}_i(k|k) = \hat{\boldsymbol{X}}_i(k|k-1) + \boldsymbol{K}_i(k)r_i(k) \tag{8.10}
$$

$$
\boldsymbol{P}_i(k|k) = [\boldsymbol{I} - \boldsymbol{K}_i(k)\boldsymbol{C}_i]\boldsymbol{P}_i(k|k-1) \tag{8.11}
$$

$$
\boldsymbol{Y}_i(k) = \boldsymbol{C}_i\boldsymbol{X}(k) + \boldsymbol{V}_i(k) \tag{8.12}
$$

The residual and estimation state vectors obtained from the respective i-th Kalman filters are defined as $\boldsymbol{r}_i(k)$ and $\hat{\boldsymbol{X}}_i(k|k)$, respectively. In the event of a fault occurring in sensor i, the vector $\boldsymbol{C}_i$ changed to be $\boldsymbol{C}_i+\boldsymbol{C}_{fi}(k)$ accordingly. At this time, the methodology employed mirrors that of the single Kalman filter approach, leading to a change in the statistical characteristics of the residuals $\mathrm{E}[\boldsymbol{r}_i(k)] \neq 0$. When sensor $j(j \neq i)$ experiences a fault, each step of the Kalman filter recursion formula corresponding to sensor i remains unaffected by the fault information $\boldsymbol{C}_{fj}(k)$. Consequently, unbiased estimation $\mathrm{E}[\boldsymbol{r}_i(k)] = 0$ is obtained.

Based on the aforementioned analysis, it becomes evident that the residual $\boldsymbol{r}_i(k)$ of a faulty system possess distinct statistical characteristics prior to and following the occurrence of a fault. By examining the statistical feature, whether a sensor has encountered a malfunction can be effectively determined. Hence, the following sensor failure detection rule is derived:

$$\begin{cases} \mathrm{E}[\boldsymbol{r}_i(k)] = 0 \text{ sensor } i \text{ is fault free;} \\ \mathrm{E}[\boldsymbol{r}_i(k)] \neq 0 \text{ sensor } i \text{ is faulty.} \end{cases} \tag{8.13}$$

8.1.2 Simulation Analysis

This section deals with how to combine the Kalman filter method mentioned in the previous section with the gap sensors, accelerometers, and current sensors in the actual levitation system to diagnose whether they are faulty in real-time.

By employing the state space methodology, the triumvirate of state variables can be linked to the gap sensor, the accelerometer, and the current sensor, individually:

$$\boldsymbol{X} = (x_1\ x_2\ x_3)^T = (z_l\ \dot{z}_l\ i_l)^T \tag{8.14}$$

The state space equation, following the linearization process of a levitation point at the state of equilibrium, is defined as follows:

$$\begin{cases} \dot{\boldsymbol{X}} = \boldsymbol{AX} + \boldsymbol{BU} \\ \boldsymbol{Y} = \boldsymbol{CX} \end{cases} \tag{8.15}$$

where the matrix are as follows:

$$\boldsymbol{A} = \begin{bmatrix} 0 & 1 & 0 \\ \frac{k_b - k_s}{m} & 0 & -\frac{k_a}{m} \\ 0 & \frac{k_d}{k_c} & -\frac{R}{k_c} \end{bmatrix}, \quad \boldsymbol{B} = \begin{bmatrix} 0 \\ 0 \\ \frac{1}{k_c} \end{bmatrix} \\ \boldsymbol{C} = \begin{bmatrix} 1\ 0\ 0 \end{bmatrix} \tag{8.16}$$

with m the mass of the electromagnet, R the resistance of the windings, k_a, k_b, k_c, k_d, and k_s the coefficients related with system structures.

Take the sampling time as $T = 0.001$ s, the state equation can be obtained as:

$$\dot{\boldsymbol{X}} = \begin{bmatrix} 1.0031 & 0.001 & 0 \\ 6.1006 & 0.9974 & -0.0028 \\ 12.1461 & 3.948 & 0.9519 \end{bmatrix} \boldsymbol{X} + \begin{bmatrix} 0 \\ 0 \\ 0.0089 \end{bmatrix} \boldsymbol{U} \\ \boldsymbol{Y} = \begin{bmatrix} 1\ 0\ 0 \end{bmatrix} \boldsymbol{X} \tag{8.17}$$

The designed Kalman filter is added to the levitation system model for simulation, and the gap deviation, speed deviation, and current deviation are monitored, respectively. Under normal operation of the levitation system sensors, the output signals obtained by the Kalman filter are residual signals with zero-mean characteristics. Figure 8.1 shows the residual signal values obtained by the levitation system under normal working conditions. The magnitude of fluctuations remains confined within tolerable limits, specifically not surpassing 0.0005, 0.005, and 0.05 distinctly. Through the statistics of the levitation system data, the residual signal thresholds are set as $\pm$ 0.001, $\pm$ 0.02, and $\pm$ 2, respectively. Consequently, such values reside well beneath the threshold indicative of sensor malfunction.

In the following, the fault diagnosis effect of the Kalman filter is tested by simulating the fault states of the gap sensor, accelerometer, and current sensor, respectively. As shown in Figs. 8.2, 8.3 and 8.4, the failure of the sensor is simulated by cutting off the power of different sensors at 0.1 s. During the simulation procedure, if the residual signal of a particular sensor undergoes substantial variations, manifesting itself as a deviation beyond the prescribed fault threshold, it can be concluded that the sensor is deemed faulty.

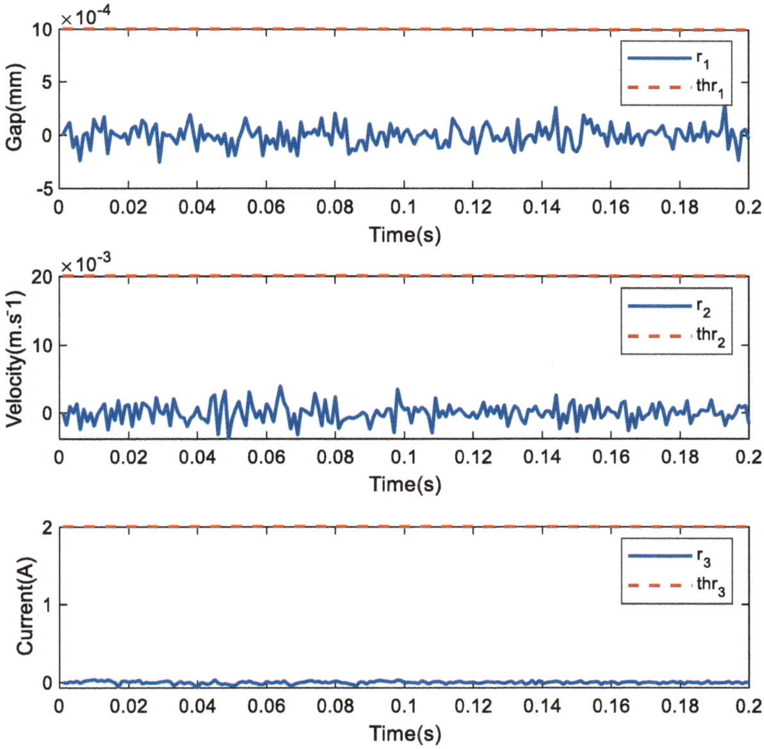

Fig. 8.1 Residual signal under fault-free condition

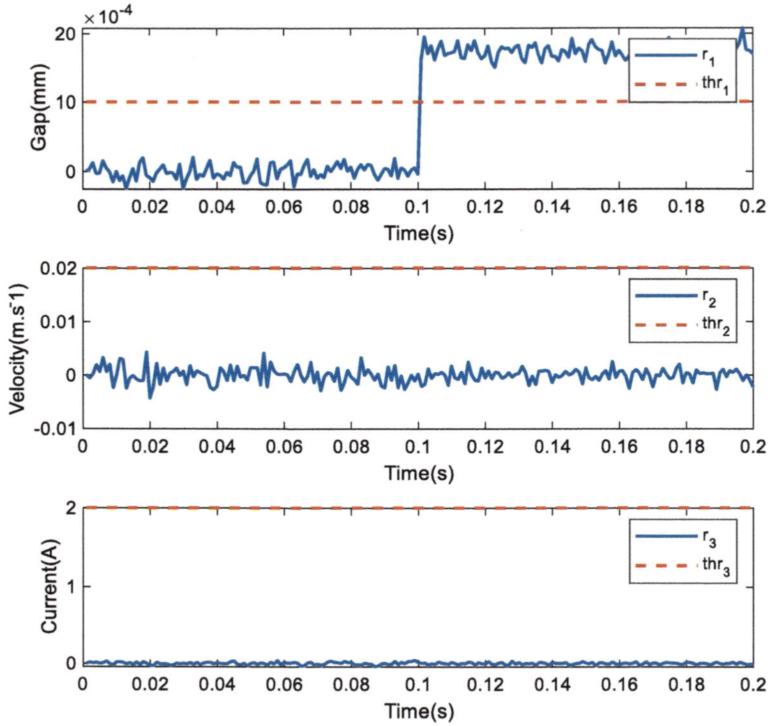

Fig. 8.2 Residual signals before and after a gap sensor failure

As shown in Fig. 8.2, the residual value of the gap deviation rapidly increases from zero mean to 0.002, which exceeds the set fault threshold, indicating that the gap sensor has failed. However, during this period, the residual values for speed deviation and current deviation remained close to their zero-mean values both before and after the occurrence of the gap sensor fault. As such, the fault did not have any impact on these residual values. Based on the previously established criteria, a faulty gap sensor can be immediately diagnosed. Similarly, it can be seen from the speed deviation residual signal in Fig. 8.3 and the current deviation residual signal in Fig. 8.4 that when the fault occurs at 0.1 s, the residual signal suddenly exceeds the fault threshold from the zero-mean value, and the corresponding sensor is immediately diagnosed as faulty.

It can be seen from the simulation results that when a single sensor fails, the sensor residual will suddenly increase from the zero-mean value state and exceed the set threshold. The Kalman filter fault diagnosis system can detect and isolate the sensor fault of the system.

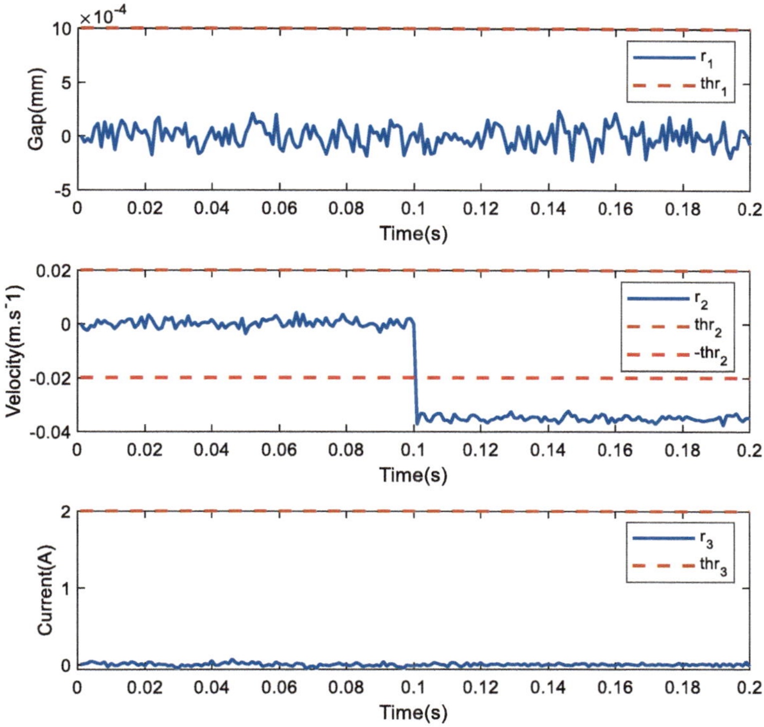

Fig. 8.3 Residual signals before and after an accelerometer failure

8.2 Data-Driven Fault Diagnosis of Levitation System

It is of great importance to monitor and diagnose faults of levitation systems in real-time for safety operation purposes. Although the mechanism of the levitation system is well studied, unavoidable noises and disturbances make it difficult to get an accurate system model. Considering the system noise and model uncertainty, a fault diagnosis method of the maglev train levitation system based on system input and output data is proposed, which combines the model-based fault diagnosis with the data-based implementation method to maximize the advantages of the model and data. Then, a secondary confirmation of fault isolation results is proposed to improve the reliability of fault diagnosis. Furthermore, a subsequent processing of fault data is proposed to enrich the fault analysis methodologies.

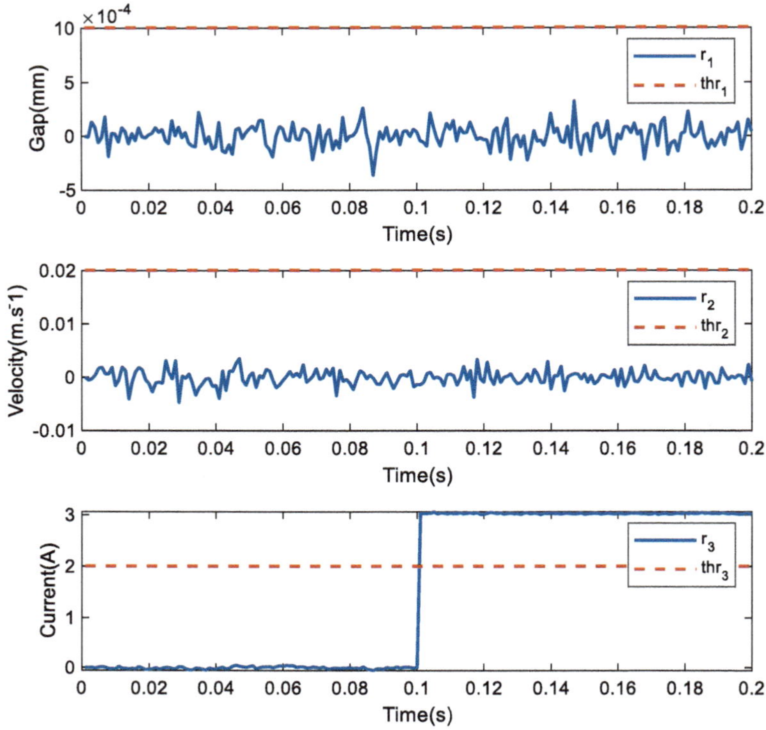

Fig. 8.4 Residual signals before and after a current sensor failure

8.2.1 Data-Driven Fault Diagnosis Architecture

Classical fault diagnosis includes two parts: fault detection and fault isolation. The function of fault detection is to judge whether the system is at fault according to the real-time status of the system, and fault isolation is to judge the type of fault (location, amplitude, etc.) after the fault is detected. The fault isolation results assume a pivotal role in the implementation of active fault-tolerant control, for any inaccuracies could potentially trigger system malfunction and jeopardize overall system safety. To bolster the reliability of the fault isolation results, an enhanced fault diagnosis architecture is introduced, leveraging the combined influence of residual signals and data integration, as is shown in Fig. 8.5. There are two processes added, the secondary confirmation of fault isolation results and the subsequent analysis and processing of fault data. On the one hand, the reliability of fault isolation results is improved; on the other hand, the analysis and evaluation methodologies of fault are enriched.
For the additional two parts in the fault diagnosis process, the problems to be addressed in each part are as follows.

Secondary confirmation of fault isolation results: after the fault occurs, the statistical characteristics of the residual also change accordingly. The core of the secondary

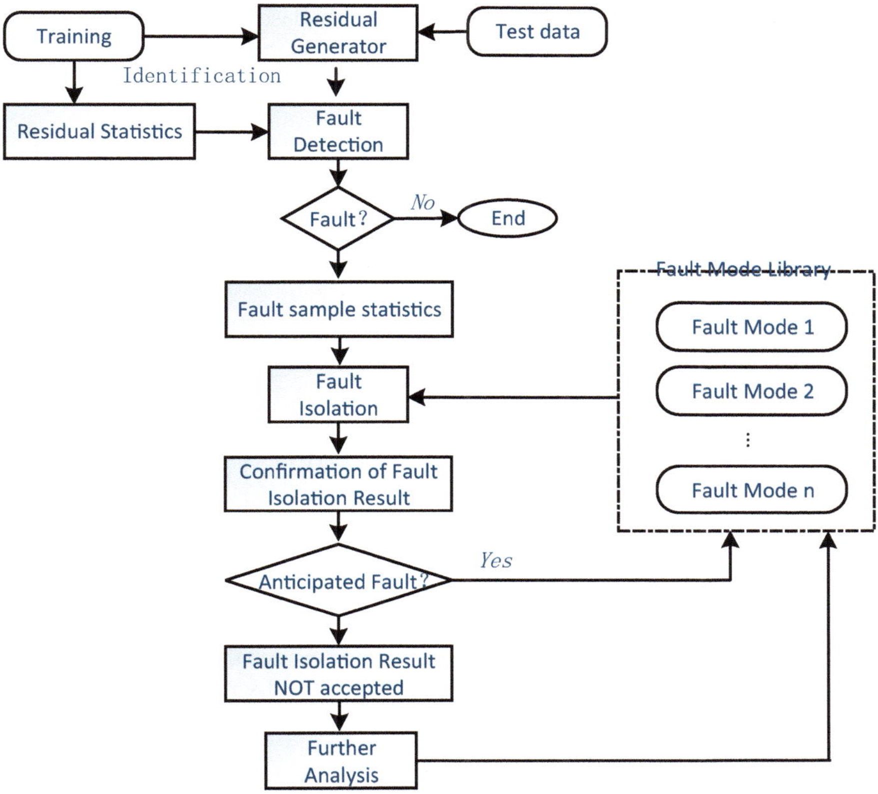

Fig. 8.5 An enhanced data-driven fault diagnosis architecture

confirmation of fault isolation results is to analyze the system characteristics after the fault and judge the fault isolation results based on it.

Subsequent processing of fault data: it includes displaying the fault data in the visual space to understand the fault intuitively and enriching the fault information library to describe the fault characteristics more accurately.

8.2.1.1 Residual Generator Identification and Fault Detection

The residuals encapsulate a multitude of fault-related insights, and acquiring the residual signal stands as a crucial step in fault diagnosis. In the subsequent segment, a detailed description is provided, outlining the procedure for identifying the residual generator based on system data in a rigorous and systematic manner.

Firstly, data structure definitions of the data-driven description of the system are introduced. For given $\boldsymbol{x}(k) \in \mathcal{R}^{n_x}$, $\boldsymbol{x}$ can be any variable in the control system, such as input variable, output variable, state variable, noise, etc., k is the discrete time index, n_x is the dimension of the variable. Defining:

$$\boldsymbol{x}_{k,s} = \begin{bmatrix} \boldsymbol{x}(k) \\ \boldsymbol{x}(k+1) \\ \vdots \\ \boldsymbol{x}(k+s-1) \end{bmatrix} \in \mathcal{R}^{n_x s} \tag{8.18}$$

$$\boldsymbol{x}_k(N) = \begin{bmatrix} \boldsymbol{x}(k) \; \boldsymbol{x}(k+1) \; \ldots \; \boldsymbol{x}(k+N-1) \end{bmatrix} \in \mathcal{R}^{n_x \times N} \tag{8.19}$$

$$\boldsymbol{x}_{k,s}(N) = \begin{bmatrix} \boldsymbol{x}_{k,s} \; \boldsymbol{x}_{k+1,s} \; \cdots \; \boldsymbol{x}_{k+N-1,s} \end{bmatrix} = \begin{bmatrix} \boldsymbol{x}_k(N) \\ \boldsymbol{x}_{k+1}(N) \\ \vdots \\ \boldsymbol{x}_{k+s-1}(N) \end{bmatrix} \in \mathcal{R}^{sn_x \times N}, \tag{8.20}$$

where k, s and N are positive integers representing data start ingress, height, and width of the data matrix.

According to Eqs. (8.18)–(8.20), the relationship between the system output data, the system input data and the state data derived from the discrete state equation can be expressed as:

$$\boldsymbol{y}_{k,s} = \boldsymbol{\Gamma}(s)x_k + \boldsymbol{H}_u(s)\boldsymbol{u}_{k,s} + \boldsymbol{H}_f(s)\boldsymbol{f}_{k,s} + \boldsymbol{H}_w(s)\boldsymbol{w}_{k,s} + \boldsymbol{v}_{k,s} \tag{8.21}$$

$$\boldsymbol{y}_{k,s}(N) = \boldsymbol{\Gamma}(s)\boldsymbol{x}_k(N) + \boldsymbol{H}_u(s)\boldsymbol{u}_{k,s}(N) + \boldsymbol{H}_f(s)\boldsymbol{f}_{k,s}(N) + \boldsymbol{H}_w(s)\boldsymbol{w}_{k,s}(N) + \boldsymbol{v}_{k,s}(N), \tag{8.22}$$

where

$$\boldsymbol{\Gamma}(s) = \begin{bmatrix} \boldsymbol{C} \\ \boldsymbol{CA} \\ \vdots \\ \boldsymbol{CA}^{s-1} \end{bmatrix} \in \mathcal{R}^{sm \times n} \tag{8.23}$$

$$\mathbf{H}_u(s) = \begin{bmatrix} \boldsymbol{D} & 0 & \cdots & 0 \\ \boldsymbol{CB} & \boldsymbol{D} & \cdots & 0 \\ \vdots & \ddots & \ddots & \vdots \\ \boldsymbol{CA}^{s-2}\boldsymbol{B} & \cdots & \boldsymbol{CB} & \boldsymbol{D} \end{bmatrix} \in \mathcal{R}^{sm \times sl} \tag{8.24}$$

$$\boldsymbol{H}_f(s) = \begin{bmatrix} \boldsymbol{F}_f & 0 & \cdots & 0 \\ \boldsymbol{C}\boldsymbol{E}_f & \boldsymbol{F}_f & \cdots & 0 \\ \vdots & \ddots & \ddots & \vdots \\ \boldsymbol{C}\boldsymbol{A}^{s-2}\boldsymbol{E}_f & \cdots & \boldsymbol{C}\boldsymbol{E}_f & \boldsymbol{F}_f \end{bmatrix} \in \mathcal{R}^{sm \times sn_f} \tag{8.25}$$

$$\boldsymbol{H}_w(s) = \begin{bmatrix} 0 & 0 & \cdots & 0 \\ \boldsymbol{C} & 0 & \cdots & 0 \\ \vdots & \ddots & \ddots & \vdots \\ \boldsymbol{C}\boldsymbol{A}^{s-2} & \cdots & \boldsymbol{C} & 0 \end{bmatrix} \in \mathcal{R}^{sm \times sn} \tag{8.26}$$

Equation (8.22) can be further written into a data-driven input and output model in the fault-free case:

$$\begin{bmatrix} \boldsymbol{u}_{k,s}(N) \\ \boldsymbol{y}_{k,s}(N) \end{bmatrix} = \boldsymbol{\Psi}(s) \begin{bmatrix} \boldsymbol{u}_{k,s}(N) \\ \boldsymbol{x}_k(N) \end{bmatrix} + \begin{bmatrix} 0 \\ \boldsymbol{H}_w(s)\boldsymbol{w}_{k,s}(N) + \boldsymbol{v}_{k,s}(N) \end{bmatrix}, \tag{8.27}$$

where

$$\boldsymbol{\Psi}(s) = \begin{bmatrix} \boldsymbol{I} & 0 \\ \boldsymbol{H}_u(s) & \boldsymbol{\Gamma}(s) \end{bmatrix} \in \mathcal{R}^{s(l+m) \times (sl+n)} \tag{8.28}$$

Define datasets as:

$$z_p = \begin{bmatrix} \boldsymbol{u}_{k-s_p,\, s_p}(N) \\ \boldsymbol{y}_{k-s_p,\, s_p}(N) \end{bmatrix} \in \mathcal{R}^{s_p(l+m) \times N} \tag{8.29}$$

$$z_f = \begin{bmatrix} \boldsymbol{u}_{k,\, s_f}(N) \\ \boldsymbol{y}_{k,\, s_f}(N) \end{bmatrix} \in \mathcal{R}^{s_f(l+m) \times N} \tag{8.30}$$

If $s_p = s_f = s$, the following QR decomposition can be obtained:

$$\begin{bmatrix} z_p \\ \boldsymbol{u}_{k,s}(N) \\ \boldsymbol{y}_{k,s}(N) \end{bmatrix} = \begin{bmatrix} \boldsymbol{R}_{11} & 0 & 0 \\ \boldsymbol{R}_{21} & \boldsymbol{R}_{22} & 0 \\ \boldsymbol{R}_{31} & \boldsymbol{R}_{32} & \boldsymbol{R}_{33} \end{bmatrix} \begin{bmatrix} \boldsymbol{Q}_1 \\ \boldsymbol{Q}_2 \\ \boldsymbol{Q}_3 \end{bmatrix} \tag{8.31}$$

which leads to:

$$\begin{bmatrix} \boldsymbol{u}_{k,s}(N) \\ \boldsymbol{y}_{k,s}(N) \end{bmatrix} = \begin{bmatrix} \boldsymbol{R}_{21} & \boldsymbol{R}_{22} \\ \boldsymbol{R}_{31} & \boldsymbol{R}_{32} \end{bmatrix} \begin{bmatrix} \boldsymbol{Q}_1 \\ \boldsymbol{Q}_2 \end{bmatrix} + \begin{bmatrix} 0 \\ \boldsymbol{R}_{33}\boldsymbol{Q}_3 \end{bmatrix} \tag{8.32}$$

Comparing (8.27) and (8.32), the following equations can be obtained as follows:

$$\boldsymbol{\Psi}(s)\begin{bmatrix} \boldsymbol{u}_{k,s}(N) \\ \boldsymbol{x}_k(N) \end{bmatrix} = \begin{bmatrix} \boldsymbol{R}_{21} & \boldsymbol{R}_{22} \\ \boldsymbol{R}_{31} & \boldsymbol{R}_{32} \end{bmatrix}\begin{bmatrix} \boldsymbol{Q}_1 \\ \boldsymbol{Q}_2 \end{bmatrix} \tag{8.33}$$

$$\boldsymbol{H}_w(s)\boldsymbol{W}_N(k, k+s) + \boldsymbol{V}_N(k, k+s) = \boldsymbol{R}_{33}\boldsymbol{Q}_3 \tag{8.34}$$

Multiply the left side of the Eq. (8.33) by $\boldsymbol{\Psi}^{\perp}(s) = \begin{bmatrix} \boldsymbol{\Psi}_u^{\perp}(s) & \boldsymbol{\Psi}_y^{\perp}(s) \end{bmatrix}$, where $\boldsymbol{\Psi}_y^{\perp}(s)$ is the orthogonal complement space of $\boldsymbol{\Psi}_y(s)$, and then the following relationship can be obtained:

$$\boldsymbol{\Psi}^{\perp}(s)\begin{bmatrix} \boldsymbol{R}_{21} & \boldsymbol{R}_{22} \\ \boldsymbol{R}_{31} & \boldsymbol{R}_{32} \end{bmatrix} = 0 \tag{8.35}$$

The singular value decomposition of $\begin{bmatrix} \boldsymbol{R}_{21} & \boldsymbol{R}_{22} \\ \boldsymbol{R}_{31} & \boldsymbol{R}_{32} \end{bmatrix}$ can be conducted as shown in Eq. (8.36), where $\boldsymbol{\Lambda}_2 \approx 0$:

$$\begin{bmatrix} \boldsymbol{R}_{21} & \boldsymbol{R}_{22} \\ \boldsymbol{R}_{31} & \boldsymbol{R}_{32} \end{bmatrix} = \begin{bmatrix} \boldsymbol{U}_1 & \boldsymbol{U}_2 \end{bmatrix}\begin{bmatrix} \boldsymbol{\Lambda}_1 & 0 \\ 0 & \boldsymbol{\Lambda}_2 \end{bmatrix}\begin{bmatrix} \boldsymbol{V}_1^T \\ \boldsymbol{V}_2^T \end{bmatrix} \tag{8.36}$$

Then, an approximate relationship can be obtained as follows:

$$\boldsymbol{\Psi}^{\perp}(s) \approx \boldsymbol{U}_2^T \tag{8.37}$$

Due to $\boldsymbol{\Psi}^{\perp}(s)\boldsymbol{\Psi}(s) = 0$, which leads to:

$$\boldsymbol{\Psi}_u^{\perp}(s) + \boldsymbol{\Psi}_y^{\perp}(s)\boldsymbol{H}_u(s) = 0 \tag{8.38}$$

$$\boldsymbol{\Psi}_y^{\perp}(s)\boldsymbol{\Gamma}(s) = 0 \tag{8.39}$$

Thus, $\boldsymbol{\alpha}_s$ and $\boldsymbol{\beta}_s$ can be obtained by:

$$\boldsymbol{\alpha}_s = \boldsymbol{\Psi}_y^{\perp}(s) \tag{8.40}$$

$$\boldsymbol{\beta}_s = \boldsymbol{\alpha}_s\boldsymbol{H}_u(s) = -\boldsymbol{\Psi}_u^{\perp}(s) \tag{8.41}$$

Until now, the residual generator can be written as follows:

$$\begin{cases} \boldsymbol{x}_o(k+1) = \boldsymbol{A}_o\boldsymbol{x}_o(k) + \boldsymbol{B}_o\boldsymbol{u}(k) + \boldsymbol{L}_o\boldsymbol{r}(k) \in \mathcal{R}^s \\ \boldsymbol{r}(k) = \boldsymbol{G}_o\boldsymbol{y}(k) - \boldsymbol{C}_o\boldsymbol{x}_o(k) - \boldsymbol{D}_o\boldsymbol{u}(k) \in \mathcal{R} \end{cases}, \tag{8.42}$$

where $\boldsymbol{A}_o$, $\boldsymbol{B}_o$, $\boldsymbol{L}_o$, $\boldsymbol{G}$, $\boldsymbol{C}_o$ and $\boldsymbol{D}_o$ are defined as Eqs. (8.43)–(8.47):

$$\boldsymbol{A}_o = \begin{bmatrix} 0 & 0 & \cdots & 0 \\ 1 & 0 & \cdots & 0 \\ \vdots & \ddots & \ddots & \vdots \\ 0 & \cdots & 1 & 0 \end{bmatrix} \tag{8.43}$$

$$\begin{bmatrix} \boldsymbol{B}_o \\ \boldsymbol{D}_o \end{bmatrix} = \begin{bmatrix} \boldsymbol{\alpha}_{s,1} & \boldsymbol{\alpha}_{s,2} & \cdots & \boldsymbol{\alpha}_{s,s} \\ \boldsymbol{\alpha}_{s,2} & \cdots & \boldsymbol{\alpha}_{s,s} & 0 \\ \vdots & \iddots & \iddots & \vdots \\ \boldsymbol{\alpha}_{s,s} & 0 & \cdots & 0 \end{bmatrix} \begin{bmatrix} \boldsymbol{D} \\ \boldsymbol{CB} \\ \vdots \\ \boldsymbol{CA}^{s-1}\boldsymbol{B} \end{bmatrix} = \mathrm{col}(\boldsymbol{\alpha}_s \boldsymbol{H}_{u,s}) \tag{8.44}$$

In (8.44), the matrix $\mathrm{col}(\boldsymbol{\alpha}_s \boldsymbol{H}_{u,s})$ represented here is a novel one, obtained by rearranging the columns of the parameter matrix $\boldsymbol{\alpha}_s \boldsymbol{H}_{u,s}$.

Suppose $\boldsymbol{\alpha}_s \boldsymbol{H}_{u,s} = \begin{bmatrix} \boldsymbol{\beta}_{s,1}^{\mathrm{T}} & \boldsymbol{\beta}_{s,2}^{\mathrm{T}} & \cdots & \boldsymbol{\beta}_{s,s}^{\mathrm{T}} \end{bmatrix}^{\mathrm{T}}$, then:

$$\boldsymbol{B}_o = \begin{bmatrix} \boldsymbol{\beta}_{s,1} \\ \boldsymbol{\beta}_{s,2} \\ \vdots \\ \boldsymbol{\beta}_{s,s-1} \end{bmatrix} \tag{8.45}$$

$$\boldsymbol{D}_o = \boldsymbol{\beta}_{s,s} \tag{8.46}$$

$$\boldsymbol{C}_o = \begin{bmatrix} 0 \; 0 \; \cdots \; 1 \end{bmatrix}, \boldsymbol{G}_o = \boldsymbol{\alpha}_{s,s} \tag{8.47}$$

In conclusion, the procedural steps involved in the identification of the data-driven residual generator are outlined as follows:

Algorithm 8.1

Step 1: Collect and organize data, then perform QR decomposition as outlined in Eq. (8.31)

Step 2: Implement singular value decomposition, replacing $\boldsymbol{\Psi}^{\perp}(s)$ with $\boldsymbol{U}_2^T$.

Step 3: Determine $\boldsymbol{\alpha}_s$ and $\boldsymbol{\beta}_s$ using Eqs. (8.40) and (8.41), respectively

Step 4: Construct the residual generator, as shown in Eq. (8.42), by employing Eqs. (8.43)–(8.47)

After the residual generator is obtained, input the test data to get the residual statistics of the test data, and then compare it with the residual statistics of the training data to detect the system fault. Suppose $\boldsymbol{x} = [x_1; x_2; \ldots; x_n] \in \mathcal{R}^{n \times N}$ is n-dimensional sample data (also known as training data) subject to the multivariate normal distribution, and $\boldsymbol{x}_t \in \mathcal{R}^n$ is the latest sampled test data. According to the statistical distribution law that the variance of residual signal is subject to χ^2 distribution, the fault detection algorithm can be designed by using the hypothesis test principle of the normal population variance.

First, generate residual signal according to the Eq. (8.42). Then, generate fault detection statistics. If the model is completely known:

$$T^2(r) = \boldsymbol{r}^T \Sigma_r^{-1} \boldsymbol{r} \tag{8.48}$$

If the noise characteristic is unknown, the variance matrix of noise should be obtained by using the sample variance approximation, then select the statistics as:

$$T^2(r) = \boldsymbol{r}^T \hat{\Sigma}_r^{-1} \boldsymbol{r} \tag{8.49}$$

Furthermore, set thresholds and make a judgment. In the case of significance level α, the threshold value can be set as $\chi^2_{1-\alpha}$. Thus, when the detection statistic (8.48) or (8.49) is less than or equal to the threshold value, the current data is considered normal; when the detection statistic is greater than the threshold value, the fault is considered to have occurred. To conclude, this fault detection process can be summarized as Algorithm 8.2.

Algorithm 8.2
Step1: Generate residual signal according to Eq. (8.42)
Step2: Generate fault detection statistics (8.48) or (8.49)
Step3: Set the threshold as $\chi^2_{1-\alpha}$ and make the judgment

8.2.1.2 Fault Isolation

If the test data $\boldsymbol{x}_t$ is identified as fault data in the fault detection stage, the next task is to determine which type of fault $\boldsymbol{x}_t$ belongs to through the fault isolation algorithm. Assuming that there are n types of fault modes and there are enough fault data for each type of fault mode, the data set consisting of the fault data marked with the fault mode is called the fault mode database. The idea of fault isolation using fault data in the fault mode database is to compare the newly detected fault data $\boldsymbol{x}_t$ with the fault data in the fault mode database, to determine the fault mode of the fault data and complete the purpose of fault isolation.

According to the comparison method of fault data, fault isolation rules can be categorized as amplitude-based method and angle-based method.

(1) Fault Isolation Based on Amplitude Feature.

Denote the deviation between current fault data $\boldsymbol{x}_t$ and fault mode i as $d(\boldsymbol{x}_t, \boldsymbol{f}_i)$, and the definition is as follows:

$$d(\boldsymbol{x}_t, \boldsymbol{f}_i) = \left\| \sigma_i^{-1} (\boldsymbol{x}_t - \overline{\boldsymbol{x}}_i) \right\|, \tag{8.50}$$

where $\overline{\boldsymbol{x}}_i$ is the average of $\boldsymbol{x}_i$, σ_i is the standard deviation of $\boldsymbol{x}_i$. The smaller $d(\boldsymbol{x}_t, \boldsymbol{f}_i)$ is, the closer $\boldsymbol{x}_t$ is to $\overline{\boldsymbol{x}}_i$, and the fault data $\boldsymbol{x}_t$ is more likely to be in fault mode i. If

$d\left(\boldsymbol{x}_t, \boldsymbol{f}_{i_0}\right)$ is the minimum, then it is reasonable to think that the current fault mode is of the fault mode type i_0. So if:

$$i_0 = \arg\min_i\left\{d\left(\boldsymbol{x}_t, \boldsymbol{f}_i\right), i = 1, \ldots, n_f\right\} \tag{8.51}$$

with n_f indicating the total number of fault modes, the fault data can be regarded as belonging to fault mode i_0.

(2) Fault Isolation Based on Angle Feature.

The fault eigenvector corresponding to the fault mode in the fault mode database can be denoted as $\left\{\vec{\boldsymbol{f}}_i\right\}_{i=1}^{n_f}$, then the angle between any two fault vectors $\vec{\boldsymbol{f}}_i$ and $\vec{\boldsymbol{f}}_j$ in the fault mode database can be computed as follows:

$$\theta\left(\boldsymbol{f}_i, \boldsymbol{f}_j\right) = \arccos\left(\left|\vec{\boldsymbol{f}}_i\vec{\boldsymbol{f}}_j\right| / \left(\left\|\vec{\boldsymbol{f}}_i\right\|\left\|\vec{\boldsymbol{f}}_j\right\|\right)\right) = \arccos\left(\left|\vec{\boldsymbol{f}}_{\mathrm{i}}^T\vec{\boldsymbol{f}}_j\right|\right) \tag{8.52}$$

The larger the angle between every two fault eigenvectors in the fault mode database is, the more advantageous it is to isolate the two faults. If the direction vector of the fault sample $\boldsymbol{x}_t$ is $\vec{\boldsymbol{f}}_t$, the angle between $\vec{\boldsymbol{f}}_t$ and $\vec{\boldsymbol{f}}_i$ is:

$$\theta\left(\boldsymbol{f}_t, \boldsymbol{f}_i\right) = \arccos\left(\left|\vec{\boldsymbol{f}}_t\vec{\boldsymbol{f}}_i\right| / \left(\left\|\vec{\boldsymbol{f}}_t\right\|\left\|\vec{\boldsymbol{f}}_i\right\|\right)\right) = \arccos\left(\left|\vec{\boldsymbol{f}}_t^T\vec{\boldsymbol{f}}_i\right|\right) \tag{8.53}$$

The fault isolation rule based on angles involves the following steps: calculate the angle between $\vec{\boldsymbol{f}}_t$ and $\vec{\boldsymbol{f}}_i$, identify the minimum angle, and assign the fault sample to the fault mode associated with the characteristic direction having the smallest angle. To elaborate further, if:

$$i_0 = \arg\min_i\left\{\theta(\boldsymbol{r}_t, \boldsymbol{r}_i), i = 1, 2, \ldots, n_f\right\} \tag{8.54}$$

then the sample fault will be considered as the fault mode i_0. One of the advantages of utilizing the angle-based fault isolation method is that it remains unaffected by variations in data amplitude between different batches, ensuring that the fault isolation outcomes remain consistent.

During the fault isolation, it is necessary to obtain the feature direction vector of the fault sample and the fault mode in the database, respectively. Considering possible faults, the system model can be extended to:

$$\begin{cases} \boldsymbol{x}(k+1) = \boldsymbol{A}x(k) + \boldsymbol{B}u(k) + \boldsymbol{E}_f\boldsymbol{f} + \boldsymbol{w}(k) \\ \boldsymbol{y}(k) = \boldsymbol{C}x(k) + \boldsymbol{D}u(k) + \boldsymbol{D}_f\boldsymbol{f} + \boldsymbol{v}(k) \end{cases} \tag{8.55}$$

Here, only sensor fault and actuator fault are considered, that is, each sensor and actuator correspond to a fault eigenvector. Besides only additive fault is considered in terms of fault form. Under this premise, setting $\boldsymbol{E}_f = \left[\boldsymbol{B}\ 0\right]$, $\boldsymbol{F}_f = \left[\boldsymbol{D}\ \boldsymbol{I}\right]$ in

Eq. (8.55), and fault vector $\boldsymbol{f} = \begin{bmatrix} \boldsymbol{f}^a \\ \boldsymbol{f}^s \end{bmatrix}$, then the system (8.55) becomes:

$$\begin{cases} \boldsymbol{x}(k+1) = \boldsymbol{A}x(k) + \boldsymbol{B}\big(\boldsymbol{u}(k) + \boldsymbol{f}^a(k)\big) + \boldsymbol{w}(k) \\ \boldsymbol{y}(k) = \boldsymbol{C}x(k) + \boldsymbol{D}\big(\boldsymbol{u}(k) + \boldsymbol{f}^a(k)\big) + \boldsymbol{f}^s(k) + \boldsymbol{v}(k) \end{cases} \tag{8.56}$$

At the same time, Eq. (8.21) becomes:

$$\boldsymbol{y}_{k,s} = \boldsymbol{\Gamma}(s)\boldsymbol{x}_k + \boldsymbol{H}_u(s)\boldsymbol{u}_{k,s} + \boldsymbol{H}_u(s)\boldsymbol{f}^a_{k,s} + \boldsymbol{H}_w(s)\boldsymbol{w}_{k,s} + \boldsymbol{f}^s_{k,s} + \boldsymbol{v}_{k,s} \tag{8.57}$$

Let $\boldsymbol{f}_{k,s} = \begin{bmatrix} -\boldsymbol{f}^a_{k,s} \\ \boldsymbol{f}^s_{k,s} \end{bmatrix}$, residual data $\boldsymbol{r}_{k,s}$ can be written as:

$$\boldsymbol{r}_{k,s} = \boldsymbol{\kappa}_s \begin{bmatrix} \boldsymbol{u}_{k,s} \\ \boldsymbol{y}_{k,s} \end{bmatrix} = \boldsymbol{\kappa}_s \boldsymbol{f}_{k,s} + \boldsymbol{\Gamma}^{\perp}(s)\big(\boldsymbol{H}_w(s)\boldsymbol{w}_{k,s} + \boldsymbol{v}_{k,s}\big) \sim N\left(\boldsymbol{\kappa}_s \boldsymbol{f}_{k,s}, \sum_{rs}\right), \tag{8.58}$$

where $\boldsymbol{\kappa}_s$ is another form of the residual generator. In Eq. (8.58), $\boldsymbol{\kappa}_s \boldsymbol{f}_{k,s}$ is related to fault vector, and $\boldsymbol{\Gamma}^{\perp}(s)\big(\boldsymbol{H}_w(s)\boldsymbol{w}_{k,s} + \boldsymbol{v}_{k,s}\big)$ is the item related to noise. The influence of fault vector $\boldsymbol{f}_{k,s}$ on residual $\boldsymbol{r}_{k,s}$ is closely related to $\boldsymbol{\kappa}_s$. Therefore, the unit direction vector of fault can be defined as Eq. (8.59), where $\boldsymbol{\kappa}(:, i)$ represents all elements of column i in the matrix $\boldsymbol{\kappa}$. Then, the angle between the direction vector of the current fault sample and the direction vector of the fault feature is calculated according to Eq. (8.53), and fault isolation is carried out according to the principle of minimum angle.

$$\vec{\boldsymbol{f}}_i = \begin{cases} \dfrac{\sum_{j=0}^{s-1} \boldsymbol{\kappa}(:, i + lj)}{\left\| \sum_{j=0}^{s-1} \boldsymbol{\kappa}(:, i + lj) \right\|}, & 1 \le i \le l \\ \dfrac{\sum_{j=0}^{s-1} \boldsymbol{\kappa}(:, i + lj + sl - l)}{\left\| \sum_{j=0}^{s-1} \boldsymbol{\kappa}(:, i + lj + sl - l) \right\|}, & l < i \le l + m \end{cases} \tag{8.59}$$

The insufficiency of fault isolation is that if there is a fault not belonging to the fault database, the fault will be classified as the fault type with the closest angle by the fault isolation rule. When the fault diagnosis results are only used as a reference, fault isolation errors do not affect system performance. When the result of fault isolation is used in fault-tolerant control, the wrong result of fault isolation will make the fault-tolerant control algorithm not perform as expected. Therefore, it is meaningful to confirm the fault isolation results for the second time.

8.2.1.3 Secondary Confirmation of Fault Isolation Results

The key to the secondary confirmation of fault isolation is to reconstruct the test statistics after the fault, to confirm the result of fault isolation according to the statistical law. According to the theory of unanticipated fault diagnosis, the following detection statistics can be constructed after a fault:

$$\chi_{cf}^2 = \tilde{\boldsymbol{r}}_s^T(k) \sum_{rs}^{-1} \tilde{\boldsymbol{r}}_s(k), \tag{8.60}$$

where

$$\tilde{\boldsymbol{r}}_s(k) = \boldsymbol{r}_s(k) - \operatorname{sign}(\boldsymbol{r}_s(k)\boldsymbol{r}_{i0})\|\boldsymbol{r}_s(k)\|\boldsymbol{r}_{i0} \tag{8.61}$$

and sign(·) is a signum function. Detection statistics χ_{cf}^2 obey the statistical law shown in Eq. (8.62). The threshold value of secondary confirmation of fault isolation is set to be $\chi_{1-\alpha}^2(n_r)$, when the false alarm rate is α.

$$\chi_{cf}^2 \sim \chi^2(n_r) \tag{8.62}$$

8.2.1.4 Subsequent Processing of Faulty Data

There are two methods to deal with fault isolation results, one is fault database training, and the other is graphical processing of diagnosis results.

(1) Database Training

The database for levitation system faults serves as a compendium of fault modes and related data. Establishing an all-encompassing fault database is a prerequisite for designing a fault-tolerant control algorithm. If the system model is known with precision, the fault mode database can be acquired accurately through the use of the model. If a system model is unavailable, experimental data can be utilized for training purposes. The methodology for training the levitation system fault mode database is as follows.

Algorithm 8.3

Step 1: Initialize the fault database, which contains known fault modes and corresponding fault samples
Step 2: Build fault detection statistics based on sample data for fault detection
Step 3: Fault isolation when the fault occurs
Step 4: Secondary confirmation of isolated fault samples. If the fault really belongs to a fault mode in the fault database, add the fault sample of the fault mode in the fault database; if the fault is not in the fault database, add the fault sample direction to the new fault feature direction

(2) Graphical Processing of Diagnosis Results

When dealing with a complex system with a fault vector of high dimensionality, it can be challenging to visualize the fault vector. In order to facilitate the comprehension and analysis of fault diagnosis results, it is possible to map the high-dimensional fault vector onto a two-dimensional or three-dimensional space through the utilization of dimension reduction techniques. Thus, a graphical illustration of the fault diagnosis results can be obtained. The essence of the graphical fault diagnosis result is to find a mapping matrix from fault vector space R^{n_f} to graphical space $R^{n_{vs}}$(two-dimensional or three-dimensional space). It is required to retain as much original fault information as possible in the mapping process. Principal component analysis (PCA), which has a good effect on the dimensionality reduction of big data, can be used to find the mapping matrix.

The specific steps of the graphical representation of fault results are as follows:

(1) Calculate the divergence matrix of fault characteristic direction $\{\boldsymbol{r}_i\}_{i=1}^{n_f}$:

$$\boldsymbol{\Phi}_{sm} = \sum_{j=1}^{n_f}\sum_{i=1}^{n_f}\left(\boldsymbol{r}_i - \boldsymbol{r}_j\right)\left(\boldsymbol{r}_i - \boldsymbol{r}_j\right)^{\mathrm{T}} \tag{8.63}$$

(2) Singular value decomposition of divergence matrix:

$$\boldsymbol{\Phi}_{sm} = \boldsymbol{\Gamma}\boldsymbol{\Lambda}\boldsymbol{\Gamma}^{\mathrm{T}} \tag{8.64}$$

(3) Determine the mapping transfer matrix:

$$\boldsymbol{T}_{pc} = \boldsymbol{\Gamma}_{1:n_{vs}}^{\mathrm{T}} \tag{8.65}$$

8.2.2 Analysis of Fault Diagnosis for Levitation System

In this part, the performances of the proposed monitoring and fault diagnosis method are testified on simulation for the maglev system. In the simulations, the control period is 200 μs, the false alarm rate is set as 0.5%, and noise variance $\sum_w = 2.5 \times 10^{-13}$ and $\sum_v = 5 \times 10^{-13}$. Chopper fault, gap sensor fault, current fault, and accelerometer fault on the left and right sides are considered. For ease of distinguishing, the fault vectors are marked as $\boldsymbol{f}_{a1}$, $\boldsymbol{f}_{a2}$, $\boldsymbol{f}_{s1}$, $\boldsymbol{f}_{s2}$, $\boldsymbol{f}_{s3}$, $\boldsymbol{f}_{s4}$, $\boldsymbol{f}_{s5}$ and $\boldsymbol{f}_{s6}$, respectively. The identification numbers corresponding to the faults are 1 to 8, respectively. The simulated faults encompass the following types: constant deviation faults with an amplitude of 8 V occur on the left and right choppers, respectively; constant deviation faults with amplitudes of 1 mm and 2 mm occur on the left and right gap sensors, respectively; furthermore, constant deviation faults with amplitudes of 20 A and 10 A appear on the left and right current sensors, respectively.

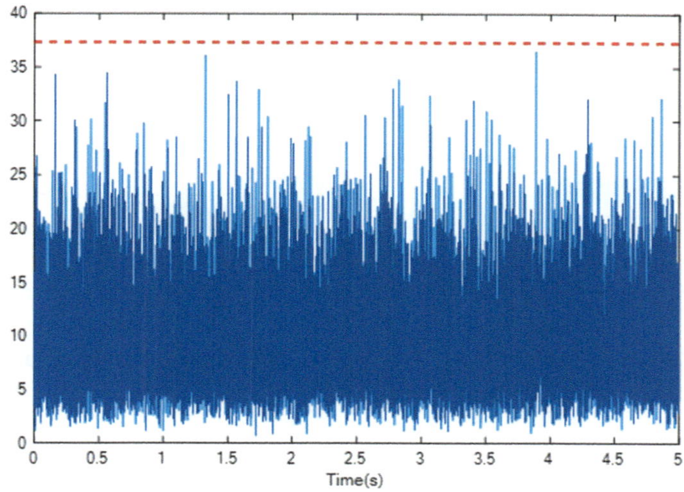

Fig. 8.6 Fault detection statistics and thresholds trained from normal data

Firstly, the residual generator is derived utilizing Algorithm 8.1. Then the fault detection statistics and threshold are obtained by using the normal data of the levitation system. According to the fault detection steps shown in Algorithm 8.2, the threshold value is $\chi^2_{99.95\%}(10) = 37.3$. The fault detection result obtained for the normal condition is shown in Fig. 8.6.

The fault detection for different types of faults, respectively, and the test results are shown in Fig. 8.7. To save space, only the test results of fault $\boldsymbol{f}_{a1}$, $\boldsymbol{f}_{s1}$, $\boldsymbol{f}_{s3}$ and $\boldsymbol{f}_{a1}\&\boldsymbol{f}_{a2}$ are given. Subsequent fault isolation and secondary confirmation of the isolation results are also shown for these fault types. From the fault detection results shown in Fig. 8.7a–d, it can be seen that all faults can be detected in a short time after the occurrence of the fault. The detection time and the fault detection rate are satisfactory.

After the detection of the fault, the fault isolation algorithm is initiated to ascertain the type of the fault. The fault isolation results are shown in Fig. 8.8. In Fig. 8.8a, the left chopper fault is correctly isolated as a type 1 fault. In Fig. 8.8b, the left gap sensor fault is isolated as the type 3 fault and the type 4 fault. There is a jump between the fault isolation results. In Fig. 8.8c, the left current sensor fault is correctly isolated as a type 5 fault. In Fig. 8.8d, the isolation result of the simultaneous faults of the left and right choppers jumps between type 1 and type 2 faults. To further analyze the cause of the jump of the fault isolation result, the fault direction vector shown in Eq. (8.66) is obtained according to the calculation method shown in Eq. (8.59).

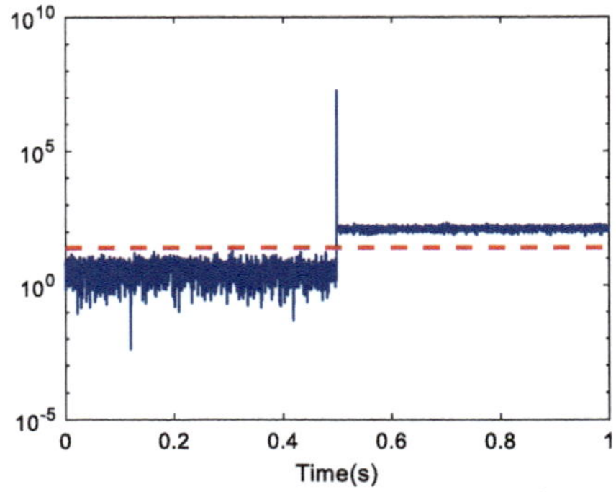

(a) Detection result of fault $\boldsymbol{f}_{a1}$

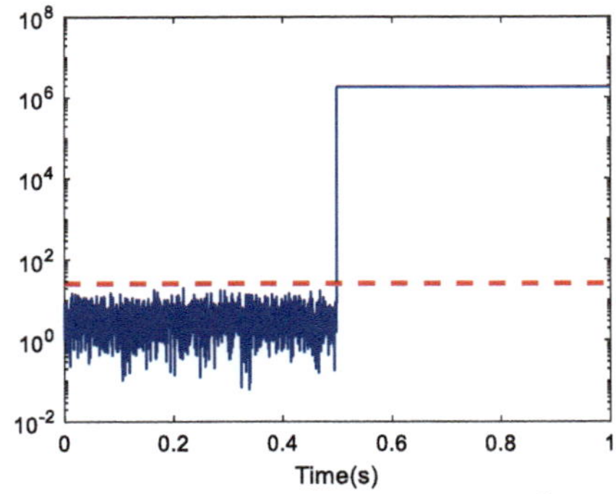

(b) Detection result of fault $\boldsymbol{f}_{s1}$

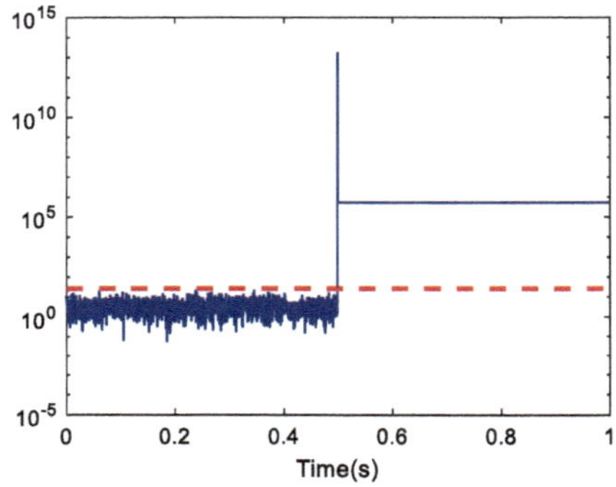

(c) Detection result of fault $\boldsymbol{f}_{s3}$

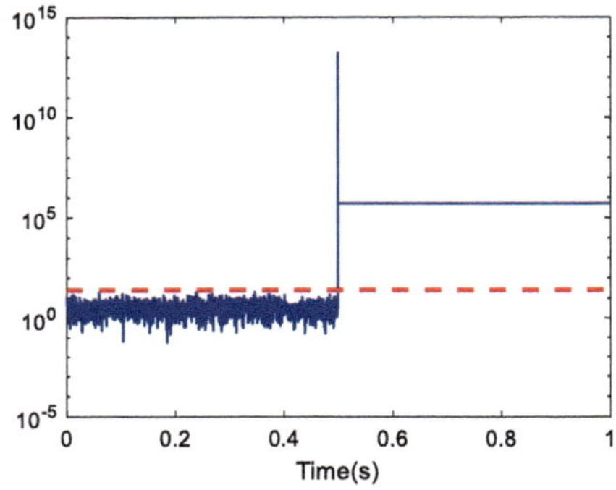

(d) Detection result of fault $\boldsymbol{f}_{a1}$ & $\boldsymbol{f}_{a2}$

Fig. 8.7 Detection results of different faults

$$\vec{r} = 10^3 \begin{bmatrix} -1.910 & 0.196 & 2.126 & -2.127 & 1.914 & -0.201 & -1.910 & 0.196 \\ 1.699 & 1.005 & 0.921 & -0.921 & -1.697 & -1.007 & 1.699 & 1.003 \\ 4.569 & 1.257 & 4.823 & -4.823 & -4.559 & -1.267 & 4.5649 & 1.260 \\ 1.648 & 6.129 & 2.701 & -2.703 & -1.643 & -6.134 & 1.652 & 6.127 \\ 0.014 & 0.009 & 3.005 & -3.005 & -0.008 & -0.015 & 0.011 & 0.009 \\ 0.127 & -0.293 & -5.155 & 5.155 & -0.137 & 0.303 & 0.127 & -0.295 \\ -7.548 & 0.501 & 3.778 & -3.778 & 7.555 & -0.508 & -7.547 & 0.501 \\ 1.910 & -2.775 & 1.090 & -1.090 & -1.908 & 2.773 & 1.911 & -2.777 \\ 2.668 & -0.944 & -0.780 & 0.780 & -2.670 & 0.946 & 2.673 & -0.946 \\ -1.456 & 7.133 & -3.520 & 3.520 & 1.449 & -7.126 & -1.460 & 7.137 \end{bmatrix} \tag{8.66}$$

From the fault direction vector shown in Eq. (8.66), the fault vector $\boldsymbol{f}_{s1}$ and $\boldsymbol{f}_{s2}$ are opposite to each other, which is why the fault isolation result of gap sensor jumps between these two fault vectors. The inverse direction of the fault vector $\boldsymbol{f}_{s1}$ and $\boldsymbol{f}_{s2}$

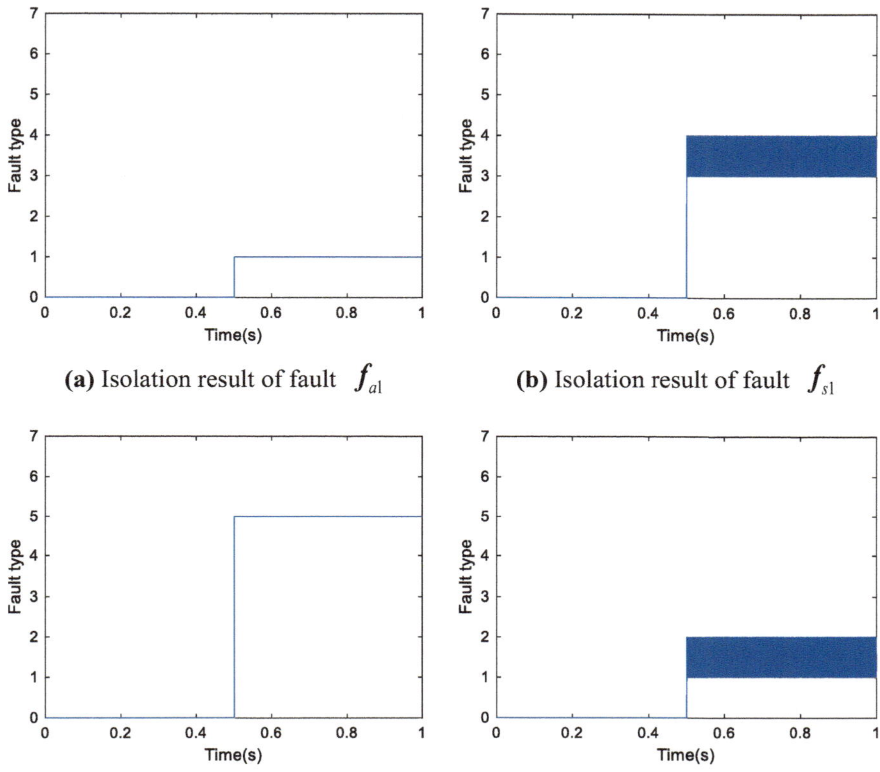

(a) Isolation result of fault f_{a1}

(b) Isolation result of fault f_{s1}

(c) Isolation result of fault f_{s3}

(d) Isolation result of fault f_{a1} & f_{a2}

Fig. 8.8 Isolation results of different faults

can be comprehended as follows: when the amplitude of the positive deviation in the signal from the left-side gap sensor is equivalent to that of the negative deviation in the signal from the right-side gap sensor, the residual signal is equally affected.

It can be seen from the fault isolation results that although there is a bounce, the bounce is only between similar faults. In another word, although the isolation result is inaccurate, it can still be used as a reference, with the isolation result helping to narrow the scope of troubleshooting.

After the fault isolation of the levitation system is completed, the secondary confirmation of the fault isolation results is obtained, as shown in Fig. 8.9. In Fig. 8.9a–c, all three faults, the fault of the left chopper, the fault of the left gap sensor, and the fault of the left current sensor, are identified as known faults in the fault database. However, the simultaneous faults of the left and right choppers are confirmed as unknown faults in the fault database, in Fig. 8.9 (d). At this time, the fault isolation results of the previous step are not accepted.

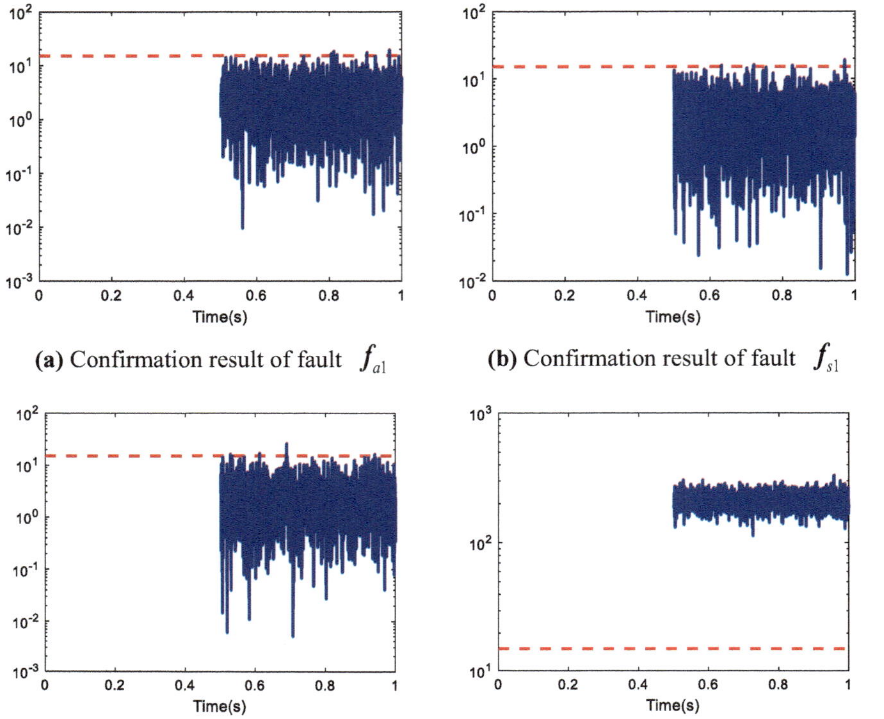

(a) Confirmation result of fault f_{a1} **(b)** Confirmation result of fault f_{s1}

(c) Confirmation result of fault f_{s3} **(d)** Confirmation result of fault f_{a1} & f_{a2}

Fig. 8.9 Secondary confirmation of fault isolation results

By introducing the data into Eq. (8.64), we will get $\boldsymbol{\Lambda} = \mathrm{diag}(50.3, 41.9, 32, 1.5 \times 10^{-5}, 1.2 \times 10^{-6}, 4.6 \times 10^{-8}, 4.1 \times 10^{-15}, 3.4 \times 10^{-15}, 9.8 \times 10^{-16}, 2.7 \times 10^{-16})$. The value of the first three elements is significantly greater than that of the last seven, so the first three elements should be retained, meaning that the residual vector is mapped to the 3D space to complete the image representation of the fault vector. In this part, the residual vector with dimension $n_f = 8$ is transferred to three-dimensional space to get more intuitive fault vector distribution results. Using the PCA method, the fault eigenvector after dimension reduction is obtained as follows:

$$\vec{r}_3 = \begin{bmatrix} 0.72 & -0.72 & 0 & 0 & -0.72 & 0.72 & 0.72 & -0.72 \\ 0.67 & 0.67 & 0 & 0 & -0.69 & -0.69 & 0.69 & 0.69 \\ 0 & 0 & -1.00 & 1.00 & 0 & 0 & 0 & 0 \end{bmatrix} \tag{8.67}$$

Considering that the fault vector corresponding to the accelerometer fault is very close to the fault vector corresponding to the actuator fault, and the method based on signal processing can be adopted for the fault diagnosis of the accelerometer.

Therefore, the result of the accelerometer fault is omitted in the graphical process. Figure 8.10 shows the graphical result of the final fault vector, where Fig. 8.10a is the complete distribution diagram of the fault eigenvector and fault data, and Fig. 8.10b shows the distribution diagram of the fault vector f_{s1}&f_{s2}. From the results shown in Fig. 8.10a, it can be seen that the normal data of the system are distributed near the origin in the figure, and the fault data of the system is distributed near the fault feature direction, while the simultaneous fault vector f_{s1} and f_{s2} are distributed between f_{s1}&f_{s2}. In Fig. 8.10b, there exists no intersection between the fault data corresponding to the two different fault types, thereby signifying a complete distinguishability of the fault data. When the angle-based fault isolation method cannot isolate faults accurately, the graphical method can be used to mark the location of the fault data in the diagram to identify the fault type by the location of the fault data, which is the advantage of the proposed method.

Fig. 8.10 Graphical fault diagnosis results

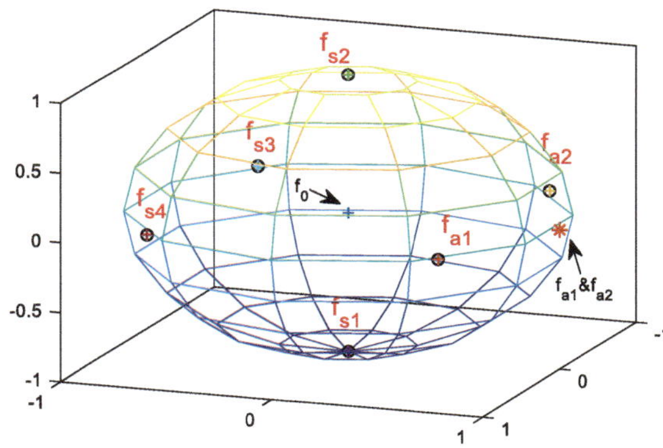

(a) Fault vector and eigenvector distribution

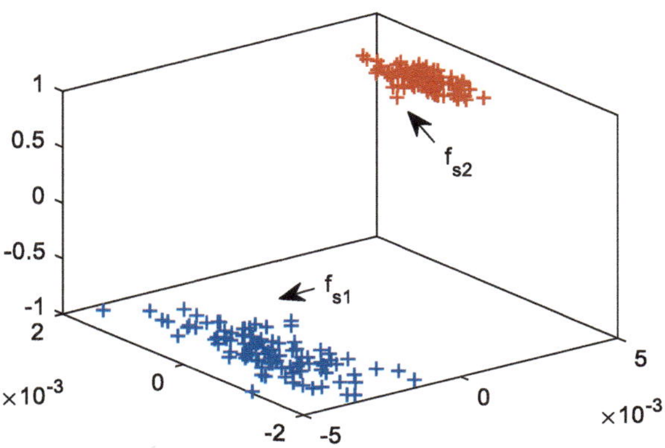

(b) Fault vectors distributions of f_{s1} & f_{s2}

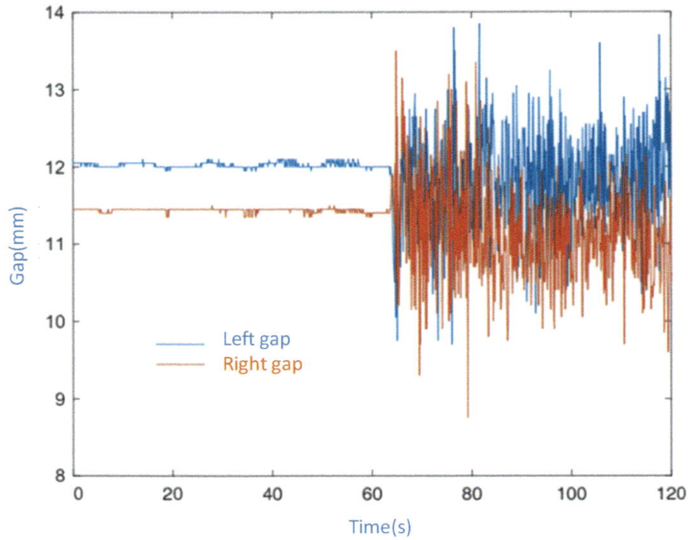

Fig. 8.11 Levitation gap under fault-free condition

The following part takes the high-speed maglev train as the research object and stores the input and output data of the system for further analysis. In the first stage, the train levitates above the track without horizontal movement. In the second stage, the maglev train moves slowly along the track. The experimental data characteristics of the levitation system under the fault-free state are shown in the following figures, in which the levitation gap is shown in Fig. 8.11, and the levitation current is shown in Fig. 8.12.

In the first 60 s, the maglev train is in a static levitating state. There are small fluctuations in the gap and current curve. After 60 s, the maglev train is in operation, and the fluctuation amplitude becomes larger. Please refer to Table 8.1 for information regarding the levitation gap, and Table 8.2 for specific statistics related to the levitation current. In the static levitating/operating state, the average values of the levitation gap and the current are the same, while the variance of the levitation gap and the current in the operating state is much larger than that in the static levitating state.

To verify the effectiveness of the method under fault conditions, the intermittent fault data in a period of operation is used as the experimental data. After executing the above procedure. The final state detection result is shown in Fig. 8.13. In this graph, the solid line is the test statistics, and the dashed line is the threshold. The first 20 s are in the static levitating state, and the last 50 s are in the operation state. Two independent test statistics and thresholds are employed for two different operating conditions. In a quiescent state, the test statistics are below the threshold, which means the system is fault-free and free of false alarms. In the running state, the test statistics value is below the threshold for the first 30 s, which also means that the

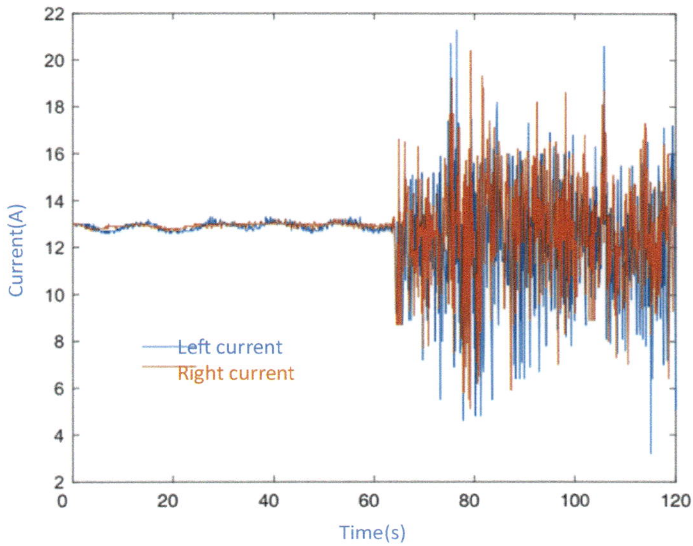

Fig. 8.12 Levitation current under fault-free condition

Table 8.1 Levitation gap statistics

State	Statistics	Value	Unit
Static levitation	Average	11.4419	mm
	Variance	0.0005	mm^2
Operation	Average	11.1046	mm
	Variance	4.776	mm^2

Table 8.2 Current statistics

State	Statistics	Value	Unit
Static levitation	Average	12.9228	A
	Variance	0.0086	A^2
Operation	Average	12.6856	A
	Variance	5.9437	A^2

system is fault-free and has no false alarms; around 50 and 70 s, the test statistics value exceeds the threshold value, which means that the fault is successfully detected.

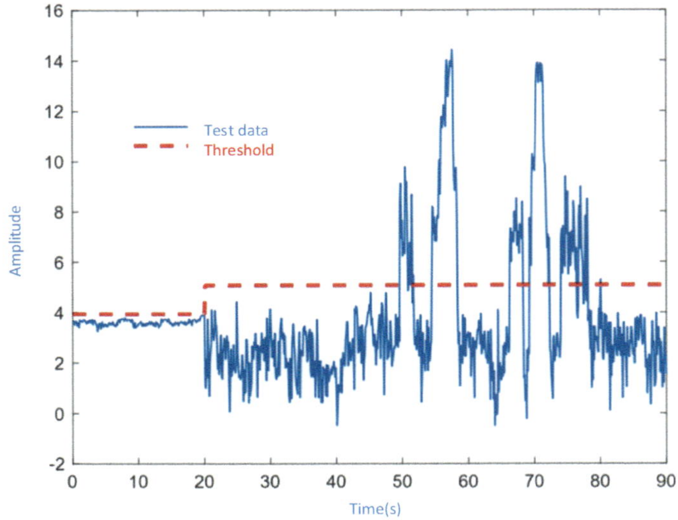

Fig. 8.13 Fault detection using system data

8.3 Signal Processing-Based Fault Diagnosis for Sensors in PEM Hybrid Levitation

8.3.1 Accelerometer Fault Diagnosis

8.3.1.1 Accelerometer Fault Diagnosis Based on Tracking-Differentiator

The levitation control algorithm utilizes the feedback of several key variables, including the electromagnet levitation gap, movement speed of the electromagnet, and the current passing through the excitation windings. The levitation gap value is obtained by a gap sensor, while the current value passing through the windings is measured using a current transformer nestled within the chassis. However, directly measuring the movement speed of the electromagnet is not feasible with a sensor.

To overcome this limitation, the movement speed of the levitation electromagnet can be estimated using one of two redundant signal acquisition methods. One approach involves differentiating the levitation gap signal. Alternatively, the movement speed can be inferred by integrating the acceleration signal. It is important to note that the gap signal may be susceptible to electromagnetic interference, resulting in the presence of noise signals. The classical method of differentiation may amplify high-frequency noise present in the gap signal, thereby reducing the signal-to-noise ratio of the gap differential signal. Therefore, under the normal operation of the train, generally, the integral of the acceleration signal is selected as the speed information for feedback.

The fault diagnosis of the accelerometer signal utilizes the redundant characteristics of the speed signal acquisition methods. Under the premise that the gap signal is fault-free, the working state of the accelerometer is determined according to the difference between the gap differential signal and the acceleration integral signal. The steps to obtain the decision quantity and the threshold are as follows:

(1) Determine the mean and variance of the residuals e, namely μ_0 and σ^2, in the normal state.

$$\mu_0 = \frac{1}{N_0}\sum_{i=1}^{N_0} e(i), \quad \sigma_0^2 = \frac{1}{N_0}\sum_{i=1}^{N_0} [e(i) - \mu_0]^2 \tag{8.68}$$

(2) Calculate the statistical properties of residuals e in real-time.

$$\mu(k) = \frac{1}{N}\sum_{i=1}^{N} e(k-i) \tag{8.69}$$

$$\sigma_1^2 = \frac{1}{N}\sum_{i=1}^{N} [e(k-i) - \mu_0]^2 \tag{8.70}$$

$$\sigma_2^2 = \frac{1}{N}\sum_{i=1}^{N} [e(k-i) - \mu(k)]^2 \tag{8.71}$$

(3) Calculate the decision quantity d and the threshold thr.

$$d(k) = \frac{\sigma_1^2(k)}{\sigma_0^2} - \ln\frac{\sigma_2^2(k)}{\sigma_0^2} - 1 \tag{8.72}$$

$$thr = 2\ln[NP_0/(1 - P_0)], \tag{8.73}$$

where P_0 is the prior probability that the system is fault-free.

When $d > thr$, the accelerometer is deemed to be malfunctioning, whereas when $d \le thr$, it is considered that the accelerometer is working normally.

In this way, the residual data of a certain system during normal operation can be obtained to calculate μ_0 and σ_0^2. When the system is working, $\mu(k)$, σ_1^2, and σ_2^2 are obtained in real-time, and then the decision quantity d and the threshold thr are calculated, and status of the accelerometer is assessed according to the relationship between d and thr.

8.3.1.2 Tracking-Differentiator Principle and Characteristics

In this section, a novel tracking-differentiator is introduced. This sophisticated mechanism possesses the remarkable ability to swiftly discern and track the input signal with utmost precision, deftly avoiding both fluttering and overshooting. Additionally, it effectively accomplishes the task of signal differentiation without amplifying the accompanying noise. Notably, this innovative approach offers a distinct advantage: the calculation process eschews any reliance on complex mathematical operations, such as radicals, exponentiation, or logarithms.

(1) Form of the Discrete Tracking-Differentiator

The general form of the discrete tracking-differentiator is:

$$\begin{cases} x_1(k+1) = x_1(k) + hx_2(k) \\ x_2(k+1) = x_2(k) + hu(k) \\ u(k) = ftd_{new}(x_1(k) - v(k), x_2(k), r, c_0h) \end{cases} \tag{8.74}$$

In this context, the state variables, represented by the symbols x_1 and x_2, can be regarded as fundamental elements that describe the system's current conditions or states. The discrete step length, denoted as h, refers to the increment or decrement applied to these state variables at each discrete time step. The control comprehensive function, denoted as u, captures the overall influence exerted on the state variables based on certain control strategies or policies. Meanwhile, the rapid coefficient and filtering coefficient, symbolized by r and c_0, respectively, represent the factors that influence the speed and accuracy of the state variable changes. Given the requisite selection of optimal values for the parameters r and c_0, along with the proficient functioning of the comprehensive control function u, the desirable outcomes of $x_1 \rightarrow v$ and $x_2 = \dot{x}_1 \rightarrow \dot{v}$ can be actualized under these circumstances.

(2) Procedure to Calculate the Control Comprehensive Function

For a continuous second-order cascaded integral system $\begin{cases} \dot{x}_1 = x_2 \\ \dot{x}_2 = u, \ |u| \leq r \end{cases}$, attaining the origin from any point situated on the phase plane necessitates the adoption of the fastest optimal control strategy $u = -r\text{sign}\left(x_1 + \frac{x_2|x_2|}{2r}\right)$. Throughout this progression, it is imperative that the control quantity undergoes, at most, a solitary change in polarity, which transpires promptly as the point converges upon the boundary line $x_1 + \frac{x_2|x_2|}{2r}=0$. Conversely, within a discrete system, it becomes challenging for the state point to effectuate an exact alteration in polarity upon reaching the boundary line due to the inherent constraints of the one-step sampling procedure. Consequently, an unsettling phenomenon known as chattering ensues, wherein the state point oscillates persistently in proximity to the boundary line. This relentless chattering not only undermines the operational efficacy of the controller but also exacerbates its overall performance. Consequently, it becomes crucial to implement tailored interventions in the vicinity of the boundary line to rectify and mitigate this undesirable behavior.

In the phase plane, there exist two boundaries: the first one is the switching line, where the control values consistently assume either $u = +r$ or $u = -r$. In this case, the state points converge precisely to the origin. The second characteristic line is formed by those initial points that converge to the origin when the control value first adopts $u = +r$ and then consistently becomes $u = -r$, or when the control value initially takes $u = -r$ and then consistently becomes $u = +r$.

Define the collection of points that converge to the origin when either $u = +r$ or $u = -r$ are consistently employed as $\{a_{+k}\}$, $\{a_{-k}\}$. Define the collection of points that converge to the origin when $u = -r$ is initially followed by $u = +r$ consistently, or $u = +r$ is initially followed by $u = -r$ consistently as $\{b_{+k}\}, \{b_{-k}\}(k \geq 2)$. Define the curve characterized by the points in $\{a_{+k}\}$ as Γ_A^+, the curve characterized by the points in $\{a_{-k}\}$ as Γ_A^-, the curve characterized by the points in $\{b_{+k}\}$ as Γ_B^+, and finally, the curve characterized by the points in $\{b_{-k}\}$ as Γ_B^-.

The mathematical description of the curve $\Gamma-A = \Gamma_A^+ \cup \Gamma_A^-$ can be derived as follows:

$$x_1 + \frac{x_2|x_2|}{2r} + \frac{1}{2}h \cdot x_2 = 0 \tag{8.75}$$

The mathematical description of the curve $\Gamma-B = \Gamma_B^+ \cup \Gamma_B^-$ is as follows:

$$x_1 - s \cdot \frac{x_2^2}{2r} + \frac{5}{2}hx_2 - s \cdot h^2 r = 0, \quad s = \operatorname{sgn}(x_1 + hx_2) \tag{8.76}$$

The region encompassing state points capable of converging to the origin within a span of two steps can be denoted as the 2-step reachable zone:

$$\Omega_r = \left\{(x_1, x_2)||x_2 + 2hx_2| \leq h^2 r, |x_1 + hx_2| \leq h^2 r\right\} \tag{8.77}$$

Define the region bounded by $\Gamma-A$ and $\Gamma-B$, excluding Ω_r, as the linear zone Ω_l. The remaining two regions can be referred to as the negative zone $\Omega-$, positioned above the region, and the positive zone $\Omega+$, located below the region. Figure 8.14 illustrates the partitioning of the phase plane into four distinct sections: Ω_r, Ω_l, $\Omega+$, and $\Omega-$.

It is obvious that in the region $\Omega-$, $u = -r$; in the region $\Omega+$, $u = r$. Different control strategies can be observed between the continuous system and discrete system exist when the state point falls within the linear region Ω_l and the 2-step reachable region Ω_r.

When the state (denoted by $P(x_1, x_2)$) falls within the linear region Ω_l, a discernible approach is to draw a trajectory line through point P on the phase plane that intersects with both Γ - A and $\Gamma-B$. By designating these intersections as points A and B, respectively, where the x-coordinate values are determined as X_A and X_B, an adjustable control variable $u = -r\frac{(X_B+X_A-2|x_1|)}{X_B+X_A}\operatorname{sign}(x_2)$ can be introduced. This

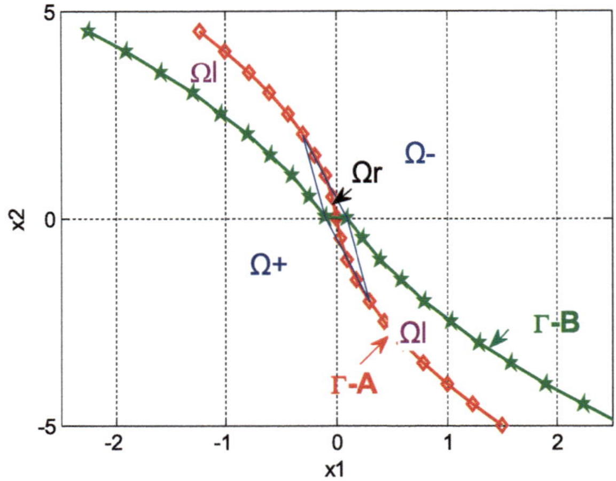

Fig. 8.14 The division of the phase plane

variable ensures that subsequent states, including the forthcoming state $P(k+1)$, gracefully navigate within the linear region and ultimately converge toward the origin along the boundaries.

When the state is within the 2-step reachable region, the control strategy can be derived from the system equation as $u = -\frac{1}{h^2}(x_1 + 2hx_2)$.

In conclusion, the basic approach of this nonlinear tracking differentiator is: given a point $P(x_1, x_2)$ in the phase plane, the value of the control comprehensive function ftd_{new} is determined so that the state point can converge to the origin as soon as possible. The strategy of calculating the control comprehensive function ftd_{new} is: make a horizontal line through the point $P(x_1, x_2)$ on the phase plane, get the value of X_A and X_B, then obtain the value of $u = ftd_{new}(x_1, x_2, r, h)$ according to which zone $P(x_1, x_2)$ belongs to.

As an instance, the trajectory for an initial point $P(x_1, x_2) = (1.1, 2)$ in the phase plane can be seen in Fig. 8.15. In this figure point $P(x_1, x_2)$ moves from zone $\Omega-$ to linear zone, entering the reachable zone along the switching line, and eventually converges to the origin in 11 steps, with no fluttering during this whole process. The locus of the state point remains remarkably proximate to the optimal convergence locus due to the proper choice of control comprehensive function ftd_{new}.

8.3.1.3 Simulation Analysis of Accelerometer Fault Diagnosis

The simulation conditions are as follows: the electromagnet is gradually stabilized at the equilibrium point, the acceleration fault is simulated at $t = 3$ s, and the change in the decision-making quantity is to be observed.

The diagnostic result when the accelerometer has a zeroing fault is shown in Fig. 8.16. Figure 8.16a is the acceleration integral value. When the accelerometer

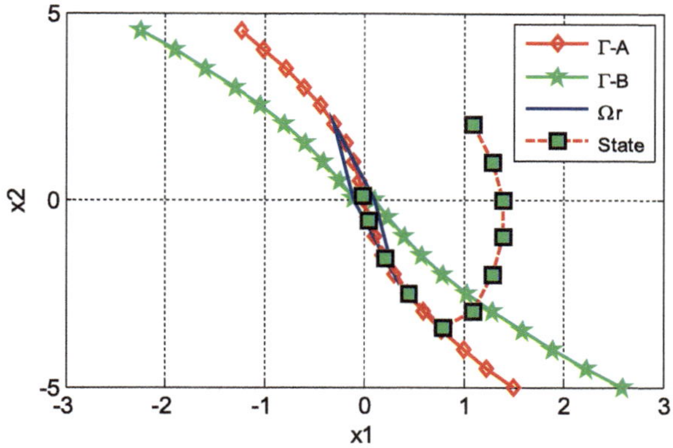

Fig. 8.15 The convergence of the state point

has a zeroing fault, the acceleration integral value remains unchanged, and the acceleration integral signal is a straight line after $t = 3$ s. Since the acceleration integral provides speed feedback information for the controller, the speed feedback information remains constant due to the failure of the accelerometer, which cannot reflect the dynamic characteristics of the system, so the system starts to oscillate. Figure 8.16b shows the residual value between the acceleration integral value and the gap differential value, and the residual value begins to diverge after the fault. Figure 8.16c is the decision quantity and the threshold value. When the decision quantity exceeds the threshold value near $t = 3.1$ s, the zeroing fault of the accelerometer is detected.

The diagnosis result when the accelerometer has a constant gain fault is shown in Fig. 8.17, and the constant gain fault coefficient is $\alpha = 4$. Due to the sudden increase in the amplitude of the accelerometer output, the acceleration integral value increases rapidly, and the residual between the acceleration integral value and the differential value of the gap also increases, and the fault is quickly detected at $t = 3.02$ s.

The simulation results show that the fault diagnosis algorithm can effectively diagnose accelerometer faults. When the accelerometer is operating normally, the decision quantity is below the threshold. When the accelerometer fails, due to the action of the integrator, the acceleration integral signal will not change abruptly, the fault diagnosis component refrains from issuing an indication of fault occurrence during this particular moment. After the fault, if the decision-making quantity exceeds the threshold, the fault diagnosis system judges the occurrence of the fault, and the value of the fault diagnosis signal jumps from 0 to 1, signifying the conclusive judgment of the accelerometer fault, and the diagnosis result is timely and effective.

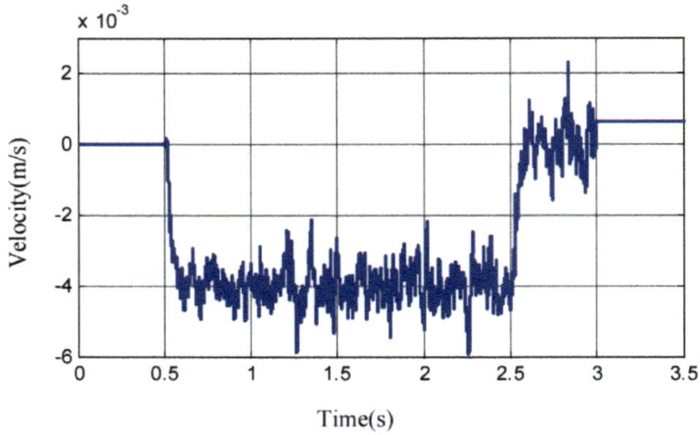

(a) Integral of accelerator signal

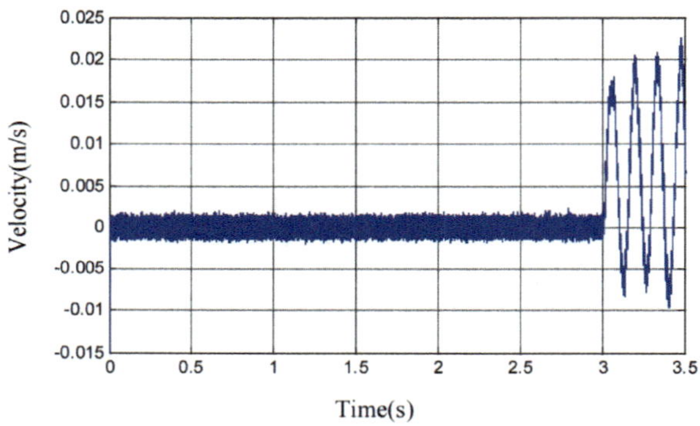

(b) Residual signal

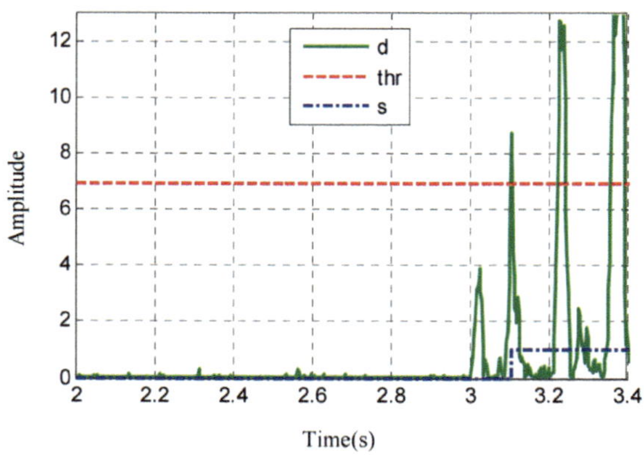

(c) Decision quantity and threshold

Fig. 8.16 Accelerator fault diagnosis result (zeroing faults)

Fig. 8.17 Accelerator fault diagnosis result (constant gain fault)

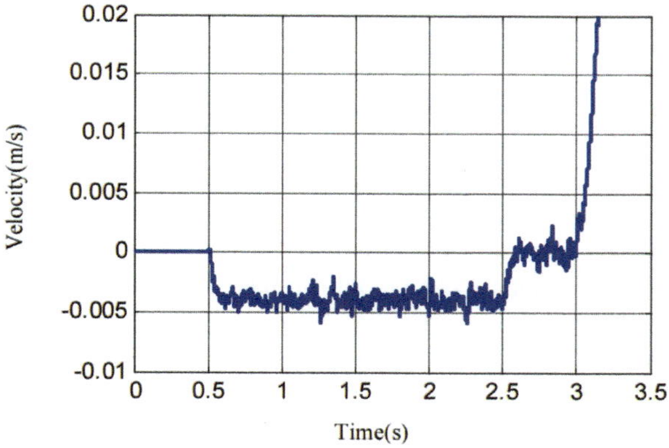

(a) Integral of accelerator signal

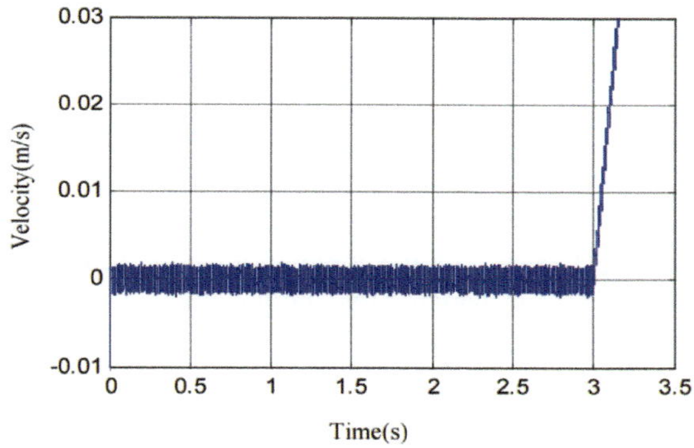

(b) Residual signal

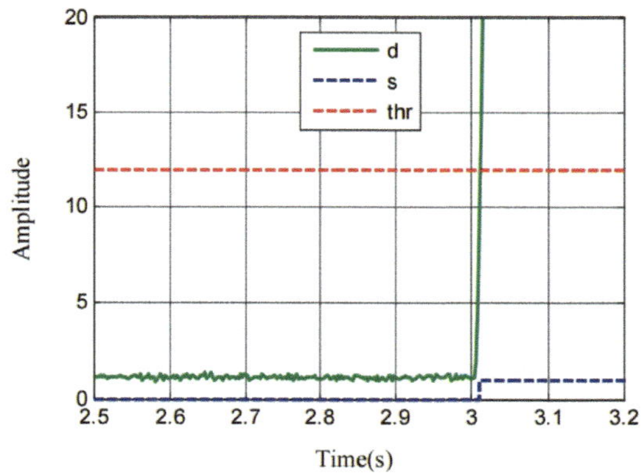

(c) Decision quantity and threshold

8.3.2 Current Sensor Fault Diagnosis

8.3.2.1 Current Sensor Fault Diagnosis Method

The current sensor is used to measure the electromagnet current, which is also an essential feedback quantity in feedback control. Two current sensors are installed inside the controller chassis to measure the output and input current of the electromagnet, respectively. Under normal operating conditions, the measured values of the two sensors are approximately equal. When the difference between the two measured values is large, it can be attributed to two potential fault conditions. One is the leakage of the electromagnet, causing the output and input currents to be unequal; the other is one of the sensors is damaged, which consequently leads to the measured value deviation. Therefore, the cause of the fault cannot be accurately located by simple comparison, and a fault diagnosis method for a single current sensor needs to be studied.

In view of the inherent inductance within the windings of the electromagnet, the current flowing within the windings assumes a continuous and gradual nature, devoid of abrupt fluctuations. The fault of the current sensor can be diagnosed according to this property. A viable approach entails harnessing the potential of a delay link to access the current magnitude at a preceding temporal instant. By comparing this historical value with the contemporaneous current measurement, a residual value is obtained, from which a threshold can be set. Should the residual value exceed this predetermined threshold, the current sensor is deemed to be faulty.

This particular method offers a means to detect faults post facto, once the current sensor has already succumbed to malfunction. To enable timely fault diagnosis, a novel approach centered on signal estimation is proposed, enabling the anticipation of potential sensor faults based on the discernible trends exhibited by the current signal. Leveraging the capabilities of a tracking differentiator, an ideal differentiation signal can be extracted, representing the rate of change of the current magnitude. Specifically, when the discrete step is denoted as h and the current differential as $\dot{i}$, the estimated current at the given step can be obtained as follows:

$$\hat{i} = i + kh \cdot \dot{i} \tag{8.78}$$

In the case of the normal operation of the current sensor, the residual between the estimated value of the current value and the actual measured value will be limited within a certain range. In this way, a threshold can be set, and the current sensor is considered to be working normally when the residual is within the threshold.

8.3.2.2 Simulation Analysis of Current Sensor Fault Diagnosis

This section simulates the output zeroing fault of the current sensor. The parameters of the differentiator in the current value estimation are $r = \pi \times 10^{10}$, $c_0 = 3$, and

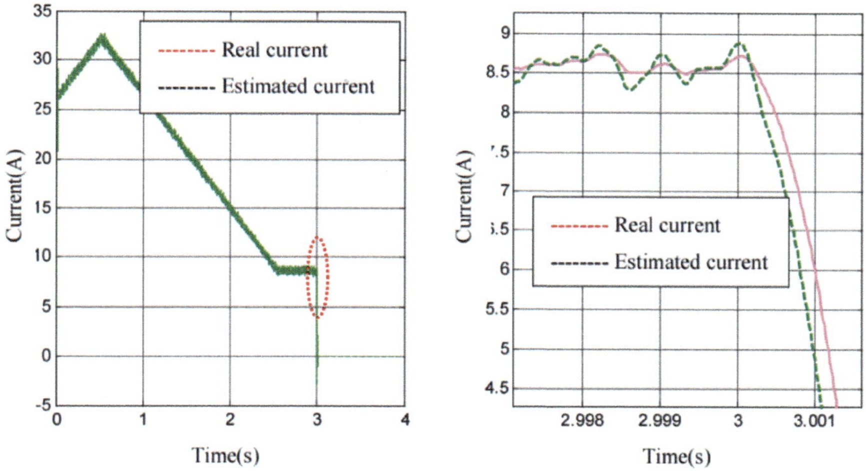

Fig. 8.18 Current measurements and estimations

the estimation step is $k = 10$, the occurrence time of the current sensor fault is $t = 3$ s. The current estimation simulation results are shown in Fig. 8.18. The figure on the left show cases the comparative trajectory of the estimated current value against the actual measured value throughout the entirety of the simulated process. From the simulation results, the estimated current value is consistent with the actual value as a whole. To offer a more focused perspective, the right panel presents an exquisite magnification of the aforementioned trajectory. Remarkably, the simulation elicits a congruence between the estimated and actual current values.

The operation of the whole process fault diagnosis is shown in Fig. 8.19. The above figure shows the residual value between the estimated current value and the measured current value. During normal operation, the residual value is within the range of 0.3 A, and the maximum value of the residual value is 2.2 A nearby $t = 3$ s. The following figure shows the decision quantity d, decision threshold thr and fault indication s obtained from Bayes decision theory, $s = 0$ indicating that the current sensor works normally and $s = 1$ indicating that the current sensor fails. The simulation results show that under normal circumstances $d \leq 4$, the decision quantity d increases to more than 400 when the current sensor fails, and s jumps from 0 to 1 when the current sensor fails.

8.3.2.3 Current Sensor Fault Diagnosis Experiment

To further verify the reliability of the algorithm, the simulation experiment of the maglev system can be carried out with the help of the maglev test bench. During the experiment, the output of the current sensor is artificially cutoff to simulate the zeroing fault of the current sensor. The output signal of the current sensor is processed

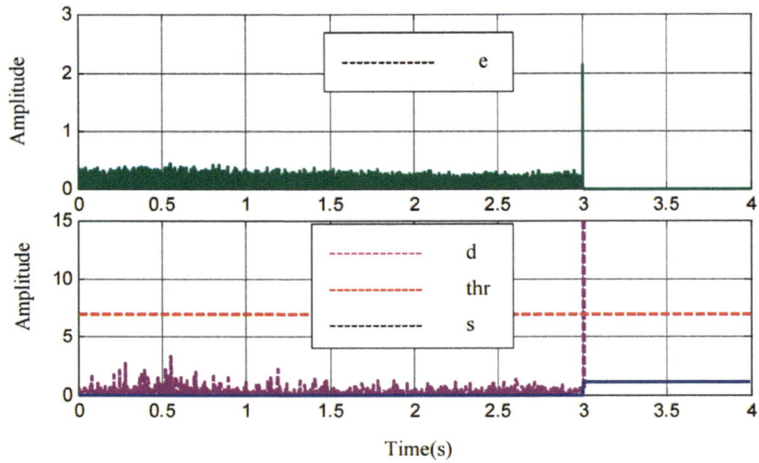

Fig. 8.19 Simulation results of current sensor fault diagnosis unit

to determine whether the current sensor has a fault. For comparison, both tracking differentiator and classical differentiator are used to estimating the current value.

The experimental result of using the tracking-differentiator to estimate the current value is shown in Fig. 8.20. The parameters of the differentiator in the current value estimation are $r = \pi \times 10^4$, $c_0 = 2$, and the estimation step is $k = 10$. Figure 8.20a is the levitation gap value of the whole process obtained by the gap sensor. When the maglev system falls on the track, the levitation gap is 10 mm. When it is stably levitated, the levitation gap is 5 mm. Then at the moment $t = 4.4$ s, the current sensor signal is cutoff to simulate the current sensor zeroing fault. Figure 8.20b shows the current measured value i and the current estimated value $\hat{i}$, the estimated value is basically coincident with the measured value. Figure 8.20c shows the decision quantity, threshold and fault indication. When the system is working normally, the decision-making quantity is less than the threshold, and the fault indication is 0, indicating that there is no fault. When the current sensor has a zeroing fault, the magnitude of the decision-making quantity increases rapidly. When the threshold is exceeded at the time $t = 4.41$ s, the fault indication jumps from 0 to 1, indicating that the current sensor is faulty. During this period, the gap value changed from 4.96 to 4.95 mm, and there is no large fluctuation, indicating that the fault diagnosis results are timely and effective.

The experimental results of the current value estimation using the classical differentiator (the parameter of the classical differentiator is $T = 0.001$) are shown in Fig. 8.21. Figure 8.21a shows the current measured value i and current estimated value $\hat{i}$, and Fig. 8.21b shows the residual value between them. On the whole, the estimated value is the same as the measured value, but compared with Fig. 8.20 (b), the estimated value is noisy, and there is a large deviation in the estimated value near $t = 1.3$, 1.9, and 3.0 s. Figure 8.21c shows the decision quantity and threshold. When

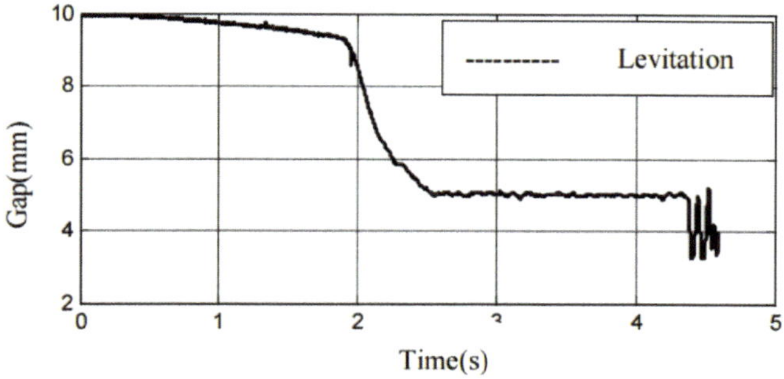

(a) Levitation gap

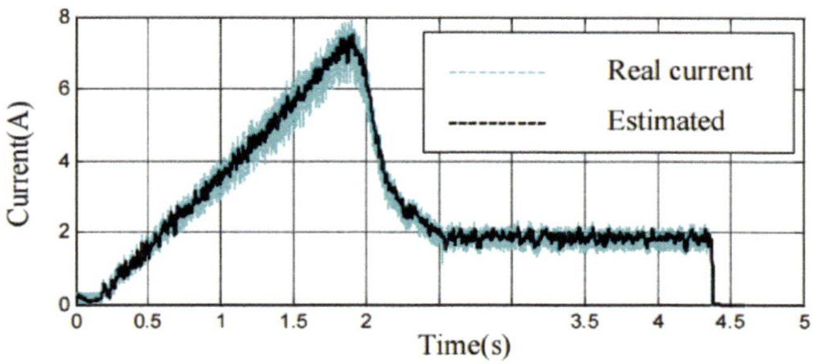

(b) Current value and current estimated value

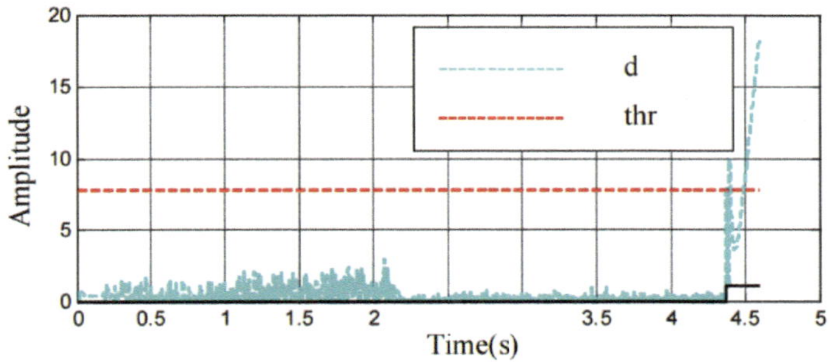

(c) Decision quantity and threshold

Fig. 8.20 Fault diagnosis with tracking-differentiator

$t = 1.9$ s, the decision quantity exceeds the threshold, and misjudgment occurs. This occurrence serves as an unequivocal testament to the superior capabilities exhibited by the tracking differentiator in its precise estimation of the current signal.

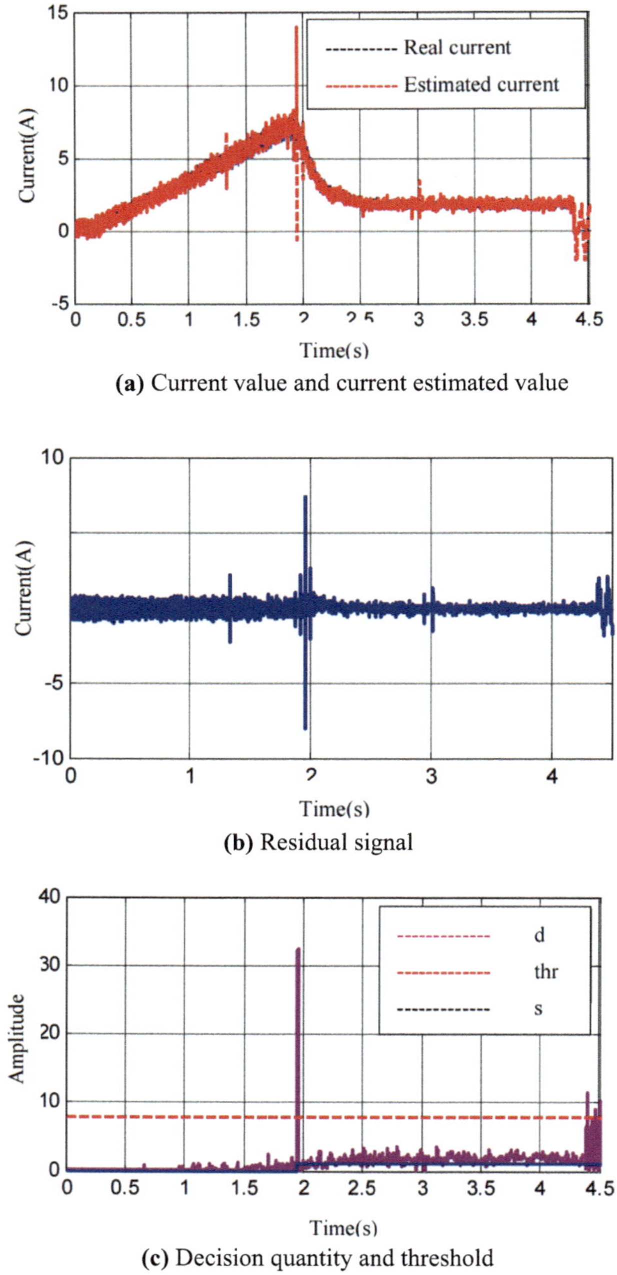

Fig. 8.21 Fault diagnosis with classical differentiator

8.4 Conclusion

This chapter entails the execution of fault diagnosis for the PEM hybrid levitation system, thus laying a foundation for subsequent fault-tolerant control implementation. The design of Kalman filter banks is employed for the purpose of diagnosing and localizing different sensor faults. Taking into account potential system model uncertainty, a fault diagnosis methodology for the levitation system is proposed, which relies on the conjoined influence of residual error and data. Building upon this framework, a subsequent processing approach for the outcomes of fault isolation is also presented. By leveraging signal analysis, a fault diagnosis method for the accelerometer and current sensor is introduced, which is based on the analysis of signals and takes into account signal redundancy.

Bibliography

1. Chen Z, Fang H, Chang Y (2016) Weighted data-driven fault detection and isolation: a subspace-based approach and algorithms. IEEE Trans Industr Electron 63(5):3290–3298
2. Chen Z, Zhang K, Ding SX et al (2016) Improved canonical correlation analysis-based fault detection methods for industrial processes. J Process Control 41:26–34
3. Ding SX (2013) Model-based fault diagnosis techniques: design schemes, algorithms, and tools. Springer, New York, pp 50–56
4. Ding SX (2014) Data-driven design of fault diagnosis and fault-tolerant control systems. Springer Science & Business Media, New York
5. Chen H, Long Z, Chang W (2006) Fault tolerant control research for high-speed maglev system with sensor failure. Sixth World Congr Intell Control Autom 3:2281–2285
6. Liu Huaqing Compile (1995) The German maglev train-Transrapid. University of Electronic Science and Technology Press, Sichuang
7. Liu Z, Long Z, Li X (2015) Maglev trains. Springer, Berlin
8. Long Z, Dou F, Wang Z et al (2023) Current status and prospects of high-speed maglev suspension and guidance control technology. Sci Technol Foresight 2(4):78–88
9. Long Z, Li X, Cheng H (2023) Research on permanent-electromagnetic levitation technology and application. Shanghai Scientific and Technical Publishers, Shanghai
10. Long Z, Li Y, He G (2010) Research on electromagnet fault diagnosis technology of suspension control system of maglev train. Control Decis 25(7):1004–1009
11. Long Z, Lv Z, Chen H et al (2003) Analysis of the reliabilities of maglev train power system with FTA method. In: Fifth international symposium on instrumentation and control technology, Beijing, China, vol 5253, pp 1001–1005
12. Long Z, Xue S, He G et al (2011) Fault-diagnosis for the accelerometer of maglev suspension system based on signal comparison. Chin J Sci Instrum 32(12):2641–2647
13. Long Z, Zhai M, Wang Z (2023) Condition monitoring, fault diagnosis and fault tolerant control of the maglev train. Shanghai Scientific and Technical Publishers, Shanghai
14. Luo H, Yin S, Liu T et al (2019) A data-driven realization of the control-performance-oriented process monitoring system. IEEE Trans Industr Electron 67(1):521–530
15. Sansan D (2022) 600 km/h high-speed maglev transportation system. Shanghai Scientific and Technical Publishers, Shanghai
16. Wang Z (2019) Fault diagnosis and tolerant control for high speed maglev train suspension system. Ph.D. dissertation, National University of Defense Technology Doctor Degree, Changsha

17. Wang Z, Li X, Xie Y, Long Z (2018) Maglev train signal processing architecture based on nonlinear discrete tracking differentiator. Sensors 18(6):1697
18. Wang Z, Long Z, Luo J et al (2020) Data driven state monitoring of maglev system with experimental analysis. IEEE Access 8:79104–79113
19. Wang Z, Long Z, Luo J et al (2023) A data-driven fault diagnosis of high speed maglev train levitation system. Int J Adapt Control Signal Process 37(10):2671–2689
20. Wang P, Long Z, Xu Y (2021) Component-level fault detection for suspension system of maglev trains based on autocorrelation length and stable kernel representation. IEEE Trans Vehic Technol 70(8):7594–7604
21. Wu X (2003) Maglev train. Shanghai Scientific and Technical Publishers, Shanghai
22. Xie Y, Long Z (2009) A high-speed nonlinear discrete tracking-differentiator with high precision. Control Theory Appl 26(2):127–132
23. Xie Y, Long Z, Li J et al (2008) Research on new nonlinear discrete-time tracking-differentiator filtering characteristic. In: The 7th World Congress on Intelligent Control and Automation, Chongqing, China, pp 6750–6755
24. Yang B, Mei Z, Long Z (2023) Online anomaly detection method integrating LSTM and MGD for suspension system of maglev trains. J Traffic Transport Eng 23(6):216–231
25. Ye J, Long Z (2007) Fault diagnosis theory based on petri net. Control Decis 22(12):1403–1407
26. Zhou X, Wen T, and Long Z (2023) Anomaly detection of suspension system in maglev train based on missed detection rate. J Southwest Jiaotong Univ 58(4):903–912

Chapter 9
Fault Tolerant Control of High-Speed PEM Hybrid Levitation System

Fault tolerance technology is an effective method to improve system reliability without changing the system structure and increasing cost. A unified fault diagnosis and tolerant control scheme for the levitation system of high-speed maglev train is proposed based on the controller parameterization theory. For incipient faults, a fault tolerant control strategy based on the online update of control parameters is proposed. Aiming at serious system failures, fault tolerant control methods based on signal reconstruction and switching strategy are proposed.

9.1 Fault Tolerant Control Architecture Design for Levitation System

The fault tolerant control based on the Youla parameterization principle is as follows: firstly, all stabilized controllers are expressed in the form only related to the parameter matrix $\boldsymbol{Q}(z)$, and then the controller output is changed by adjusting the value of $\boldsymbol{Q}(z)$, to achieve the purpose of fault tolerant control. After completing the parameterization of the controller, the next task is to adjust the value of the parameter matrix according to the fault condition, to achieve the purpose of fault tolerant control.

A realization form of Youla parameterization for fault tolerant control is proposed as: suppose the nominal system can be stabilized by $u_n(z)$, then all proper controllers achieving internal stability can be parameterized by:

$$u(z) = u_n(z) + \boldsymbol{Q}(z)\boldsymbol{r}(z) \tag{9.1}$$

where $\boldsymbol{r}(z)$ is the residual of plant, $\boldsymbol{Q}(z)$ is the parameterization matrix which can be updated iteratively online for tolerant purpose concerning possible faults.

The idea of fault tolerant control of the system under this control structure is to improve the response of the system under fault conditions by adjusting the values in

Z. Long et al., *High-Speed Maglev Train's Levitation and Guidance Control*, Advances in High-speed Rail Technology, https://doi.org/10.1007/978-981-97-2309-6_9

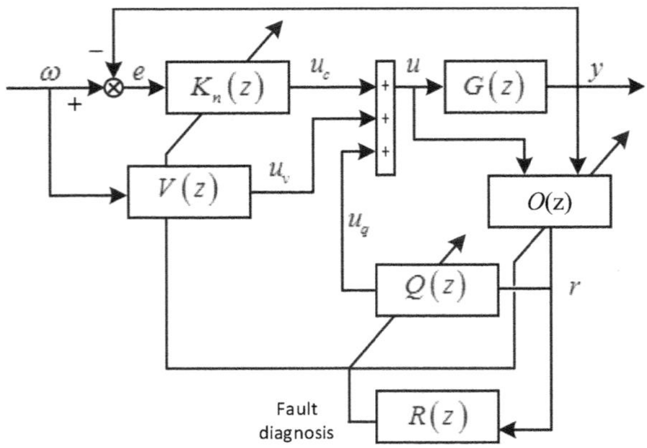

Fig. 9.1 Fault tolerant control based on Youla parameterization

the Youla parameterized controller. Under minor faults, the value $\boldsymbol{Q}(z)$ can be directly fine-tuned to improve the system response after the fault. In the case of serious faults, the main controller $K_n(z)$, residual generator $\boldsymbol{O}(z)$, and compensation controller $\boldsymbol{Q}(z)$ need to be redesigned. In addition, a residual signal processing post-filter $\boldsymbol{R}(z)$ can be added for fault diagnosis purpose. Based on the above analysis, the fault tolerant control structure of the system can be represented by the frame shown in Fig. 9.1.

In the actual system, many factors affect the performance of the levitation system, such as the change of the equilibrium point and the working environment of the levitation system, the aging or even failure of the electronic components in the system during the operation process, the wear and deformation of the mechanical structure, and the maintenance and replacement. The possible impact of these factors is that the initially designed controller cannot guarantee that the levitation system always maintains the optimal control effect. Therefore, it is necessary to design a fault tolerant control method that can update the controller parameters and even the controller structure online, so that the system can still maintain the optimal control effect in the presence of faults. The fault mentioned here is a generalized fault, that is, it includes not only the complete failure of the components in the system, but also the minor faults of the components in the system before the complete failure. Minor faults are characterized by small fault amplitudes, which may be intermittent and time-varying. Considering that minor faults are usually time-varying and uncertain, it is more difficult to diagnose minor faults, and it is more appropriate to use a structure that does not require a fault diagnosis unit for fault tolerant control.

Considering the influence of different faults on the performance of the levitation system, faults are divided into serious faults, general faults, and minor faults.

Serious faults refers to the complete failure of a single levitation point, that is, a levitation point in the joint structure cannot output the control voltage, resulting in the levitation point being completely uncontrollable. Potential causes of this type of failure include drive circuit failure, power failure, IGBT failure, etc. At this time, the performance of the levitation system of the entire joint structure can only be improved by adjusting the working state of the normal working levitation points in the joint structure. In addition, in the case of some unknown faults, to avoid further damage to the system, the controller will take protective measures to close the output of the single levitating point output voltage. This situation is also classified as a single point of complete failure and can be handled by the same fault tolerant control strategy.

The general fault mainly refers to the failure of the sensor, including the fault of the gap sensor, the fault of the accelerometer, the fault of the current sensor, etc., also known as the controllable single-point partial fault. For some sensor faults, the signal reconstruction method can be used to build an observer by using the analytical redundancy relationship between the internal variables of the system or directly using the redundant relationship between the sensor signals to reconstruct the fault signal.

Minor faults mainly refer to the performance degradation of system components before complete failure, such as sensor signal drift, electromagnet characteristics changes, simulated device characteristics changes, mainly caused by mechanical friction, material deformation, and device aging accumulated in the long-term operation of the levitation system and other factors. Minor faults have the characteristics of small fault amplitude, uncertain fault trend, and time-varying, which will not affect system stability but affect system performance. Ideally, the nominal controller can make the levitation system achieve the optimal performance. In contrast, the minor fault makes the levitation system deviate from optimal performance, so the controller parameters need to be adjusted. Given the time-varying and uncertain nature of such faults, the ideal solution is to make the levitation system maintain the optimal performance through the online parameter adjustment strategy.

A complete fault tolerant control architecture can be seen in Fig. 9.2. Residual $\boldsymbol{r}(z)$ can be obtained through model-based or data-driven observers design approaches. For the single levitation subsystem zero output fault, observers should be designed respectively for the normal system and the faulty system to generate residual signals $\boldsymbol{r}_n$ and $\boldsymbol{r}_f$, which carry the incipient faults information. The corresponding parameterization matrix $\boldsymbol{Q}_n$ and $\boldsymbol{Q}_f$ are online updated to make the system robust to the potential incipient faults. Then with the fault information about the single levitation subsystem zero output fault, the final control output is chosen from the normal control value $u_n + u_{qn}$ and the fault tolerant control value $u_f + u_{qf}$.

9.2 Fault Tolerant Control of Incipient Fault

The key to fault tolerant control for minor faults lies in how to design the online update strategy of the parameterized controller, so that the parameterized controller can improve the tolerant performance of the system under the premise of maintaining system stability. Focusing on the details of the parameter online update, the fault tolerant control ability of the levitation system under the action of the parameter update method is analyzed. Firstly, the method of converting the Youla parameter matrix into the canonical type is given to realize the one-to-one correspondence

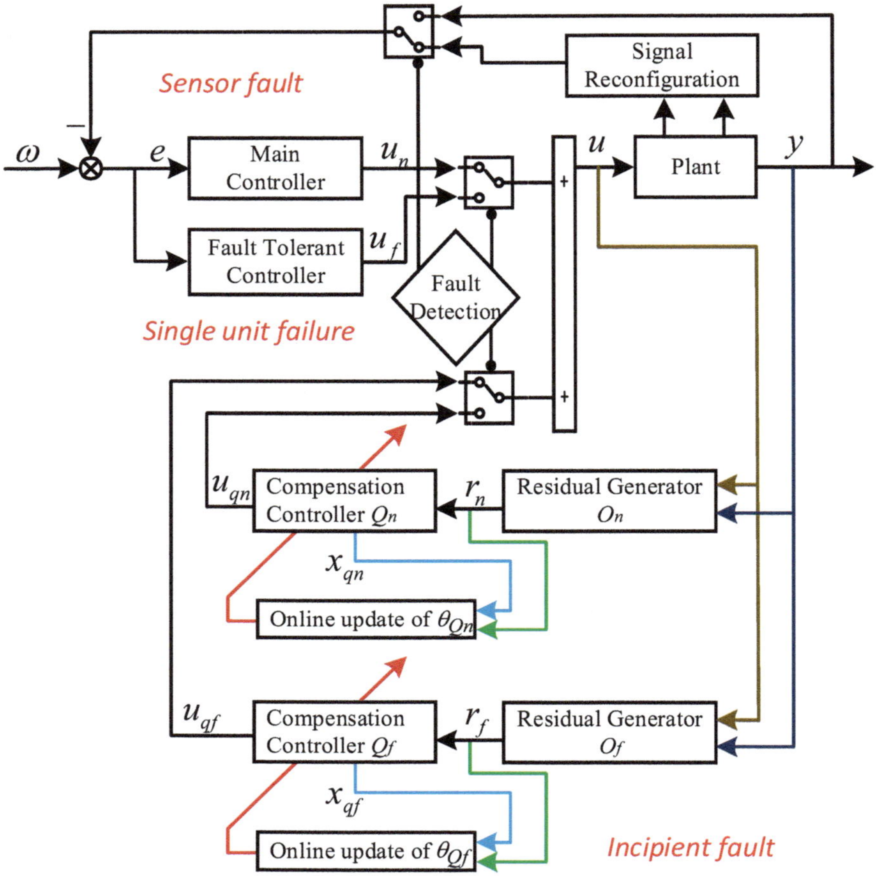

Fig. 9.2 Unified fault diagnosis and fault tolerant control architecture

between the Youla parameter matrix and the canonical parameter. Then the relationship between the performance index and the canonical parameter is given, and the parameter update law based on gradient descent is given.

9.2.1 Canonical Representation of Youla Parameter Matrix

The input–output canonical type was first proposed by Hanzon et al. for the continuous-time state-space model, and then extended to the multiple-input multiple-output discrete-time state-space model. The advantage is that the parametric model can be guaranteed asymptotically stable without additional constraints on the parameter space. Compared with the observable standard type and other forms, the input standard type is less sensitive to parameters, and the small-scale adjustment of parameters will not have much impact on the system performance, so it is more suitable for parameter adjustment. For the state-space model defined by the system matrix $\boldsymbol{A} \in \mathcal{R}^{n\times n}$, $\boldsymbol{B} \in \mathcal{R}^{n\times l}$, $\boldsymbol{C} \in \mathcal{R}^{m\times n}$, $\boldsymbol{D} \in \mathcal{R}^{m\times l}$, the basic definition of the input canonical form corresponding to the matrix pair $(\boldsymbol{A}, \boldsymbol{B})$ is:

$$\begin{bmatrix} \boldsymbol{A}^{\mathrm{T}} \\ \boldsymbol{B}^{\mathrm{T}} \end{bmatrix} = \prod_{i=1}^{n} \boldsymbol{T}_i(\theta_{AB}(i)) \begin{bmatrix} \boldsymbol{I}_n \\ 0_{l\times n} \end{bmatrix} \tag{9.2}$$

where $\boldsymbol{\theta}_{AB} \in \mathcal{R}^{nl}$ is the parameter vector after the input canonical transformation of the matrix pair $(\boldsymbol{A}, \boldsymbol{B})$, and its components satisfy $\theta_{AB}(i) \in [0, 1]$, and $\boldsymbol{T}_i(\theta_{AB}(i)) \in \mathcal{R}^{(n+l)\times(n+l)}$ is the transformation matrix in the process of the input canonical transformation. The value of the transformation matrix $\boldsymbol{T}_i(\theta_{AB}(i))$ is given by:

$$\boldsymbol{T}_1(\theta_{AB}(1)) = \begin{bmatrix} \boldsymbol{I}_{n-1} & 0 & 0 \\ 0 & \boldsymbol{U}(\theta_{AB}(1)) & 0 \\ 0 & 0 & \boldsymbol{I}_{l-1} \end{bmatrix}$$

$$\vdots$$

$$\boldsymbol{T}_l(\theta_{AB}(l)) = \begin{bmatrix} \boldsymbol{I}_{n+l-2} & 0 \\ 0 & \boldsymbol{U}(\theta_{AB}(l)) \end{bmatrix}$$

$$\boldsymbol{T}_{l+1}(\theta_{AB}(l+1)) = \begin{bmatrix} I_{n-2} & 0 & 0 \\ 0 & U(\theta_{AB}(l)) & 0 \\ 0 & 0 & I_l \end{bmatrix}$$

$$\vdots$$

$$\boldsymbol{T}_{2l}(\theta_{AB}(2l)) = \begin{bmatrix} \boldsymbol{I}_{n+l-3} & 0 & 0 \\ 0 & \boldsymbol{U}(\theta_{AB}(2l)) & 0 \\ 0 & 0 & 1 \end{bmatrix}$$

$$\vdots$$

$$\boldsymbol{T}_{(n-1)l+1}(\theta_{AB}((n-1)l+1)) = \begin{bmatrix} \boldsymbol{U}(\theta_{AB}((n-1)l+1)) & 0 \\ 0 & \boldsymbol{I}_{n+l-2} \end{bmatrix}$$

$$\vdots$$

$$\boldsymbol{T}_{nl}(\theta_{AB}(nl)) = \begin{bmatrix} \boldsymbol{I}_{l-1} & 0 & 0 \\ 0 & \boldsymbol{U}(\theta_{AB}(nl)) & 0 \\ 0 & 0 & \boldsymbol{I}_{n-1} \end{bmatrix}$$

where $\boldsymbol{U}(\theta_{AB}(i))$ is the matrix consisting of the parameter components $\theta_{AB}(i)$ shown in (9.3), which satisfies the normalization condition $\boldsymbol{U}^{\mathrm{T}}(\theta_{AB}(i))\boldsymbol{U}(\theta_{AB}(i)) = \boldsymbol{I}$.

$$U(\theta_{AB}(i)) = \begin{bmatrix} -\theta_{AB}(i) & \sqrt{1-\theta_{AB}^2(i)} \\ \sqrt{1-\theta_{AB}^2(i)} & \theta_{AB}(i) \end{bmatrix}, \quad i = 1, 2, \ldots, nl \tag{9.3}$$

The canonical transformation shown in Eq. (9.2) completes the mapping from the parameter $\mathcal{R}^{n(n+l)}$ space to $\mathcal{R}^{nl}$, reduces the required parameters, and ensures that the matrix pair $(\boldsymbol{A}, \boldsymbol{B})$ is stable when $\theta_{AB}(i) \in [0, 1]$.

9.2.2 Online Update of Youla Parameters Based on Gradient Descent

The initial stage of performance optimization, utilizing the gradient descent technique, involves acquiring the gradient pertaining to the parameter with respect to the performance index. Subsequently, the parameter undergoes successive iterations, updating its value in the direction of the gradient descent.

Set the quadratic performance index as:

$$J(i) = \sum_{k=k_0+N(i-1)+1}^{k_0+Ni} \left[\boldsymbol{e}(k)^{\mathrm{T}}\boldsymbol{W}_e(k)\boldsymbol{e}(k) + \boldsymbol{u}(k)^{\mathrm{T}}\boldsymbol{W}_u(k)\boldsymbol{u}(k)\right] \tag{9.4}$$

where i is the number of iterations, k_0 is the initial value of the window, N is the width of the window, $\boldsymbol{W}_e(k)$ is the weight value of the error independent of the parameter θ, and $\boldsymbol{W}_u(k)$ is the weight value of the control variable independent of the parameter θ. The value of the performance index has a close relationship with the compensation controller, and the control goal is to reduce the quadratic performance index by adjusting the compensation controller online. Due to the one-to-one correspondence between the parameter θ and the compensation controller $\boldsymbol{Q}(z)$, the control objective is equivalent to adjusting θ to make J decrease rapidly. A possible strategy is to change θ along the direction of the negative gradient of J:

$$\theta(i+1) = \theta(i) - \lambda \nabla J(i) \tag{9.5}$$

where λ is the parameter update step size, and $\nabla J(i)$ represents the gradient of the indicator $J(i)$:

$$\nabla J(i) = 2 \sum_{k=k_0+N(i-1)}^{k_0+Ni} \left[\boldsymbol{e}(k)^{\mathrm{T}}\boldsymbol{W}_e(k)\frac{\partial \boldsymbol{e}(k)}{\partial \theta} + \boldsymbol{u}(k)^{\mathrm{T}}\boldsymbol{W}_u(k)\frac{\partial \boldsymbol{u}(k)}{\partial \theta}\right] \tag{9.6}$$

It can be seen from Eq. (9.6) that when calculating the gradient $\nabla J(i)$, it is necessary to obtain the partial differential of the tracking error $\boldsymbol{e}(k)$ and the control variable $\boldsymbol{u}(k)$ with respect to the parameter θ.

9.3 Fault Tolerant Control of Levitation System Based on Switching

The key to fault tolerant control for sensor fault is reconstructing the signal value by using the analytical redundancy relationship between variables in the levitation system and the interaction relationship between two levitation points in the joint structure. The key to fault tolerant control for a single levitation point fault is establishing the joint structure levitation system model and designing the corresponding fault tolerant controller for the system model after the fault.

9.3.1 Active Fault Tolerant Control Based on Signal Reconstruction

The fault tolerant control strategy based on signal reconstruction is mainly aimed at the sensors directly related to the calculation of control quantity in the levitation system. The most important and most prone to failure are the levitation gap sensor and acceleration sensor. In the "two in one" sensor, the gap sensor and acceleration sensor are installed in the same shell and have an independent analog signal conditioning circuit. The premise of using signal reconstruction for fault tolerant control is that there is analytical redundancy in the system, and there is a direct redundancy relationship between acceleration signal and levitation gap signal. The sensor distribution of the joint structure levitation system is shown in Fig. 9.3. Similarly, the sensor signals in the left and right single-point levitation systems are distinguished by affixing 1 and 2 respectively in the figure. The sensor configurations on the left and right sides in the joint structure are the same, in which the current sensor, voltage sensor, and temperature sensor are located in the controller cabinet, and the gap sensor and accelerometer are integrated into a sensor shell and installed on the upper surface of the electromagnet. Among them, the current sensor signal, gap sensor signal, and accelerometer signal are directly related to the calculation of control quantity. The main reason why each single levitation point is equipped with two gap sensors is that there are joints between two tracks. The continuous arrangement of two gap sensors can ensure that when any gap sensor faces the track seam, the other gap sensor can provide the levitation gap signal. The levitation gap sensor and accelerometer are installed on the surface of the electromagnet module. Because the sensor experiences long-time oscillation with the electromagnet module, there is a complex leakage magnetic field between the electromagnet module and the track, and the sensor will be affected by the change of electromagnet pole temperature, the gap sensor and accelerometer have a higher failure rate than other sensors in the actual levitation system, the corresponding fault tolerant control problem should be paid more attention.

The fault tolerant control structure based on the signal reconstruction strategy is shown in Fig. 9.4. Under normal conditions, the normal sensor value of the system

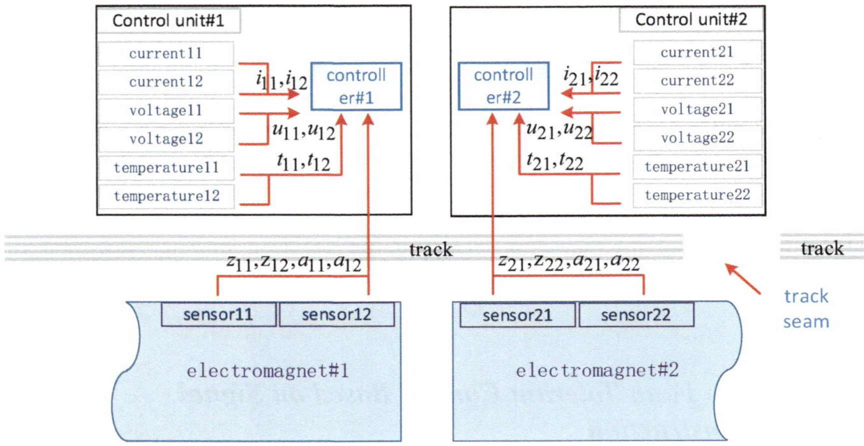

Fig. 9.3 Schematic diagram of sensor distribution of joint structure levitation system

is selected for feedback control purpose. When the system fails, the signal source is switched by the signal reconstruction method to achieve the purpose of fault tolerant control. This method also needs a fault detection mechanism to judge the switching time between the original signal and the reconstructed signal. The key to the fault tolerant control method is that the fault diagnosis mechanism can make accurate and timely switching. The reconstructed signal needs to reflect the actual change or change trend of the system as the original signal. The reconstruction signal can be obtained by constructing an observer or using the simple redundant relationship between system variables and sensors.

(1) Reconstruction method of gap differential signal under accelerometer fault

In the calculation of the control quantity, the required system variables are the left and right levitation gaps, and the derivatives of the left and right levitation gaps.

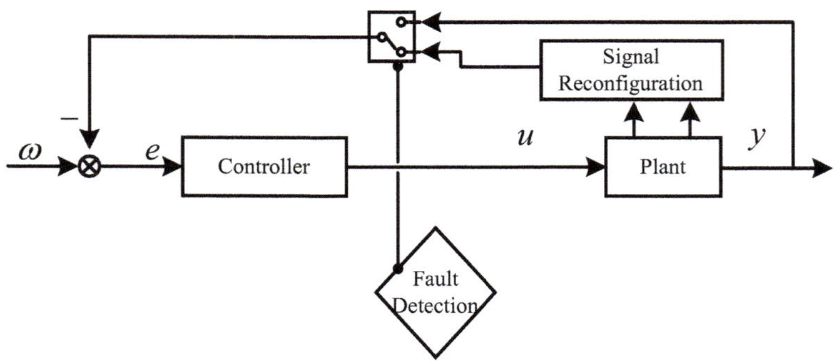

Fig. 9.4 Fault tolerant control structure based on signal reconstruction

According to the sensor configuration shown in Fig. 9.1, the levitation gaps on the left and right sides are obtained by the gap sensors, and the derivative of the levitation gaps are equivalently obtained by integrating the accelerometer signals on the left and right sides. It is easy to differentiate the signal in the analog circuit, but in the digital controller of the magnetic levitation system, the gap signal obtained by the controller is a digital quantity. Differentiating the digitized gap signal will seriously amplify the high-frequency noise within the gap signal. Therefore, only by finding a method that can effectively obtain the differential signal can we reconstruct the differential signal of the gap, to realize the fault tolerant control of the accelerometer failure.

The advantage of the tracking-differentiator is that it can effectively extract the differential signal of the gap while suppressing the amplification of high-frequency noise. Figure 9.5 shows the schematic diagram of the fault tolerant control structure of the accelerometer based on gap differential signal reconstruction. After the gap signal is processed by the tracking-differentiator, two outputs can be obtained, which are the filtered value and the differential value of the gap signal, and the differential output is used as the backup signal of the acceleration integral signal. Under normal circumstances, the levitating acceleration integral is selected to calculate the control amount. When the fault detection unit detects that the accelerometer is faulty, this signal is switched to the backup signal obtained by the tracking-differentiator.

(2) Reconstruction method of levitation gap signal in case of single gap sensor failure

In high-speed maglev tracks, track seams are inevitable (Fig. 9.6). When the levitation gap sensor faces the track seam, the levitation gap sensor outputs its maximum value, which is 20 mm. In the high-speed maglev train levitation system, each controller chassis corresponds to two gap sensors, and the two gap sensors are arranged on the electromagnet module in tandem along the track direction. At the same time, only one gap sensor faces the seam, so the track seam will not affect the levitation system

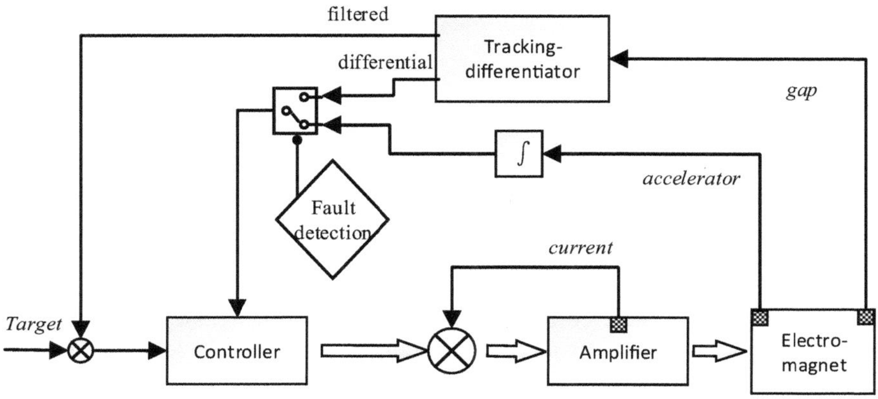

Fig. 9.5 Accelerometer fault tolerant control architecture

Fig. 9.6 High speed maglev train track seam

when the system is normal, that is, there is always a sensor that can feedback the real levitation gap value at any time. When one of the two gap sensors fails, for example, its output value is stuck at 20 mm, the comprehensive gap value at this time is equal to the value of one of the gap sensors, so it cannot deal with the problem caused by the track seam.

Since accelerometers are available in a levitation system and it is well known that the double integral of acceleration is equal to the position of the electromagnet, an easy signal reconfiguration method is prompted as:

$$\hat{z}_{rcf}(t) = \int_0^t \int_0^t a(t)\mathrm{d}t\mathrm{d}t \tag{9.7}$$

where $a(t)$ is the acceleration signal from the accelerometer. However, since there exist deviations of acceleration signals due to the accuracy limit of the accelerometer, like $a(t) = a_{\text{real}}(t) + \varepsilon$, where $a_{\text{real}}(t)$ is the real acceleration and ε is the sensor deviation. Then (9.7) equals to:

$$\hat{z}_{rcf}(t) = \int_0^t \int_0^t a(t)\mathrm{d}t\mathrm{d}t = \int_0^t \int_0^t a_{real}(t)\mathrm{d}t\mathrm{d}t + \int_0^t \int_0^t \varepsilon\mathrm{d}t\mathrm{d}t \tag{9.8}$$

The term of the double integral of ε in (9.8) can accumulate this sensor deviation to a pretty big error, so some measures are to be adopted to correct this error. In the maglev system, the equilibrium current is closely related to the equilibrium levitation gap. Suppose there is a positive error in the accelerometer signal. In that case, the double integral of the accelerometer signal is larger than the real levitation gap, and a larger than needed output voltage is generated to minimize the levitation gap. To rectify this issue, the current difference between the two single subsystems can be used to reduce the output to a proper degree. Considering this, a modified signal reconfiguration method is proposed as:

$$\hat{z}'_{rcf}(t) = \int_0^t \int_0^t a_1(t)\mathrm{d}t\mathrm{d}t - k_{\mathrm{cmp}}[i_1(t) - i_2(t)]\mathrm{d}t \tag{9.9}$$

where k_{cmp} is the compensation coefficient.

To conclude, under normal conditions, the sensor signals are used directly as feedback; the reconfigured gap signal is used instead when a fault is detected.

9.3.2 Active Fault Tolerant Control Based on Control Law Switching

The active fault tolerant control structure based on the control law switching strategy is shown in Fig. 9.7. In this structure, a main controller and a fault tolerant controller are designed, respectively. The main controller is the nominal controller mentioned above, which ensures that the system has optimal performance under no faults. In contrast, the fault tolerant controller is designed for fault conditions. When the system is working normally, the main controller is used for control, and when the system fails, it is switched to the fault tolerant controller. The switching action depends on the fault indication signal given by the fault detection unit. When the fault indication signal indicates a fault, the switching action occurs. The fault tolerant control method based on the switching strategy relies on the fault detection accuracy and timeliness of the fault detection unit. If there is a missed detection, the system loses its fault tolerant control ability; if a false detection occurs, it will affect the performance of the system during normal operation; if the detection time is too long, it will also weaken the tolerant ability of the fault tolerant control algorithm.

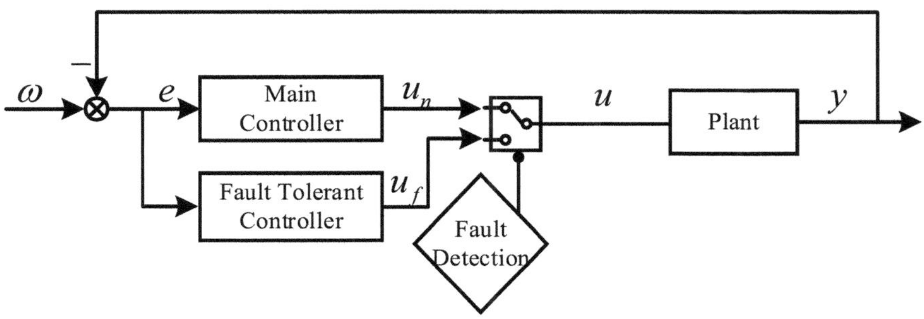

Fig. 9.7 Fault tolerant control structure based on switching strategy

For the joint structure levitation system, two single levitation points in the joint structure send working status signals to each other. Every single levitating point possesses an inherent self-detection capacity. Upon discerning its own operational inadequacies leading to an inability to achieve the desired voltage output, it promptly modifies the value pertaining to its operational state and disseminates this information to its corresponding counterpart within the same joint structure. In tandem, the aforementioned counterpart proactively checks the real-time value of the operational status bit to ascertain the presence of a potential single-point fault, which facilitates the identification of any single-point fault. When the received working status bit shows that the other levitation point is faulty, or the life signal sent by the other levitation point cannot be received, it is considered that the other levitation point is faulty, and the control output is immediately switched to the fault tolerant controller.

The design steps of the fault tolerant controller are first to obtain the post-fault system model, and then to design the controller according to the post-fault system model. Assuming that the left single levitating point fails, the system model after the failure can be obtained based on the simplified normal system model:

$$\begin{cases} m_l\ddot{\delta}_l = m_l g - F_{pel}^f + \frac{1}{2}k_s(\delta_r - \delta_l) + \frac{1}{2}(m_b + m_L)g \\ m_r\ddot{\delta}_r = m_r g - F_{per} + \frac{1}{2}k_s(\delta_l - \delta_r) + \frac{1}{2}(m_b + m_L)g \end{cases} \tag{9.10}$$

where $F_{pel}^f = \frac{\mu_0 N^2 S\cdot(H_c\delta_m/2N)^2}{\delta_1+\delta_m S/2\mu_r S_m^2} = k_{pe}\frac{\alpha^2}{(\delta_l+\beta)^2}$, $F_{per} = \frac{\mu_0 N^2 S(i_r+H_c\delta_m/2N)^2}{(\delta_r+\delta_m S/2\mu_r S_m)^2} = k_{pe}\frac{(i_r+\alpha)^2}{(\delta_r+\beta)^2}$. The state quantities of the system during rebalance after the fault are obtained by simulation: $\delta_{l0}^f = 12.48$ mm, $\delta_{r0}^f = \delta_{r0} = 12$ mm, $u_{r0}^f = 19.38$ V, $i_{r0}^f = 19.38$ A. After the occurrence of a single-point fault, the state variable and output variable remain unaltered, thus enabling the selection of a new control variable $u = u_r$. The linearized system model can then be derived as follows:

$$\begin{cases} \boldsymbol{x}(k+1) = \boldsymbol{A}_d^f x(k) + \boldsymbol{B}_d^f u(k) \\ \boldsymbol{y}(k) = \boldsymbol{C}_d^f x(k) + \boldsymbol{D}_d^f u(k) \end{cases} \tag{9.11}$$

The specific form of each parameter matrix in the linear model (9.11) is:

$$\boldsymbol{A}_d^f = \begin{bmatrix} 0 & 1 & 0 & 0 \\ \frac{2k_{bl}^f - k_s}{2m} & 0 & \frac{k_s}{2m} & 0 \\ 0 & 0 & 0 & 1 \\ \frac{k_s}{2m} & 0 & \frac{2k_{br}^f - k_s}{2m} & 0 \end{bmatrix}, \quad \boldsymbol{B}_d^f = \begin{bmatrix} 0 \\ 0 \\ 0 \\ -\frac{k_a^f}{m} \end{bmatrix}$$

$$\boldsymbol{C}_d^f = \begin{bmatrix} 1\,0\,0\,0 \\ 0\,0\,1\,0 \end{bmatrix}, \quad \boldsymbol{D}_d^f = 0 \tag{9.12}$$

where, $k_{bl}^{f} = 2k_{pe}\frac{\alpha^2}{(\delta_{l0}+\beta)^3}$, $k_{br}^{f} = 2k_{pe}\frac{\left(i_{r0}^{f}+\alpha\right)^2}{(\delta_{r0}+\beta)^3}$, $k_{a}^{f} = 2k_{pe}\frac{\left(i_{r0}^{f}+\alpha\right)}{(\delta_{r0}+\beta)^2}$. By substituting the parameters of the joint structure's levitation system and discretizing its model, the system matrix of the corresponding discrete-time state-space model can be obtained as:

$$A_d^f = \begin{bmatrix} 0.9995 & 0.0002 & 0.0006 & 0.0000 \\ -5.1816 & 0.9995 & 5.7124 & 0.0006 \\ 0.0006 & 0.0000 & 0.9995 & 0.0002 \\ 5.7124 & 0.0006 & -4.5865 & 0.9995 \end{bmatrix} \quad B_d^f = 10^{-4}\begin{bmatrix} 0 \\ -0.001 \\ 0 \\ -3.129 \end{bmatrix} \tag{9.13}$$

In the quadratic performance index, the value of the weighting coefficient is selected as $\tilde{W}_x^f = \text{diag}([1000\ 100\ 1000\ 100\ 80\ 80])$, $\tilde{W}_u^f = 100$, and the standard LQR controller design steps are also used to obtain the control law after the fault. The feedback control coefficient of the right levitation point is:

$$F^f = \begin{bmatrix} F_P^f & F_I^f \end{bmatrix} = \begin{bmatrix} 5054.3 & 78.1 & 5491.9 & 83.5 & -65.38 & -59.31 \end{bmatrix} \tag{9.14}$$

When the communication between the two controllers in the joint structure fails, the left sensor signal is unavailable, and a control method based on a single-point system model is required. At this time, the model needs to be further simplified, and the left electromagnet is regarded as a fixed load. For the electromagnetic levitation system, the force generated by the relative movement of the two sides through the levitation frame during the dynamic adjustment process is ignored; for the permanent magnet electromagnetic hybrid levitation system, in addition to the ignored interaction force generated by the levitation frame, the force generated by the hybrid electromagnet is also assumed to be a fixed attraction force, and the value is the residual magnetic force generated by the hybrid electromagnet at the equilibrium point.

9.4 Verification of Fault Tolerant Control of Levitation System

9.4.1 Aanlysis of Online Optimization of Youla Parameters

Suppose the system model is discretized at a sampling time of 1 ms, the matrix of the discretized system model is obtained as:

$$A = \begin{bmatrix} 1.0021 & 0.001 \\ 4.1799 & 1.0021 \end{bmatrix}, \quad B = \begin{bmatrix} 0 \\ 0.0013 \end{bmatrix}, \quad C = \begin{bmatrix} 1 & 0 \\ 0 & 1 \end{bmatrix}, \quad D = \begin{bmatrix} 0 \\ 0 \end{bmatrix} \tag{9.15}$$

The state space expression of the control law designed for the model is:

$$\begin{cases} \boldsymbol{x}_c(k+1) = \boldsymbol{x}_c(k) + \begin{bmatrix} 0.0391 & 0.0391 \end{bmatrix} \boldsymbol{e}(k) \\ \boldsymbol{u}_c(k) = 128\boldsymbol{x}_c(k) + \begin{bmatrix} 6493.8 & 105.3351 \end{bmatrix} \boldsymbol{e}(k) \\ \boldsymbol{x}_c(0) = \boldsymbol{e}(0) = 0 \end{cases} \tag{9.16}$$

The specific form of the residual generator is:

$$\begin{cases} \boldsymbol{x}_o(k+1) = \begin{bmatrix} 1.0021 & 0.001 \\ 4.1799 & 1.0021 \end{bmatrix} \boldsymbol{x}_o(k) + \begin{bmatrix} 0 \\ 0.0013 \end{bmatrix} \boldsymbol{u}(k) + \begin{bmatrix} 1.0021 & 0.0010 \\ 4.1799 & 1.0021 \end{bmatrix} \boldsymbol{r}(k) \\ \boldsymbol{y}_o(k) = \begin{bmatrix} 1 & 0 \\ 0 & 1 \end{bmatrix} \boldsymbol{x}_o(k) + \begin{bmatrix} 0 \\ 0 \end{bmatrix} \boldsymbol{u}(k) \\ \boldsymbol{r}(k) = \boldsymbol{y}(k) - \boldsymbol{y}_o(k), \boldsymbol{x}_o(0) = \boldsymbol{u}(0) = \boldsymbol{r}(0) = \begin{bmatrix} 0 \\ 0 \end{bmatrix} \end{cases} \tag{9.17}$$

The order of the parameterized matrix is taken as $n_q = 1$, and the initial value is:

$$\boldsymbol{A}_q = 0, \quad \boldsymbol{B}_q = [1, 1], \quad \boldsymbol{C}_q = 0, \quad \boldsymbol{D}_q = [0, 0] \tag{9.18}$$

The online iterative process is carried out according to the steps described above. Set the window width in the performance index (9.4) as $N = 1000$, and during the simulation process, $\boldsymbol{A}_q$, $\boldsymbol{B}_q$, $\boldsymbol{C}_q$, $\boldsymbol{D}_q$ are adjusted every 1 s.

(1) Simulation analysis of sensor fault situation

Upon initiating the simulation, a square wave signal with a 1 mm amplitude and a duration of 5 s is periodically introduced to the levitation gap signal beginning from the 11th second. This addition serves to simulate a sensor additive fault situation, akin to that experienced by a maglev train in motion encountering a 1 mm track irregularity. For the convenience of the display, the data in the 10-s window is displayed in the same abscissa, and different time periods are distinguished by the legend. The simulation results are shown in Fig. 9.8. Among them y_0 is the levitation gap response when the nominal controller is used in the interval of 10–20 s, y_1–y_4 are iteratively updated online using the controller parameters in the interval of 10–20 s, 50–60 s, 80–90 s, and 480–490 s, respectively. It can be seen from the simulation results that with the iterative updating of parameters, the performance of the levitation system under fault conditions is gradually improved. Since the iteration of parameters requires a certain processing time, the compensation situation needs to be strengthened at the moment when the disturbance suddenly occurs.

During this process, the change of the output control amount of the compensation controller is shown in Fig. 9.9. The value of the compensation control variable gradually increases during each disturbance application process, and finally stabilizes to a square wave shape compensation output corresponding to the disturbance. There

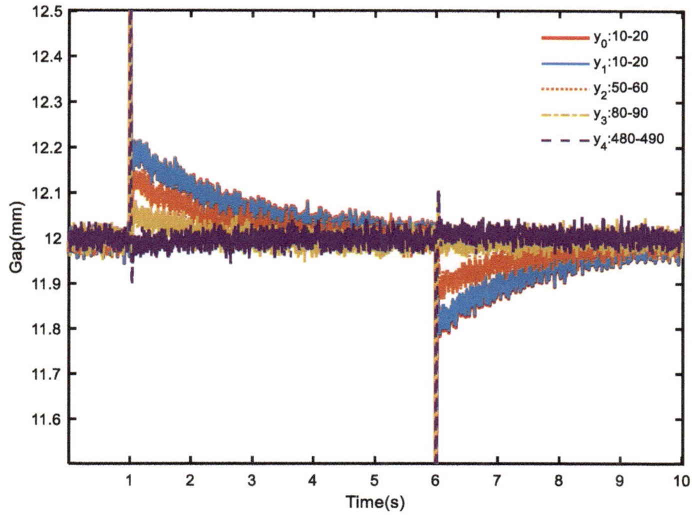

Fig. 9.8 Levitation gap response during iterative update process of controller parameters when gap sensor fails

are sharp peaks in the waveform at the moment when the disturbance is applied and canceled, which is conducive to the rapid stabilization of the system. When the disturbance is stable, it is a constant value deviation, which is beneficial to balance the external disturbance. It can also be seen from the response curve of the compensation control amount that the compensation control amount plays a role only when a fault occurs, and the value of the compensation control amount is approximately zero when there is no fault. This ensures that the fault tolerant controller can improve the performance under faults without affecting the normal operation of the system when there is no fault.

The online update situation of the controller parameter iterative update process is shown in Fig. 9.10. As the number of iterations increases, the control amount gradually approaches the step-type output. The maximum amplitudes of the main control quantity fluctuation at the moment when the fault appears and disappears are 7 V and 12 V, respectively.

(2) Simulation analysis of actuator fault conditions

In this part of the simulation, failure of the actuator is simulated. After the simulation starts, the current value is periodically increased, and the fault amplitude is set to 1 A, and the duration is 0.1 s. In this fault situation, the response of the levitation gap under the parameter online update strategy is shown in Fig. 9.11. Figure 9.12 and Fig. 9.13 show the online update of u_q and u_c under actuator failure.

In terms of the order of the parameter matrix, it is also possible to choose a form that is easier to implement, such as $n_q = 0$, and the parameter matrix at this time is simplified to a proportional link. When $n_q = 0$, the response of the levitation

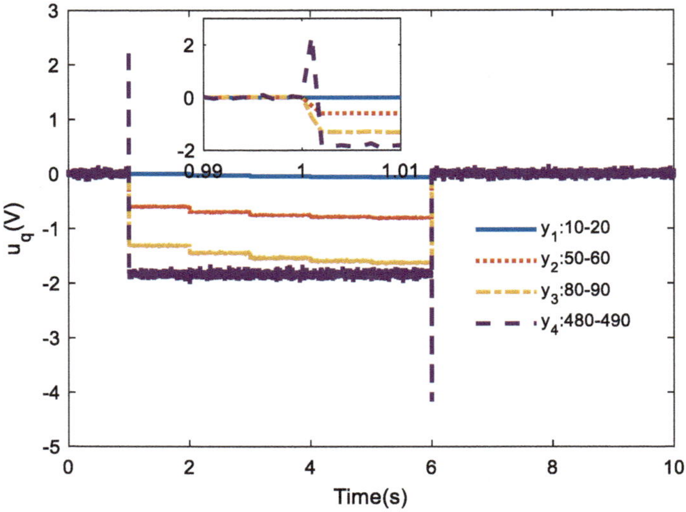

Fig. 9.9 Online update of compensation control variable u_q in the iterative update process of controller parameters

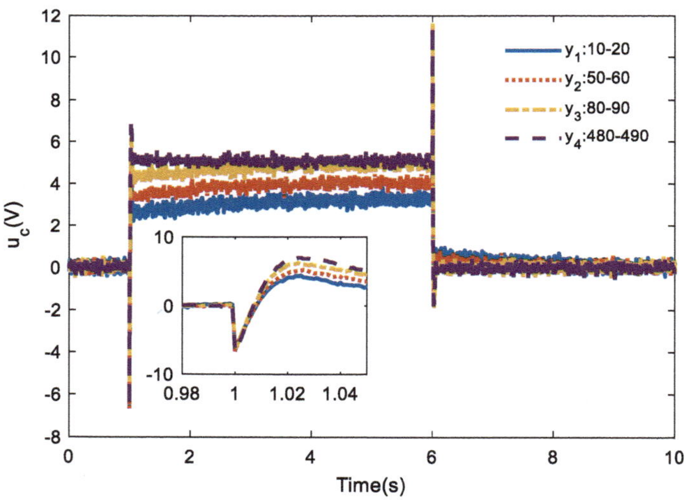

Fig. 9.10 Online updates of u_c

gap before and after parameter update in the case of actuator failure is shown in Fig. 9.14. In Fig. 9.14, only the levitation gap response before and after the parameter iterative update is given. It can be seen from the simulation results that even a simple proportional coefficient can improve the performance of the system after failure.

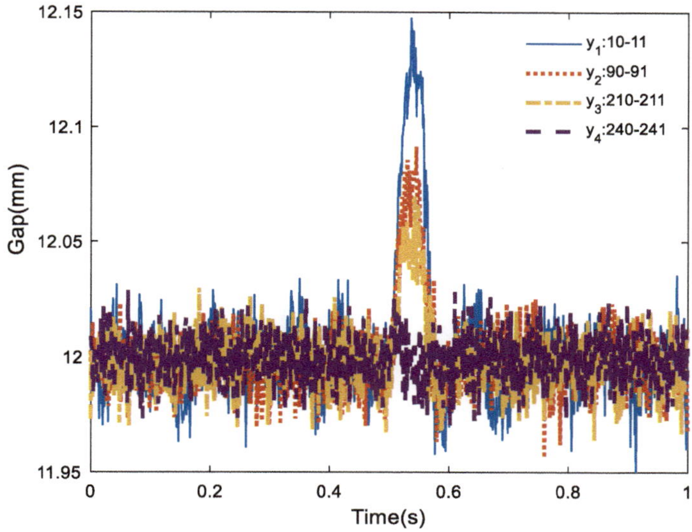

Fig. 9.11 Response of levitation gap during parameter update in case of actuator failure

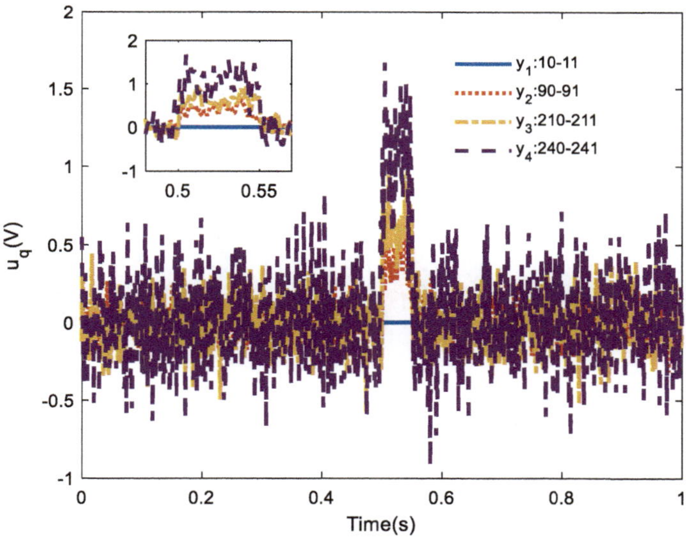

Fig. 9.12 Online update of compensation control variable u_q in case of actuator failure

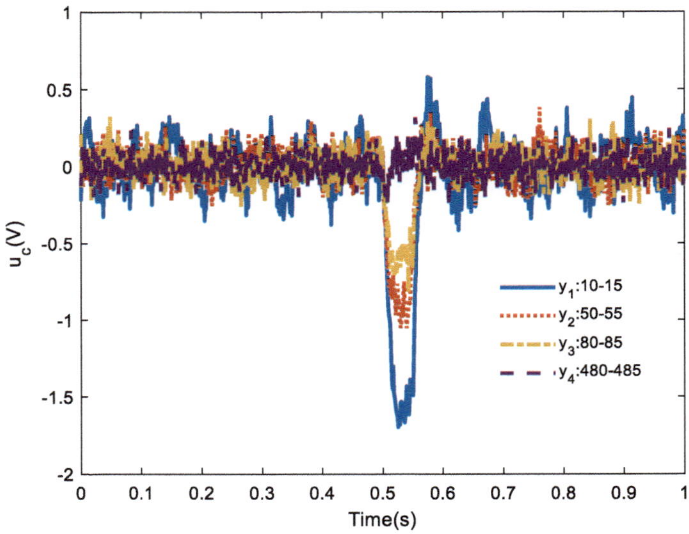

Fig. 9.13 Online update of main control variable u_c in case of actuator failure

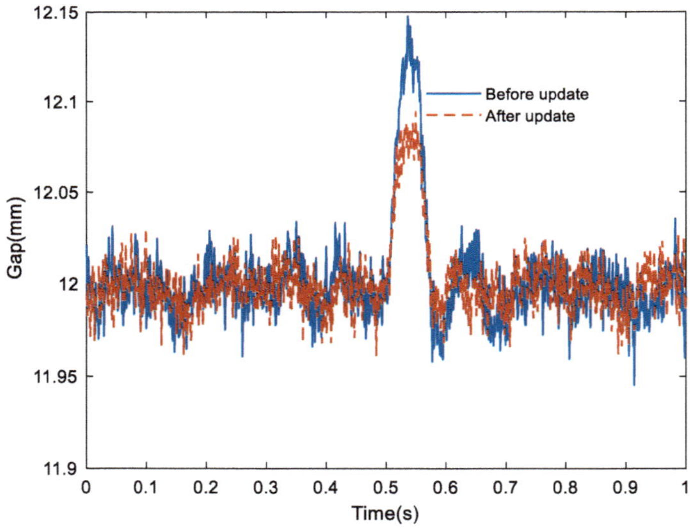

Fig. 9.14 Response of levitation gap before and after parameter update in case of actuator failure ($n_q = 0$)

9.4.2 Analysis of Fault Tolerant Control Based on Signal Reconstruction

In this part, the influence of a single gap sensor failure on the maglev train passing the track seam is illustrated by simulation. After the simulation starts, the electromagnet is stably levitated at the target gap value $\delta_0 = 12$ mm. Then, it is assumed that a single gap sensor failure occurs $t = 10$ seconds, at which time the combined gap obtained from the two gap sensor signals is equal to the normal operating sensor signal. To simulate faults and track seams, starting from $t = 10$ seconds, the gap sensor signal is set to 20 mm, and a pulse with an amplitude of 20 mm, and a duration of 0.015 s is added to the gap sensor signal. The fault response during this process is shown in Fig. 9.15. Figure 9.15a is the real levitation gap on both sides. The levitation gap is reduced by about 4 mm during this process due to the effect of the track seam. Figure 9.15b shows the integrated gap value of the left and right sides in this process. In the presence of a track seam, an upward pulse manifests in the left integrated gap signal, leading the controller to erroneously conclude that the electromagnet is descending. Consequently, a higher output voltage is applied, generating an increased current that drives the electromagnet upwards. Additionally, the permanent magnet attraction in the hybrid electromagnet exacerbates this upward movement trend.

Next, the simulation analysis of the signal reconstruction processing method in the case of single gap sensor failure is carried out. The modified gap signal reconstruction method is taken as an example, and the resulting system response is shown in Fig. 9.16.

For the corrected gap signal reconstruction (9.9), the fluctuation amplitude of the reconstructed gap signal is reduced to 0.8 mm (Fig. 9.16d), and the fluctuation amplitude of the levitation gap is a maximum of 0.2 mm during this process (Fig. 9.16a).

9.4.3 Analysis of Fault Tolerant Control Based on Control Law Switching

(1) Response of the levitation system under the failure of a single levitation unit

Assuming that a single-point levitation system failure occurs at the left levitation point in the 10th second of the simulation, and other conditions remain unchanged, the response of the levitation system before and after the failure is obtained, as shown in Fig. 9.17. Figure 9.17a shows the levitation gap of the left and right levitation electromagnets and the displacement of the levitation frame in the process of single-point failure. After the fault occurs, due to the reduction of the electromagnetic force on the left side, the electromagnetic force is smaller than the load pressure transmitted by the metal rubber spring, causing the left electromagnet to move downward rapidly, which in turn causes the levitation frame and the right electromagnet to

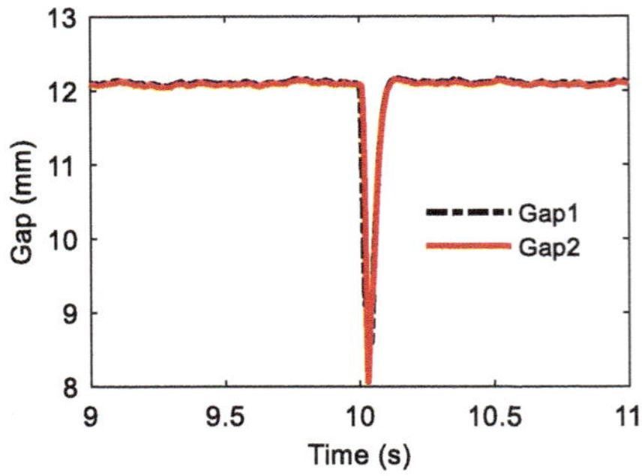

(a) Gap

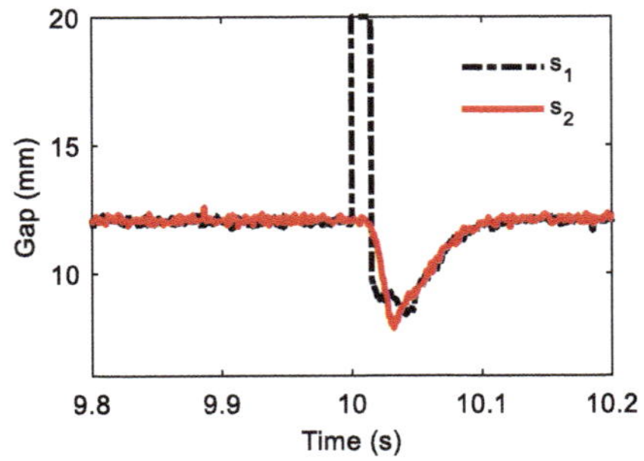

(b) Comprehensive gap value

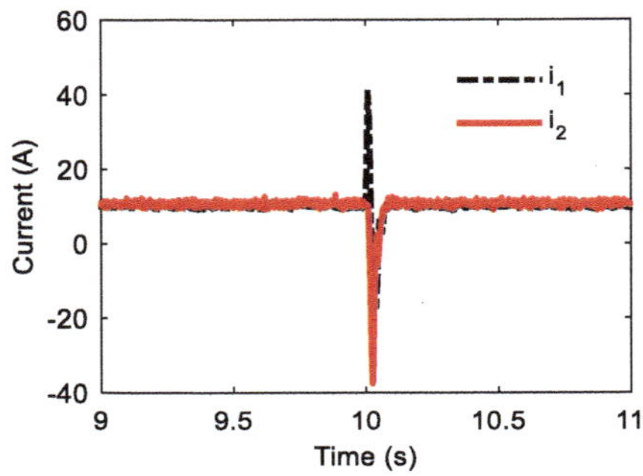

(c) Current in the solenoid coil

Fig. 9.15 Levitation system response when a single gap sensor fails over a track seam

move downward successively. After that, the right electromagnet gradually returns to the target position under the action of the control law, and the magnetic force loss of the left electromagnet is gradually compensated by the integral link. During the fluctuation of the levitation gap, the maximum fluctuation amplitude of the left electromagnet is 5.48 mm, and the maximum fluctuation amplitude of the right electromagnet is 4.65 mm. After the right levitation gap is re-stabilized, the levitation gap of the left electromagnet is 12.5 mm. Figure 9.17b shows the response of the control variable, the voltage across the electromagnet, and the current in the coil winding of the electromagnet obtained during a single-point fault. After the fault occurs, the voltage on the left side is zero, and the current in the coil winding then rapidly drops to zero. Figure 9.17c shows the electromagnetic force generated by the left and right electromagnets during a single-point fault. After the fault occurs, the current in the coil winding of the left electromagnet is zero, and only the residual magnetic force generated by the permanent magnet remains. Its size is not controlled, only related to the levitation gap, and decreases with the increase of the levitation gap, and increases with the decrease of the levitation gap. The electromagnetic force generated by the right electromagnet increases rapidly after the failure to allow the

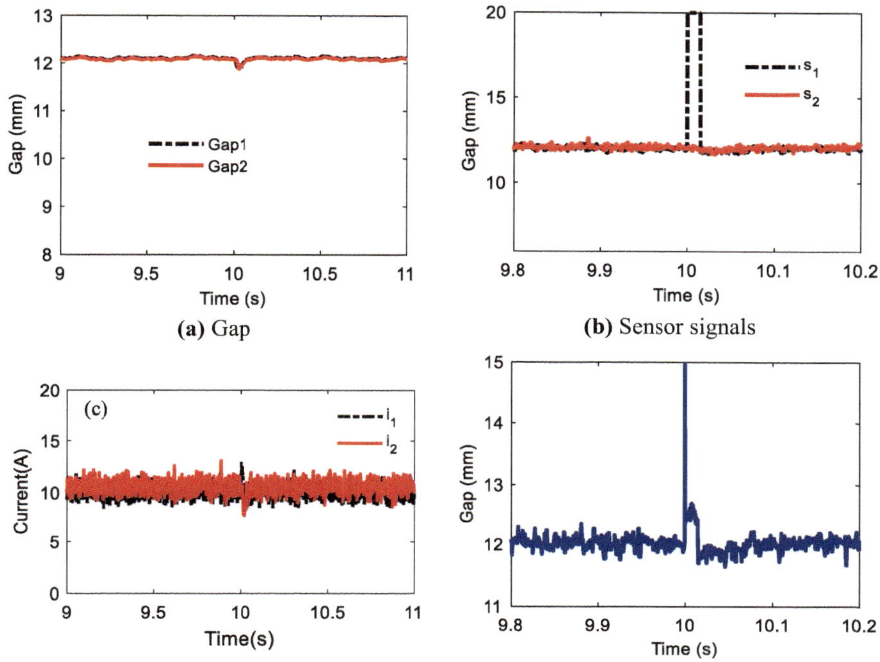

(a) Gap **(b)** Sensor signals **(c)** Current **(d)** Reconstructed gap signal

Fig. 9.16 The response of the levitation system when the train passes through the track joint under the improved gap signal reconstruction algorithm

levitation electromagnet to return to the target position. After stabilization, the resultant force generated by the left and right electromagnets is the same as before the failure. The left levitation electromagnet provides a magnetic force of 8950 N, and the right levitation electromagnet provides a magnetic force of 18,490 N. Figure 9.17d shows the pressure generated by the metal rubber springs on the left and right sides during a single-point failure. After the failure, the pressure of the left metal rubber spring decreases rapidly, and the pressure generated by the right metal rubber spring increases rapidly, with a maximum value of 18,900 N. After re-stabilization, the average pressure of the left metal rubber spring is 5510 N, and the average pressure of the right metal rubber spring is 15070 N. Based on the simulation results shown in Fig. 9.17, after the single-point levitation system fails, the joint structure levitation system can still recover stability after the fluctuation when the controller remains unchanged. However, the maximum fluctuation amplitude of the gap exceeds 5 mm, and the fluctuation time of the gap exceeds 1 s. It is necessary to design a fault tolerant control algorithm to improve the system response after a fault.

(2) Fault tolerant control effect of single-point fault

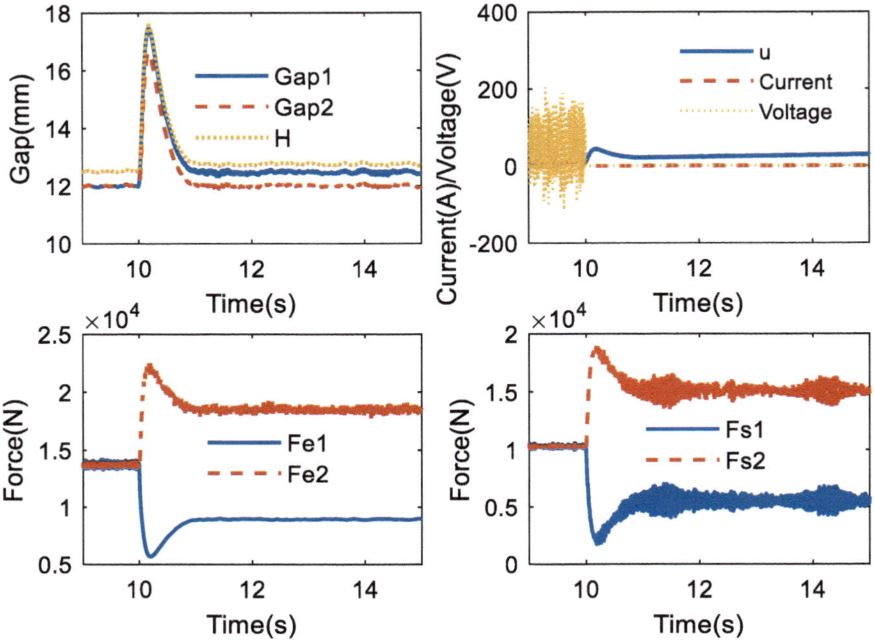

Fig. 9.17 Response of the joint structure levitation system in the case of a single levitation point fault

In this part, the fault tolerant control strategy based on the switching strategy is simulated. Without loss of generality, the simulation object is a permanent magnetic-electromagnetic hybrid high-speed maglev joint structure levitation system.

Assuming the left levitation point fails completely in the 10th second of the simulation during the normal levitation process, however, the signals of the levitation sensors on both sides can still be obtained normally after the failure. The fault diagnosis unit gives the fault diagnosis result after 0.02 s, and the controller switches to the fault tolerant control strategy. The response of the levitation gap of the levitation system in this switching process is shown in Fig. 9.18. During the process, the control amount and current, and the voltage response are shown in Fig. 9.19a, b. For 0.02 s after the failure, the controller remains unchanged. The system responds identically, showing a rapid increase in the levitation gap. After the control law is switched, the control amount increases, and the levitation gap gradually returns to the vicinity of the target position. The duration of this violent fluctuation process is about 0.5 s. The maximum amplitude of the fluctuation of the left levitation gap is 1.78 mm, the maximum amplitude of the fluctuation of the right levitation gap is 0.86 mm, and the maximum amplitude of the current fluctuation of the right electromagnet is 25.7 A. After that, under the action of the integral term in the controller, the levitation gap gradually stabilizes to the target gap accurately. After stabilization, the left levitation gap is 12.21 mm, the right levitation gap is 11.76 mm, and the right current is 19.07 A.

To verify the effectiveness of the fault tolerant control algorithm based on the switching strategy, a test is carried out on the maglev train levitation system (Fig. 1. 7). The test object is a PEM hybrid high-speed maglev train, and one of the joint

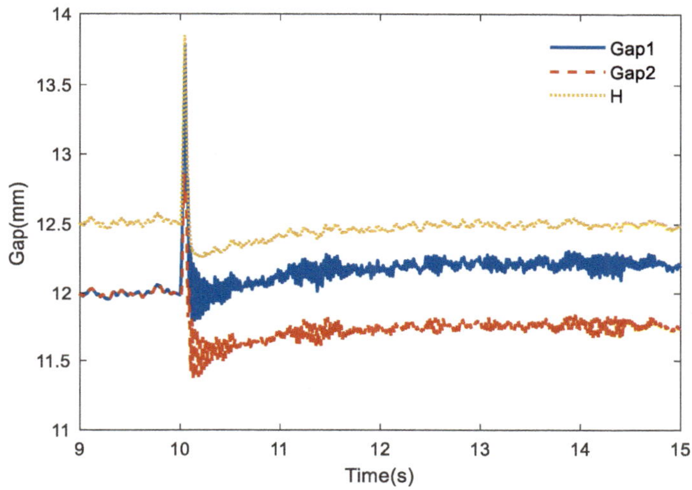

Fig. 9.18 Levitation gap response under control law switching strategy

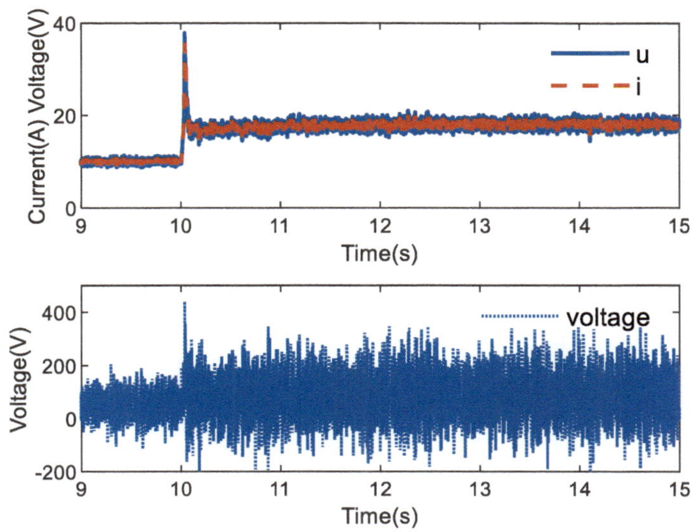

Fig. 9.19 Response of voltage and current control quantities under control law switching strategy

structures is used for testing. During the test, the maglev train first floated statically and then towed. During the operation, the output voltage of the right electromagnet was set to zero. The test results are shown in Fig. 9.20.

Figure 9.20a are the levitation gaps on the left and right sides, and Fig. 9.20b are the currents in the left and right electromagnet windings during this process. When the static levitation is started, the fluctuation amplitude of the levitation gap on both

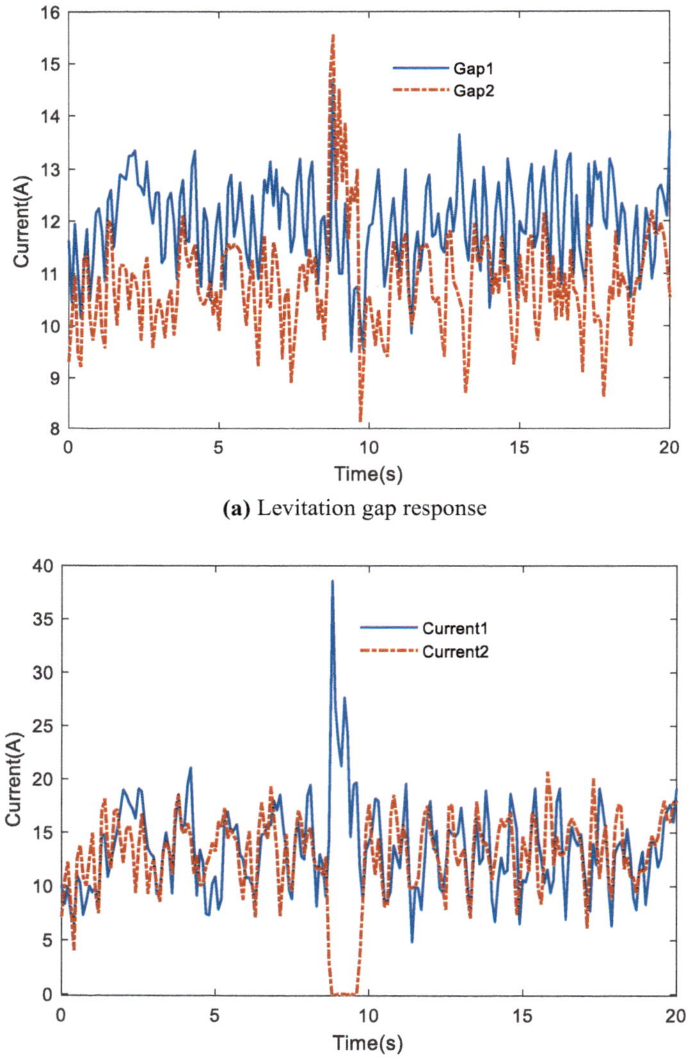

(a) Levitation gap response

(b) Current response

Fig. 9.20 Fault tolerant control effect under single levitating point zero output fault sides is within 0.5 mm. When the maglev train moves, the fluctuation amplitude of the levitation gap increases to 2 mm. When $t = 9.5$ s, a zero output fault occurs at the right single levitation point, causing the right electromagnet to lose active control. After the fault has been successfully detected, the control output is switched to the fault tolerant controller so that the left-hand gap returns to the equilibrium position after fluctuations of up to 3 mm. The experimental test results show that the fault tolerant control method based on the switching strategy can make the electromagnet return to the equilibrium position after a small fluctuation oscillation, so that the entire joint structure remains stable. Judging from the current response in the coil windings of the left and right levitation electromagnets, the maximum fluctuation amplitude of the coil winding current of the normal working electromagnet during the dynamic adjustment process is about 29.5 A. The experimental test results are aligned with the simulation. The experimental results show that the fault tolerant control algorithm based on the switching strategy can ensure the stable levitation of the entire joint structure when a single levitation point fault occurs during the train operation.

9.5 Conclusion

In this chapter, a systematic fault tolerant control scheme is proposed for the PEM hybrid levitation system of high-speed maglev train. For the fault tolerant control method of incipient faults, the online updating rule of parameters based on gradient descent method is given. A fault tolerant control method based on signal reconstruction is proposed for accelerometer fault and single gap sensor fault. This method is practical and does not depend on the system model. The second differential relationship between levitation gap and acceleration is used for signal reconstruction of accelerometer signal. The signal reconstruction of single gap sensor fault utilizes the interaction between two single-point levitation systems in the joint structure. For single levitation point fault, an active fault tolerant control strategy based on control law switching is proposed. This method can quickly adjust the control law and complete the fault tolerant control task in a short time.

Bibliography

1. Ding S (2022) 600 km/h high-speed maglev transportation system. Shanghai Scientific and Technical Publishers, Shanghai
2. Ding SX, Yang G, Zhang P et al (2010) Feedback control structures, embedded residual signals, and feedback control schemes with an integrated residual access. IEEE Trans Control Syst Technol 18(2):352–367
3. Huang C, Yang Y, Dai C et al (2023) Tracking-differentiator-based position and acceleration feedback control in active vibration isolation with electromagnetic actuator. Actuators 12(271):271

4. Liang S, Long Z, Wang P et al (2023) An anti-interference method and system for detecting shallow-ground metal based on hyper-sphere model. Measure Control 56(7–8):1202–1216
5. Lin G, Liu W, Xu J et al (2023) Opportunities and challenges for the development of high-speed maglev transportation in China. Sci Technol Foresight 2(4):7–18
6. Liu Huaqing Compile (1995) The German Maglev train-transrapid. University of Electronic Science and Technology Press, Sichuang
7. Liu Z, Long Z, Li X (2015) Maglev trains. Springer, Berlin
8. Long Z, Cai Y, Xu X (2009) Comprehensive fault evaluation on maglev train based on estimation of distribution algorithms. Control Decis 24(4):551–556
9. Long Z, Chen H, Chang W (2007) Fault tolerant control on single suspension module of Maglev train with electromagnet failure. Control Theory Appl 24(6):1033–1037
10. Long Z, Dou F, Wang Z et al (2023) Current status and prospects of high-speed maglev suspension and guidance control technology. Sci Technol Foresight 2(4):78–88
11. Long Z, Xue S, He G et al (2011) Fault-diagnosis for the accelerometer of maglev suspension system based on signal comparison. Chin J Sci Instrum 32(12):2641–2647
12. Long Z, Xue S, She L et al (2009) Distributed active fault-tolerant control for module suspension system of maglev train. J China Railway Soc 31(2):38–43
13. Long Z, Zhai M, Wang Z (2023) Condition monitoring, fault diagnosis and fault tolerant control of the maglev train. Shanghai Scientific and Technical Publishers, Shanghai
14. Long Z, Zhang Z, Chang W (2007) Simulation research of fault tolerant control for maglev system with sensor's faults. J Syst Simul 19(19):4469–4472
15. Luo H (2017) Plug-and-Play monitoring and performance optimization for industrial automation processes. Springer Fachmedien, Wiesbaden
16. Sun Y, Li F, Lin G et al (2023) Adaptive fault-tolerant control of high-speed maglev train suspension system with partial actuator failure: design and experiments. J Zhejiang Univ Science A 24(3):272–283
17. Wang Z (2019) Fault diagnosis and tolerant control for high speed maglev train suspension system. Ph.D. dissertation, National University of Defense Technology Doctor Degree, Changsha
18. Wang Z, Long Z, Li X (2021) Fault tolerant control for joint structure in PEMS high speed Maglev train. Asian J Control, 2021, 23(1): 486-498
19. Wang Z, Long Z, Li X (2019) Fault analysis and tolerant control for high speed PEMS maglev train end joint structure with disturbance rejection. J Electr Eng Technol 14(3):1357–1366
20. Wang Z, Long Z, Li X (2019) Track irregularity disturbance rejection for maglev train based on online optimization of PnP control architecture. IEEE Access 7:12610–12619
21. Wu X (2003) Maglev train. Shanghai Scientific and Technical Publishers, Shanghai
22. Xu Y (2018) Camera-based real-time control of highly dynamic systems with performance recovery. Shaker Verlag, Germany
23. Yang Y, Xie Y, Dai C et al (2023) An improved nonlinear extended state observer with adaptive variable gain. Asian J Control 25(3):2388–2397
24. Yin S, Luo H, Ding SX (2014) Real-time implementation of fault-tolerant control systems with performance optimization. IEEE Trans Industr Electron 61(5):2402–2411
25. Zhang Z (2006) Research on fault-tolerant control issues of high-speed maglev train single-rail suspension system. Ph.D. dissertation, National University of Defense Technology Doctor Degree, Changsha
26. Zheng Q, Ni F (2023) High-speed maglev transportation: from Germany to China. Sci Technol Foresight 2(4):31–39
27. Zhou K, Ren Z (2001) A new controller architecture for high performance, robust, and fault-tolerant control. IEEE Trans Autom Control 46(10):1613–1618

GPSR Compliance

The European Union's (EU) General Product Safety Regulation (GPSR) is a set of rules that requires consumer products to be safe and our obligations to ensure this.

If you have any concerns about our products, you can contact us on ProductSafety@springernature.com

In case Publisher is established outside the EU, the EU authorized representative is:

Springer Nature Customer Service Center GmbH
Europaplatz 3
69115 Heidelberg, Germany

Batch number: 10280114

Printed by Printforce, the Netherlands